Environmental Management

for Cambridge International AS Level

COURSEBOOK

Tana Scott

CAMBRIDGE
UNIVERSITY PRESS

Shaftesbury Road, Cambridge CB2 8EA, United Kingdom

One Liberty Plaza, 20th Floor, New York, NY 10006, USA

477 Williamstown Road, Port Melbourne, VIC 3207, Australia

314–321, 3rd Floor, Plot 3, Splendor Forum, Jasola District Centre, New Delhi – 110025, India

103 Penang Road, #05–06/07, Visioncrest Commercial, Singapore 238467

Cambridge University Press is part of the University of Cambridge.

It furthers the University's mission by disseminating knowledge in the pursuit of education, learning and research at the highest international levels of excellence.

www.cambridge.org
Information on this title: www.cambridge.org/9781009306256

© Cambridge University Press & Assessment 2023

This publication is in copyright. Subject to statutory exception and to the provisions of relevant collective licensing agreements, no reproduction of any part may take place without the written permission of Cambridge University Press.

First published 2023

20 19 18 17 16 15 14 13 12 11 10 9 8 7 6 5 4 3 2 1

Printed in Malaysia by Vivar Printing

ISBN 978-1-009-30625-6 Print Coursebook with Digital Access
ISBN 978-1-009-30622-5 Digital Coursebook
ISBN 978-1-009-30623-2 Coursebook eBook

Additional resources for this publication at www.cambridge.org/9781009306256

Cambridge University Press has no responsibility for the persistence or accuracy of URLs for external or third-party internet websites referred to in this publication, and does not guarantee that any content on such websites is, or will remain, accurate or appropriate. Information regarding prices, travel timetables, and other factual information given in this work is correct at the time of first printing but Cambridge University Press does not guarantee the accuracy of such information thereafter.

Cambridge International copyright material in this publication is reproduced under licence and remains the intellectual property of Cambridge Assessment International Education.

Third-party websites and resources referred to in this publication have not been endorsed by Cambridge Assessment International Education.

....

NOTICE TO TEACHERS IN THE UK
It is illegal to reproduce any part of this work in material form (including photocopying and electronic storage) except under the following circumstances:
(i) where you are abiding by a licence granted to your school or institution by the Copyright Licensing Agency;
(ii) where no such licence exists, or where you wish to exceed the terms of a licence, and you have gained the written permission of Cambridge University Press;
(iii) where you are allowed to reproduce without permission under the provisions of Chapter 3 of the Copyright, Designs and Patents Act 1988, which covers, for example, the reproduction of short passages within certain types of educational anthology and reproduction for the purposes of setting examination questions.

....

NOTICE TO TEACHERS
The photocopy masters in this publication may be photocopied or distributed [electronically] free of charge for classroom use within the school or institution that purchased the publication. Worksheets and copies of them remain in the copyright of Cambridge University Press, and such copies may not be distributed or used in any way outside the purchasing institution.

Endorsement statement

Endorsement indicates that a resource has passed Cambridge International's rigorous quality-assurance process and is suitable to support the delivery of a Cambridge International syllabus. However, endorsed resources are not the only suitable materials available to support teaching and learning, and are not essential to be used to achieve the qualification. Resource lists found on the Cambridge International website will include this resource and other endorsed resources.

Any example answers to questions taken from past question papers, practice questions, accompanying marks and mark schemes included in this resource have been written by the authors and are for guidance only. They do not replicate examination papers. In examinations the way marks are awarded may be different. Any references to assessment and/or assessment preparation are the publisher's interpretation of the syllabus requirements. Examiners will not use endorsed resources as a source of material for any assessment set by Cambridge International.

While the publishers have made every attempt to ensure that advice on the qualification and its assessment is accurate, the official syllabus, specimen assessment materials and any associated assessment guidance materials produced by the awarding body are the only authoritative source of information and should always be referred to for definitive guidance. Cambridge International recommends that teachers consider using a range of teaching and learning resources based on their own professional judgement of their students' needs.

Cambridge International has not paid for the production of this resource, nor does Cambridge International receive any royalties from its sale. For more information about the endorsement process, please visit www.cambridgeinternational.org/endorsed-resources

CAMBRIDGE DEDICATED TEACHER AWARDS 2022

Teachers play an important part in shaping futures. Our Dedicated Teacher Awards recognise the hard work that teachers put in every day.

Thank you to everyone who nominated this year; we have been inspired and moved by all of your stories. Well done to all of our nominees for your dedication to learning and for inspiring the next generation of thinkers, leaders and innovators.

Congratulations to our incredible winners!

WINNER — Regional Winner, Australia, New Zealand & South-East Asia
Mohd Al Khalifa Bin Mohd Affnan
Keningau Vocational College, Malaysia

Regional Winner, Europe
Dr. Mary Shiny Ponparambil Paul
Little Flower English School, Italy

Regional Winner, North & South America
Noemi Falcon
Zora Neale Hurston Elementary School, United States

Regional Winner, Central & Southern Africa
Temitope Adewuyi
Fountain Heights Secondary School, Nigeria

Regional Winner, Middle East & North Africa
Uroosa Imran
Beaconhouse School System KG-1 branch, Pakistan

Regional Winner, East & South Asia
Jeenath Akther
Chittagong Grammar School, Bangladesh

For more information about our dedicated teachers and their stories, go to
dedicatedteacher.cambridge.org

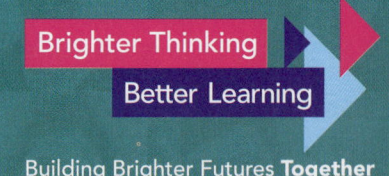

Building Brighter Futures **Together**

Contents

How to use this series	vi
How to use this book	vii
Introduction	ix
Key Skills in Environmental Management	1
1 Introduction to Environmental Management	19
2 Environmental Research and Data Collection	58
3 Managing Human Population	108
4 Managing Ecosystems and Biodiversity	140
5 Managing Resources	187
6 Managing Water Supplies	234
7 Managing the Atmosphere	273
8 Managing Climate Change	301
Glossary	335
Acknowledgements	345
Index	347

CAMBRIDGE INTERNATIONAL AS LEVEL ENVIRONMENTAL MANAGEMENT: COURSEBOOK

> How to use this series

This suite of resources supports learners and teachers following the Cambridge AS Level Environmental Management syllabus (8291). The components in the series are designed to work together and help learners develop the necessary knowledge and skills for studying Environmental Management.

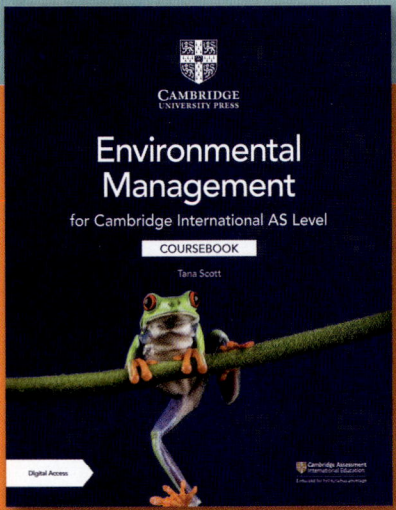

The Coursebook is designed for learners to use in the classroom with guidance from teachers. It offers full coverage of Cambridge AS Level Environmental Management (8291). Each of the eight chapters covers the corresponding units in the syllabus. The chapters contain in-depth explanations, definitions, questions, case studies, worked examples, and a range of other features to engage learners. A variety of activities allow learners to practice their skills and demonstrate their knowledge. Learners have plenty of opportunities to engage in meaningful debate, pair work, and group work.

The Teacher's Resource offers countless inspiring plenary, lesson and homework ideas for teachers of Cambridge AS Level Environmental Management. It also provides teachers with worksheets for students to use in class, practice tests, answers, and a bank of case studies from prominent third party organisations such as the World Wildlife Fund for Nature, which can be used in class or as homework tasks. Lesson planning guidance and pedagogical support is also available in the Teacher's Resource. Teachers are encouraged to use a blend of Coursebook and Teacher's Resource activities depending on the needs of their class.

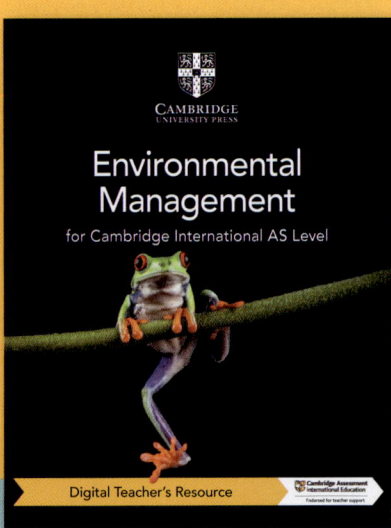

> How to use this book

LEARNING INTENTIONS

These set the scene for each chapter, help you to navigate through the coursebook and highlight the most important learning points in each topic.

GETTING STARTED

At the beginning of each chapter there will be a getting started activity. These are pair, group or class activities designed to introduce the chapter and provide the opportunity for you to show how much you already know about the topic you will be learning.

ENVIRONMENTAL MANAGEMENT IN CONTEXT

This feature presents real-world examples and applications of the content in a chapter, encouraging you to look further into topics. There are discussion questions at the end which will ask you to think about the benefits and problems of these applications.

KEY TERMS

Key vocabulary for the syllabus is highlighted in the text when it is first introduced. There are also key terms that do not relate to the syllabus – these are marked with an asterix (*). Definitions are given in the margin and can also be found in the Glossary at the back of this book. Any key words marked with an asterix (*) do not appear in the syllabus and are therefore not required knowledge, but are useful terms for the subject.

COMMAND WORDS

In the early chapters, command words from the syllabus have been pulled out and will appear in feature boxes. These will remind you of what each command word means and which skills you should apply when answering different types of questions. As you become more familiar with the command words, the boxes will appear less. You can still find the definitions for each command word in the Glossary at the back of the book.

Questions

Regular sets of questions throughout the book provide you with the chance to check your knowledge and understanding of what you have learned.

ACTIVITIES / PRACTICAL ACTIVITIES

You will find a variety of activities and practical activities throughout this coursebook. These provide a good opportunity for you to think about what you have learned and take part in discussions, answer questions or complete a practical task. Activities will help you to develop different learning styles and provide an opportunity for you to produce your own work either individually, or in pairs or groups.

TIP

These are helpful reminders or notes that will give advice on skills or methodology. You will find them most often near activities or questions, where they will be directly relevant to the task.

INVESTIGATIVE SKILLS

This feature provides an opportunity for you to develop your scientific enquiry skills. This might involve carrying out an experiment, handling data, completing a research task or predicting outcomes.

WORKED EXAMPLE

Worked examples show you step-by-step how to work through a particular process or question and then provide you with an opportunity for you to try it for yourself. You will find this feature helpful for questions which require a mathematical approach to work out the answer.

REFLECTION

These activities ask you to think about the approach that you take to your work, and how you might improve this in the future.

CASE STUDIES

The case studies and the accompanying questions allow you to actively explore real environmental management scenarios. You are provided with opportunities to produce your own work either as an individual, in pairs or in groups.

SELF / PEER ASSESSMENT

At the end of some Activities and Practical tasks, you will find opportunities to help you assess your own work, or the work of your peers and consider how you can improve the way you learn.

SUMMARY

This feature contains a series of statements which summarise the key learning points you will have covered in the chapter.

EXTENDED CASE STUDY

In each chapter you will find an extended case study which looks in detail at a particular issue or situation in a real-word setting. Extended case studies encourage you to think about a particular issue in more depth and will have embedded activities, questions or projects for you to complete.

PRACTICE QUESTIONS

At the end of each chapter, you will find a set of practice questions that use the command words from the syllabus. To answer some of these, you may need to apply what you have learned in previous chapters as well as the current chapter you are studying.

SUMMARY CHECKLIST

At the end of each chapter, you will find a series of statements outlining the content that you should now understand. You might find it helpful to rate how confident you are for each of these statements when you are revising. You should revisit any topics that you rated 'Needs more work' or 'Almost there'.

> Introduction

Environmental management is the study of how to manage human impact on the environment. Some people study it because they are interested in how humans interact with the natural environment. Others want to find out how to protect the interests of both humanity and the environment. It is difficult to provide just one definition of environmental management. This is because it operates on so many scales and impacts every aspect of our lives, from how much water an individual uses to the pollution of a large river.

Environmental managers study every aspect of our natural world, and how these interact. They do this to understand the functioning of ecosystems, from the smallest puddle to the planet as a whole. There are many different fields of environmental management. However, every field is underpinned by foundation knowledge. The knowledge you will gain in this syllabus can be applied across a wide range of future careers.

Environmental scientists often work in remote areas, sampling environmental conditions. Here, a researcher collects water quality data.

Thinking environmental management

One of the skills that an environmental manager needs is the ability to think critically. An environmental manager should consider the multiple factors that could impact a specific environment, and how these factors could cause change within it. They need to develop analytical skills in order to assess data that has been collected. They also need to employ problem solving skills in order to develop strategies which will prevent or repair environmental damage. An environmental manager is a solution seeker in a world where no two problems are ever the same. As this course progresses, you too will learn to think in a solutions-oriented way.

Using environmental management

Environmental management can be applied in a variety of practical ways to a wide range of careers. Environmental managers are specialists in industry, nature conservationists, engineers, architects, lawyers, journalists, environmental health specialists, government employees and non-governmental organisation workers. Environmental management teams are often multi-disciplinary, made up of specialists from many different fields. Team members work together to find solutions to potential or identified environmental concerns, each sharing their own field of expertise. As an environmental manager, you will work as part of a team that is working towards a common goal.

Joining in

Your knowledge of the environment, and your wish to influence how humans interact with the physical world can result in you making a change for the better. When you study environmental management, you:

- learn more about the planet
- consider how our actions as a species impact the planet
- think critically, considering strategies to manage how we, as a species, impact our world
- develop critical academic skills.

Whatever field of study you pursue, the environmental knowledge and skills you learn on this course will remain with you. Both will help you make sense of the rapid changes that we are seeing in the world today.

Note on maps: The boundaries and names shown, the designations used and the presentation of material on any maps contained in this resource do not imply official endorsement or acceptance by Cambridge University Press concerning the legal status of any country, territory, or area or any of its authorities, or of the delimitation of its frontiers or boundaries.

Key Skills in Environmental Management

LEARNING INTENTIONS

In this chapter you will:

- learn about the experimental skills you will develop throughout this book
- find out how to present data in the form of graphs and charts
- employ skills which are relevant to research
- develop evaluative essay-writing skills.

0.1 Introduction

In this course, you will learn about the issues facing the environment today. You will also develop a range of important academic skills, such as:

- Experimental skills
- Data handling, analysis, and representation
- Research skills
- Evaluative writing
- Critical thinking
- Problem-solving
- Analysis.

Some of these skills, such as problem-solving, will evolve as you move through the course. Others require context and knowledge before you can put them into practice. This short chapter will look at the latter types of skill.

0.2 Experimental skills

Chapter 2 of this Coursebook covers the scientific method, strategies and methods for collecting environmental data and analysis of that data. You will also develop the following skills in this chapter:

Unit 2.1: Understanding the structure and steps of the scientific method, including:	Unit 2.3: Collection of environmental data including:

• Setting the research question
• The formulation of hypotheses
• Types of data
• Testing the hypothesis and managing variables
• Data interpretation
• Limitations in scientific method
	• Sampling strategies
	• Sample size
	• Sampling techniques including:
	  • Point sampling
	  • Line sampling
	  • Area or quadrat sampling
	  • Pitfall traps, sweep nets, beating trays, kick sampling, light traps, capture-mark release, water turbidity and questionnaires and interviews

Unit 2.4: Data analysis using statistical tools including:	Unit 2.5: Using technology in data collection including:
• The Lincoln Index • The Simpson Index • Percentage and frequency calculations • Estimated abundance using the ACFOR scale	• Computer models • Geospatial systems • Satellite sensors • Radio tracking • Crowd sourcing

Table 0.1: Skills included in Chapter 2 of this Coursebook.

Key Skills in Environmental Management

0.3 Graphs and charts

You need to present the **numerical data** collected during investigations in the form of graphs or charts. This helps you to visualise and interpret your data. It can also enable you to see if your data contains consistent patterns, trends or **anomalies**.

When creating graphs, ensure to clearly label axes and include units where needed. The axes scale should allow for the graph to fill more than half of the grid in both directions. Do not compress the graph into one corner of the grid. In addition to this, a sensible linear scale should be used.

There are many chart types, and it is important to select the correct one. Doing so will ensure that your readers clearly understand what your data represents.

When collecting data for use in graphs, you should create clear tables to record the data in. The tables should include units in row or column headings and not in individual cells. Table 2.4 in chapter two gives an example of how a data table should be set up.

Before you choose a graph, consider the type of data you are collecting and its purpose:

	Reasons for collecting data		The type of data that has been collected
1	Find **trends**	1	Continuous
2	Compare different options	2	Categories
3	Show distribution	3	Primary category with secondary categories
4	Observe relationships between different data sets (e.g. time, size, type, ranking, correlation)	4	Data that requires two different types of graph on one display (e.g. a bar chart and line graph).

Table 0.2: The type of data and its purpose will help you to select the most appropriate graph.

TIP

The graphs in this book are typically in colour. However, in any exam situation where you might need to draw a graph, remember to use shading or patterns, as you may not have access to colours.

Types of graphs

You can represent data using different types of graph. Certain data works well with particular types of graph, as outlined below.

Bar chart

Bar charts are used to compare variables. Bar charts work well with data that has different categories, particularly with bigger sets of data, when highlighting different categories or when showing changes over time.

When creating a bar chart:

- leave spaces between the bars (this shows there is not a link between the information, which are independent variables)
- make the bars equal width
- ensure the bars are shaded differently and use a key to indicate the different categories.

For example, you could use a bar chart when collecting data that counts the number of animals of different species on a reserve.

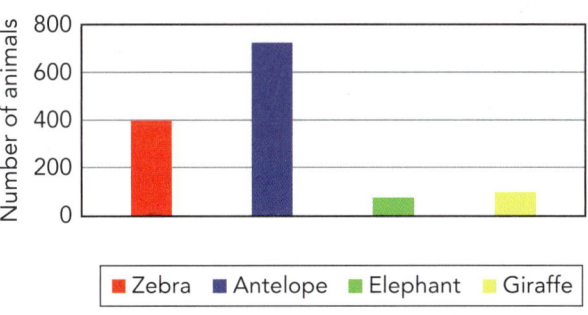

Graph 0.1: A bar chart.

KEY TERMS

numerical data: information that is expressed as numbers

anomalies: data that is unusual, and which deviates from the patterns and trends that the rest of the data indicates

trends: the general relationship between two sets of data

Stacked bar chart

Stacked bar charts are used when a second level of data is collected for a number of primary categories. They are mainly used to compare sub-divisions of data between different categories, so the columns are divided into a number of sub-bars stacked on top of each other. The bars show the frequency of secondary categories in each primary category.

For example, you could use a stacked bar chart to show data collected from four different game reserves (primary category) on the number of antelope, zebra, elephant and giraffe in each reserve (secondary categories).

Primary category	Secondary categories			
Reserve	Antelope	Zebra	Elephant	Giraffe
1	425	392	22	55
2	329	210	15	41
3	391	187	23	63
4	210	305	27	50

The primary category is the reserve, and information about each reserve is being compared. Reserve 1 has the greatest number of animals and Reserve 4 the fewest. Each category is subdivided, based on the wildlife found in the reserve. In most reserves, antelope have the highest numbers, while the elephants have the lowest.

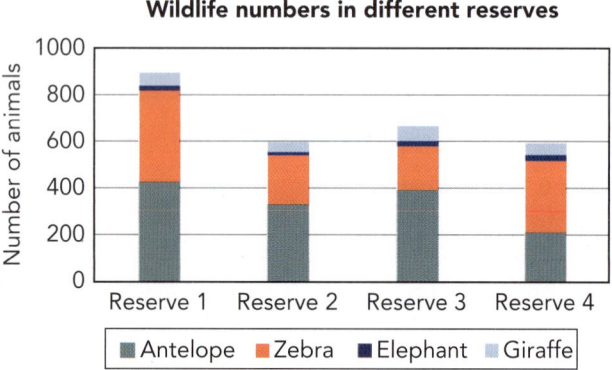

Graph 0.2: A stacked bar chart.

Divided bar chart

A divided bar chart is a rectangle divided into smaller sections along its length in proportion to the data. The value of the exact whole must be known, because each piece of the bar represents a part of the total (a percentage or a ratio). Divided bar charts are useful for showing data that is split into categories, and when comparing percentages of data that form subsets of a larger category.

For example, you could use a divided bar chart to show the labour force in a population by occupation, such as the percentages working in agriculture, services and industry. You could then create a similar chart for another country and do a direct comparison.

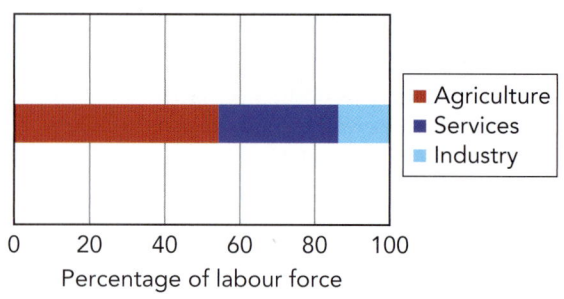

Graph 0.3: A divided bar chart.

Pie chart

Pie charts are similar to divided bar charts, except that they are circular and divided into sections. Each piece of a pie chart represents the relative size of each category as part of a whole. Pie charts are best used with small data sets. They are effective when comparing the effect of one factor on up to six different categories.

For example, you could use a pie chart to compare data collected from four different game reserves on the number of baby bison in each bison herd. Pie charts should be drawn with the sectors in rank order, with the largest first, beginning at 'noon' and proceeding clockwise. Pie charts should preferably contain no more than six sectors.

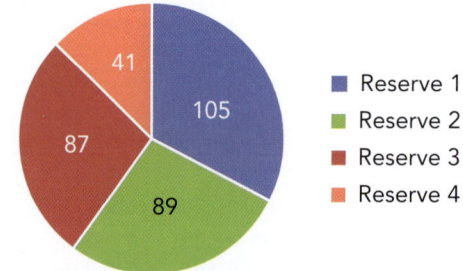

Graph 0.4: A pie chart.

Key Skills in Environmental Management

Line graph

Line graphs are used to represent continuous data sets that change over time. They are particularly useful:

- if your dataset is too big for a bar chart
- for displaying multiple sets of data for the same time period
- for identifying trends in data.

For example, you could use a line graph to show the daily midday temperature at two sites over a two-week period.

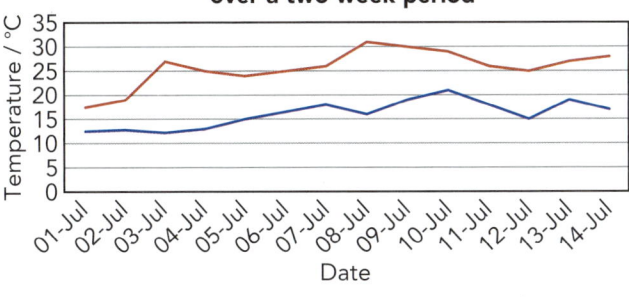

Graph 0.5: A line graph.

Scatter graph

A scatter graph uses dots to represent values for two different variables. Scatter graphs are useful for showing **correlations** and clustering in big data sets, in situations where the order of the points in the data set is not important.

For example, you could use a scatter graph to see if there is a correlation between the height and diameter of a species of tree. In Graph 0.6, a **trend line** has been added to show the relationship between the two variables.

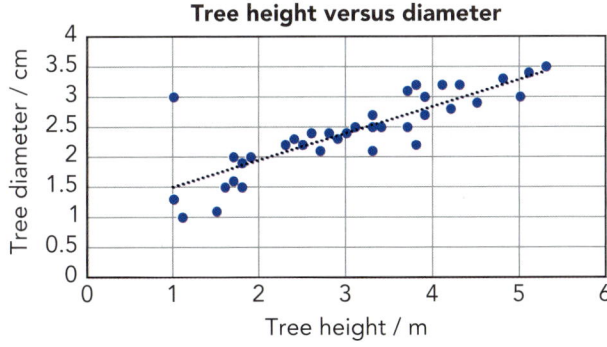

Graph 0.6: A scatter graph showing a positive linear relationship.

> **KEY TERMS**
>
> **correlation:** a relationship or link between two sets of data
>
> **trend line:** a line indicating the general relationship between two sets of data

Scatter graphs are also used to observe the relationship between two sets of variables. This relationship can be described in a number of ways:

- Positive linear relationship: As one variable increases, so does the other (Graph 0.6).
- Negative linear relationship: As one variable increases, the other decreases (Graph 0.7).
- No relationship: Where no trend line can fit to the spread of the data (Graph 0.8).
- Non-linear relationship: Where the trend line curves to fit the scatter graph (Graph 0.9).

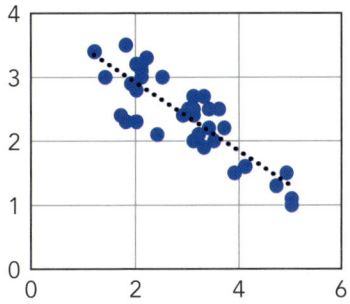

Graph 0.7: Negative linear relationship.

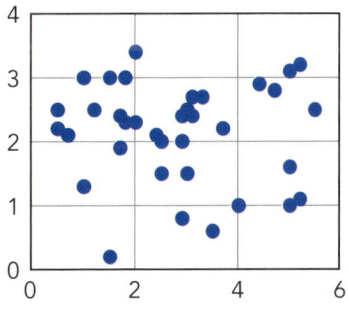

Graph 0.8: No relationship.

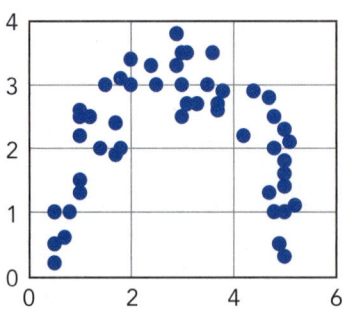

Graph 0.9: Non-linear relationship.

Combined graph

Combined graphs use two different types of graphs on one chart area, such as a bar chart and a line graph. Combined graphs are used when the data sets collected show information that can be related to each other, but have different graphing requirements.

For eample, you could use a combined graph to show climate in the form of twemperature and rainfall. Temperature is continuous data, requiring a line graph, while rainfall is non-continuous, requiring a bar chart to show average rainfall per month.

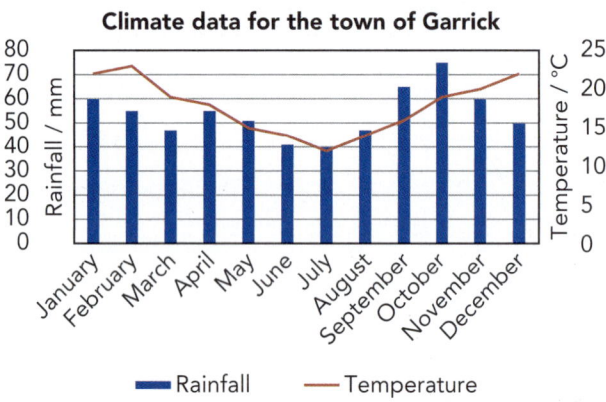

Graph 0.10: A combined graph.

> **TIP**
>
> - Use a bar chart for data which has six or more categories and a pie chart for data with less than six categories.
> - If your data is continuous, use a line graph.
> - To compare values, use a pie chart. For a more precise comparison, use a bar chart.

> **CONTINUED**
>
> - To show trends or patterns in your data, use a line graph, bar chart or scatter graph.
> - If you have large data sets, use the line graphs or scatter graphs, scatter graphs. For parts of a large data set, use bar charts.
> - If you have small data sets, use pie charts or bar charts.

Analysis of graphs and charts

Every graph tells a story. Now that you are familiar with some of the types of graphs, you need to understand how to interpret them.

There are six main steps to interpreting a graph:

1. **Read the basics:** Read the labels and the legend on the graph. What information does the graph give you? For example, Graph 0.6 tells you about the relationship between the height (metres) and diameter (cm) of a specific type of tree.

2. **Read the important numbers:** Important points are peaks, lows, clusters, points where data meet or cross over, anomalies.

3. **Define the trends:** What trends do you notice in the data? For example, in Graph 0.6, you might identify a strong positive relationship in the data, with the diameter of the tree increasing as height increases.

4. **Compare the trends:** Where data sets are plotted together on the same graph, you can compare the information or trends. For example, in Graph 0.10 it is possible to say how rainfall and temperature change and how their patterns are similar or differ.

5. **Analyse the trends:** The data may either prove or disprove a hypothesis, or it may set a new hypothesis. For example, in Graph 0.5 the researcher may have predicted an increase in temperature in July for both sites. In this case, the researcher would have been correct, as both sets of data show a gradual increase over the two weeks.

6. **Draw conclusions:** Use the data in the graph to draw conclusions and use the data from the graph to support your conclusion. For example, looking at Graph 0.3, you could say that, at approximately 55%, the agricultural sector makes up the bulk of the labour for this country. The service industry

makes up the second largest sector, at approximately 30%, with the industrial sector making up approximately 15% of the labour force. You could therefore conclude that labour in this country is predominantly within the farming sector.

Terminology that you can use to describe the change in a graph includes:

- **Going up:** increase, rise, climb, increasing trend
- **Going down:** decrease, drop, decline, fall
- **No change:** remain stable, constant
- **Frequently changing:** fluctuating
- **At the top:** peak, highest point
- **At the bottom:** lowest point.

Make sure you clearly describe what is happening in the graph, and support your description with data from it. If giving the highest temperature from Graph 0.10, for example, you could say:

The highest temperature in the year is in February, which reaches an average of 22°C, while the coolest month is in July at 12°C.

ACTIVITY 0.1

Graphing activity

1   Copy out Table A below. Complete the column with the type of graph or chart you would use to represent the data.

Description of data set	Type of graph used to represent the data set
Small sets of data collected to determine how much water is used during various activities. There are 5 categories: Shower, Toilet, Leaks, Washing clothes, Watering the garden.	
A graph showing the snowfall and temperature for a site in the Swiss Alps.	
A set of data with 12 categories you want to compare.	
A set of data showing the growth of an animal over time.	
A set of data with primary and secondary categories you want to compare.	
Two sets of data you want to analyse for correlations in the data sets.	

Table A: Matching graphs to data sets.

Snack	Monday	Tuesday	Wednesday	Thursday	Friday
Chocolate bar	5	6	4	5	9
Potato crisps	3	2	1	1	5
Fruit	4	5	4	3	6
Sweets	0	1	2	1	4
Nuts	3	1	2	3	2
Cheese	5	4	5	6	7
Popcorn	1	2	1	3	3

Table B: School snack shop sales.

> CAMBRIDGE INTERNATIONAL AS LEVEL ENVIRONMENTAL MANAGEMENT: COURSEBOOK

CONTINUED

2   With reference to Graph 0.2, **suggest** two reasons for why there are fewer animals in reserve 4 than there are in reserve 1.

3   Identify the point in Graph 0.6 that is an **anomaly**. Explain why it is an anomaly.

4   With reference to Graph 0.10, describe the climate in the town of Garrick. Include information which shows the coolest or hottest temperatures and the months that they occur in. **State** when the highest or lowest rainfall is recorded. Good data analysis of graphs always includes data from the graphs.

5   A group of students recorded the number of different types of snacks sold each day over a full week in the school shop. Table B shows the data that they collected.

Using the information about different types of graphs as guidance, decide which type of graph or chart would work best to show the snack shop sales. Create your chosen graph and then answer these questions:

a   Look at your graph. Which snack had the highest sales?

b   Look at your graph again. Which day had the highest sales?

c   Consider how else you could analyse the data to determine highest and lowest sales.

COMMAND WORDS

suggest: apply knowledge and understanding to situations where there are a range of valid responses in order to make proposals / put forward considerations

state: express in clear terms

KEY TERM

anomaly: data that deviates from what is expected

TIP

When drawing a graph, use a sharp pencil and a ruler. Using a pencil means you can easily erase any errors. A ruler will help you to be precise.

0.4 Research skills

Research skills are the ability to search for, locate and evaluate the relevance of information on a particular topic. Well-developed research skills will help you find an answer to a question or a solution to a problem. These skills will also enable you to collect information on a topic, review it and then evaluate and interpret the information.

Environmental management is a subject that lends itself to research, as so much information exists on any given environmental topic. There are six key research skills (Table 0.3).

Key Skills in Environmental Management

1	Task definition	• Identify the information you want to find out. • Develop an outline, including a plan for the question(s) you need to research. • Decide what you want to include in your research to help you identify the type of information you are going to look for.
2	Information-seeking strategies	• Identify potential sources of information on the topic. • Identify potential key words or key terms to use when conducting internet searches.
3	Finding and accessing resources	• Find information sources and locate the relevant information within each resource. • Try different key words and terms if the ones you have previously selected do not give you the information you need. • Evaluate how reliable your source is to ensure you are gathering accurate information. • Verify information from one source by finding a similar argument or set of data in another. • Locate the original information to check its reliability.
4	Using the information	• Extract the relevant information from the resources.
5	Synthesising the information	• Organise the information from the various sources into a structured essay or report. • Decide how the various sources relate to each other and the topic.
6	Evaluating the information	• Evaluate the strengths and weaknesses of the information collected.

Table 0.3: The six key research skills.

ACTIVITY 0.2

Practising your research skills

1. In pairs or small groups plan a vacation. Decide on your destination and use different websites to look for:
   - flights/bus/train/boat routes to get there
   - accommodation at your destination, and different prices available
   - different activities you can do when you get there.

2. Create a brief report comparing the different transport options and prices, the various accommodation options and prices, and any activities that you would be interested in doing. Your research should give you an insight into the wide variety of options available for a holiday.

   The details and number of options that you find will reflect the depth of the research that you carried out.

Figure 0.1: Students carrying out research in a library.

0.5 Evaluative writing

The purpose of a scientific evaluative essay is to present an opinion or a viewpoint on a subject using knowledge to support the argument. With this style of essay, evidence is used to support a conclusion. To write an effective evaluative essay, you need sound knowledge of the different factors that make up the argument. This style of writing requires **critical thinking** from the writer.

The general structure of an essay is as follows.

1. The introduction, which clearly states what you will be evaluating and the **criteria** you will be using to evaluate it.

   The introduction should have a **thesis statement**, which clearly states what your argument is and how you are going to explore it.

2. The body of the text.

   This is where you will present your argument, with evidence and explanations, relating each point back to your thesis.

3. The conclusion and evaluation.

   This should briefly reiterate your thesis statement and the evidence you have provided to support your argument.

Throughout this section of the Coursebook, the following essay question will be used to give supporting examples:

> **Evaluate** the extent to which the rapidly growing population is the main cause of food insecurities in different parts of the world.*

* This essay topic is relevant to Chapter 5 in this Coursebook. When you have covered Chapter 5, you can try writing it.

To answer a question like this, you first need to determine what the question is asking. This will help you to decide what to include in the essay and how you want to structure it. You then need to plan your essay. Table 0.4 gives more detail on what you should do in each of these steps to achieve the best results in your writing.

COMMAND WORD

evaluate: judge or calculate the quality, importance, amount, or value of something

KEY TERMS

*critical thinking: the objective analysis and evaluation of an issue in order to form a judgement

*criteria: a characteristic that is considered important, and by which something may be judged or decided

*thesis statement: a sentence in a paper or essay which introduces the main topic or argument to the reader

Figure 0.2: A learner practising their writing skills.

Planning an evaluative essay

Key points	What to consider or include
The question or statement	Read the question or statement carefully. Make sure you understand exactly what it is asking. Some questions may have more than one part. Make sure you are clear on exactly what you need to answer.
The 'hidden question'	Look for the 'hidden question' within the essay topic. If the topic asks if one thing is the cause, then it is **implying** that something else is not the cause, or is less important. *In this example,* population is being given as the main cause for food insecurity. You need to ask yourself 'in comparison to what?' This is the hidden question, and is essential in structuring your essay.
The **counter-argument** or additional argument	The hidden question will reveal the counter or additional arguments that your essay must explore. You cannot argue for or against a statement or question unless you clarify what you are assessing it against. In this example, the answer to 'in comparison to what?' includes factors such as: • climatic causes (floods, droughts, storms) • pests and diseases • lack of capital investment in agriculture or lack of infrastructure • land degradation • poor water management • war or conflict • poverty. These points are all, arguably, causes of food insecurity.
Planning	• Planning is key to developing a strong evaluative essay. Once you have determined what the question is asking and the factors that you need to discuss, take a few minutes to plan your essay's structure. You could do this in a series of bullet points, a spider diagram, flowchart, or whatever works for you. • Note down any case studies or examples relevant to the topics you want to cover in your essay.

Table 0.4: Key essay-writing points to consider.

KEY TERMS

*imply: to suggest or express something indirectly or without saying it

*counter-argument: an argument that is given in response to another argument, in order to support a viewpoint

The structure of an evaluative essay

Essay structure	
Introduction	**Introductory paragraph:** A clear thesis statement followed by what you will be evaluating and the criteria you will be using to evaluate it.
	An example of an introduction for the question being evaluated here is: Food insecurity is when a population does not have reliable access to sufficient nutritious food that supports an active and productive lifestyle. A variety of factors threaten food supply or production. These include farming failures, interruptions to the food supply chain or instances of the population exceeding the carrying capacity of the region. Such factors are caused by rapid population growth, climate change, pests and diseases, land degradation, pandemics and war. The extent to which rapid population growth is the key factor in food insecurities in various parts of the world will be evaluated here. Ultimately, this essay will suggest that there is more than one factor responsible for food insecurity in different parts of the world.
Body	**Paragraphs:** break the body of the text up into paragraphs to separate your ideas. This engages the reader and helps them to follow your argument. Use linking words or phrases, such as 'however', 'despite this', 'another factor', 'as well as', to connect your paragraphs. This helps your writing flow.
Conclusion and evaluation	Include the following steps in your conclusion: 1 Close the loop: revisit the question and the statement you made in the introduction. 2 Clearly state what your final position is. Do you agree, partially agree or disagree with the question or statement you are evaluating? If you have expressed any doubts in your argument, include them in your conclusion. 3 Clarify: explain how your final position is relevant to the essay question or statement.
	An example of a conclusion and evaluation for this question might be: The examples outlined within this essay show that, even though rapid population growth has played a significant role in putting pressure on food resources, other factors have played a part. Drought and conflict in East Africa have added to pressure in that region, for example. In addition, recent global food insecurities were triggered not by population growth, but by the pandemic and the resulting disruption to food supply chains. Therefore, although our rapidly growing population is a key cause of food insecurity in different parts of the world, it is not the only one.

Table 0.5: Essay stucture.

Paragraph structure

Your essay needs a clear structure, but each paragraph must have its own structure, too. You might be familiar with Point, Evidence, Explanation, and Evaluation paragraphs from English lessons (Table 0.6). Use this technique in every piece of academic writing you do. This will ensure that each of your paragraphs has a key point relevant to your thesis statement.

Key Skills in Environmental Management

Point, Evidence, Explanation and Evaluation paragraph structure (PEEE)			
Point	Evidence	Explanation	Evaluation
1 Assess the student lunch services at your school.			
e.g. food quality	Wednesday's great burger lunch	Explain how the lunch on Wednesday was better than usual.	This positively impacts your view on school lunches because it was a good meal.
e.g. temperature of the food	Long waiting times while students move through the lines	Explain how long waiting times and lack of heating for the food means it is cold when you get to eat it.	This negatively impacts your view on food services at school because you like warm food.
2 Using the example essay question: Evaluate the extent to which the rapidly growing population is the main cause of food insecurity in different parts of the world.			
e.g. drought	Prolonged drought in the Sahel (south of the Sahara desert)	Explain how the drought is causing food shortages in the region.	Drought is a significant contributor to food shortages in the Sahel region.
e.g. war	An example of an area experiencing unrest	Explain how war impacts food insecurity in the region (farmers displaced, farming decreases).	War is a significant contributor to food insecurity in the region discussed.

Table 0.6: PEEE structure.

Another factor to consider when writing an evaluative essay is the type of language you use and how you express yourself. Table 0.7 provides guidance on what to avoid when writing evaluative essays.

Good practice in academic writing	
Evaluative essay writing in science is formal, objective (impersonal) and technical. Avoid using casual conversation or language.	Use the term 'approximately' instead of 'just about'. Use 'insufficient' instead of 'not quite enough'.
Develop your terminology and use it correctly in your writing.	**Try:** The precipitation was intercepted by the vegetation. **Avoid:** The trees blocked the rain.
Avoid direct reference to people or feelings. Focus on facts, ideas and objects.	**Try:** Food shortages resulted in the malnutrition and ultimate starvation of the villagers. **Avoid:** Food shortages caused the tragic death of the man and his family in the village.
Avoid contractions	**Try:** Did not **Avoid:** Didn't

Good practice in academic writing	
Avoid evaluative words that are based on non-technical feelings.	**Try:** The data shows a significant increase in temperature in the volcano. **Avoid:** The data shows an amazing change in the heat of the volcano.
Avoid using first person pronouns such as I, me and my, except if it is necessary for your thesis statement. Use the third person point of view throughout the essay.	**Try:** These findings indicate that the scientist's prediction was correct. **Avoid:** I believe that the data they collected shows that prediction was correct.
Vary the length of your sentences. Shorter sentences are easier to understand. However, they can make your writing seem choppy. Varying the length of your sentences engages the reader. • Use shorter sentences to emphasise important points. • Use longer sentences to add rhythm to your writing, and convey complex points. • Check your grammar, spelling and punctuation to ensure that your essay is fluent and clear. • Use appropriate words and phrases to link your paragraphs. Examples include: 'Although', 'Because', 'Despite', 'However', 'Another factor', 'Regardless of'.	Short sentence: Snacks are popular at school. Long sentence: Every student in the survey reported buying at least one snack a week, whereas 41% of respondents reported buying one snack at least every day, citing the excellent selection of snacks the school had to offer as their main reason for doing this. Linking words in action: 'Because of food shortages . . .' 'Although Japan has an ageing population . . .' 'Another factor in the population growth in . . .' 'Despite growing concerns about the ageing population . . .'

Table 0.7: Good practice tips for academic writing.

> **TIP**
>
> **Check your sentence length**
>
> Read your work aloud to yourself. This will help you notice where your essay needs work.
>
> - If you find yourself out of breath, you may need more punctuation or to shorten your sentences.
> - If you stumble over your words, you may need to restructure your sentences so they flow better.
> - If a sentence or paragraph does not make sense, see how you could make yourself clearer.
>
> Take this approach with different types of writing you do. You will start to notice patterns in your work, and this will help you to identify areas for improvement.

Figure 0.3: Taking it in turns to read out your writing in a group is a good way to get feedback and improve.

Key Skills in Environmental Management

ACTIVITY 0.3

Writing clearly

Read the example sentences A to E:

A Phones are used by everyone. Phones are good because they mean you can talk to people this proves that technology is a force for good.

B 96% of people enjoy going to the beach. Meaning 4% of people don't like it. This shows that most people prefer the beach to snowy holidays.

C I think the most important reason for poverty is no jobs.

D Many people think climate change isn't real. Why? Well, some people point out that winters are still cold, using observation to make a sweeping statement.

E Many species are endangered which is a bad thing because they have a special job in the ecosystem which means that when they aren't there any more there's a hole in the ecosystem which has a bad impact on other animals living in the same area such as no wolves meaning there is no predator to eat deer which means there are too many deer. Which means that they eat more grass and there isn't enough for other animals that eat grass so they die.

Tasks

1 Give one strength and one weakness for each sentence. Consider: language, structure, evidence, knowledge, logic, and clarity.

2 Rewrite the sentences for clarity, or to make them more suitable for an academic essay. Then take one sentence and develop it into a paragraph.

PEER ASSESSMENT

Swap your amended sentences from Activity 0.3 with a classmate.

Rank each of your classmate's amended sentences between 1 and 3:

1 This sentence does not make sense or is not appropriate for an academic essay.

2 This sentence is OK, but could be improved.

3 This sentence makes sense and is appropriate for an academic essay.

If you have given your partner a score of 1 or 2 for any of their sentences, suggest 3 ways to help them improve.

What to avoid

- Only using descriptive writing

  Descriptive writing provides information about 'who', 'what', 'where' and 'when', but it does not include insight, analysis or consequence. Evaluative writing seeks to answer the questions of 'why', 'so what' and 'what next'. Expand your ideas to say why they are important.

- Going off topic

  Refer to your plan and the question as you write the essay. This ensures that you are answering the question and have not gone off topic.

- Repeating yourself

  Do not try to lengthen your essay by repeating something you have already said.

- Plagiarism

  Plagiarism is when work of another writer is copied and presented as your own. Make sure the essay is in your own words.

- Not having a thesis statement

  Make sure you have an argument. Do not just describe and explain factors. You should have a stance on the topic: your essay should be an argument with a purpose, e.g. 'This essay will argue that cats are the best pets' is stronger than 'Some people think cats are the best pets, whereas others think fish are the best pets. I will explore both sides.'

- Giving reasons or examples in the introduction

  Your introduction should not be a mini essay. It should contain your thesis statement, and show how you will explore your thesis. It is a statement of intent.

- Asking rhetorical questions

  In English academic style, very few **rhetorical questions** are used. This is appropriate in other styles of writing, but you should write with authority in academic writing, not persuasively.

- Not explaining how data or facts support the thesis

  Be explicit. Link your point to your thesis statement and do not assume the reader will make that jump for you. Ask yourself, 'So what?' when you make a point. For example, if you insert a graph, don't simply say 'the graph shows X'. Ask yourself, 'So what? What does X tell us?'

- Not using facts, evidence or data and making leaps in logic

  Make sure that you do not make claims without evidence. Ask yourself, 'How do I know this?' when making a point. If you have no evidence, you should not make the point. Similarly, avoid making leaps in logic: the fact that cats are the most popular pet does not mean they are the best pet. Prove that there is an undeniable link between the two concepts, or do not make the leap.

- Bringing new evidence into the conclusion and making the conclusion too long

  A conclusion should *concisely* summarise the evidence you have presented. Provide a clear answer to the question and briefly reiterate why you have given the answer you have. Do not introduce new information.

- The body paragraphs are not connected through use of transition words and phrases

  Use phrases such as 'However', 'Despite this', 'As well as', 'Another factor', 'Although', to link your paragraphs. This makes the essay feel well-connected and coherent.

> ### KEY TERM
>
> ***rhetorical question:** a question that is asked in order to create a dramatic effect, or to make a point, rather than to get an answer

> ### ACTIVITY 0.4
>
> **Evaluative essay and research activity**
>
> 1. Write an evaluative essay of around 600 words on one of the following topics:
>   - Compare iOS and Android: which one is better and why or for whom?
>   - The challenges of living in a country with a different culture.
>   - The best animal to have as a pet.
>
> 2. Carry out research to develop your argument. Take time to plan and develop a well-structured essay. Think of your paragraph and sentence structure when working through the task.
>
> 3. Once you have written your essay, complete a self-evaluation, or swap your essay with another learner to give and receive constructive feedback on how to improve essay structure and evaluative skills.
>
>
>
> **Figure 0.4:** You could write an evaluative essay on why cats are the perfect pet.

Key Skills in Environmental Management

EVALUATIVE ESSAY WRITING CHECKLIST

Use this checklist to help remind you how to structure your essay and rate how confident you are with each aspect.

Essay checklist	Confidently completed	Getting there	Needs more work	Forgot to do this
I made a brief plan first.				
My introduction includes what I was evaluating and the criteria.				
I identified the 'hidden question'.				
I clearly stated the counter or alternative arguments.				
I developed a paragraph for each counter argument with a brief evaluation of the importance of each factor in relation to the question.				
I used examples to support the evaluation.				
I stayed on topic throughout the essay.				
My conclusion clearly evaluates the validity of the statement or question being answered.				
The outcome of the evaluation is well supported by the information in the essay.				
I used formal language and relevant terminology.				

REFLECTION

Do you find evaluative essays challenging? What do you struggle with the most?
How do you think you can improve?

SUMMARY

Different types of graphs are used for different types of data. The type of graph used depends on what the researcher wants to show.
Graphs must be clearly drawn and include all the relevant labels.
Analysis of graphs must include supporting data, indicate or compare trends, and draw conclusions.
When conducting research, you need to define your task. This will enable you to use the correct search terms to find the information you need.
Research skills include logical use of the material sourced, and an evaluation of the information within it.
Academic writing should always contain a thesis statement.
You must use data and evidence in academic writing.
Academic writing should follow a clear structure overall, and at the level of individual paragraphs.
Academic writing requires a strong conclusion which links back to the question and the thesis.

SELF-EVALUATION CHECKLIST

After studying this chapter, think about how confident you are with the different topics.
This will help you to see any gaps in your knowledge and help you to learn more effectively.

I can	Needs more work	Getting there	Confident to move on	See Section
Use appropriate graphs and charts to present data.				0.1
Employ relevant research skills.				0.2
Practise good academic writing.				0.3
Structure an evaluative essay				0.3

Chapter 1
Introduction to Environmental Management

LEARNING INTENTIONS

In this chapter you will:

- learn about the world's continents and major oceans
- identify the income groups that the World Bank uses to classify countries
- define and understand the term sustainable management
- describe the water cycle, and draw and interpret diagrams of the water cycle
- state the major components of the atmosphere and the natural greenhouse effect
- define an ecosystem, and describe the interaction of biotic and abiotic components along with their related energy flows and terminology
- describe the carbon cycle, and draw and interpret diagrams of the carbon cycle

CAMBRIDGE INTERNATIONAL AS LEVEL ENVIRONMENTAL MANAGEMENT: COURSEBOOK

GETTING STARTED

1 Work with a partner to answer the following questions. **Discuss** your answers with the class.

   a Consider the economies of different countries. Discuss how the quality of living varies between them. Give reasons for why some countries are more developed than others.

   b How does temperature change as you get further away from Earth's surface?

   c **Define** the meaning of sustainability. Give two examples of resources that could be used sustainably.

   d Draw the water cycle using as many of the correct terms as possible (you should have at least six key terms).

2 In small groups discuss:

   a three changes you could make that would make your classroom or school more sustainable

   b how those changes support sustainability.

COMMAND WORDS

discuss: write about issue(s) or topic(s) in depth in a structured way

define: give precise meaning

ENVIRONMENTAL MANAGEMENT IN CONTEXT

Species distribution: how does environment affect species behaviour and adaptation?

The **environment** on planet Earth is perfectly set up to support life. Rather than individual sections making up a whole, it is a single system made up of interacting parts. Changes to one part may result in changes to another.

Organisms live in a wide variety of habitats on Earth. Some organisms, such as the Namibian desert elephant, live in habitats that do not seem to suit their needs, but they have adapted and evolved in order to survive.

The Namibian desert elephant is an example of a **species** that has adapted to live in an unusual habitat. The dry, barren landscape of a desert, with little water or vegetation and high day-time temperatures, is an extreme environment. This would present challenges for the smallest of organisms let alone a large mammal like an elephant.

Figure 1.1: Namibian desert elephants have adapted to their habitat by remembering where to find food and water.

1 Introduction to Environmental Management

CONTINUED

For years, scientists thought that desert-dwelling elephants were a separate species from the other elephants in Southern Africa. However, there is no genetic difference between desert-dwelling elephants and the ones found in savanna grassland ranges. The differences between the two types of elephants come down to the adaptations made by the desert-dwellers.

Physically, desert-adapted elephants have bigger feet than their grassland-dwelling counterparts. Their greater surface area helps them move over thick sand as it prevents them from sinking. They have adapted to the lack of food in the desert by having a smaller body mass, giving them the appearance of longer legs. This makes them seem taller, although they are not. The desert elephant can tolerate longer periods without drinking water, giving it time to move between the water sources in the desert, which are great distances apart. It travels at night time, and has learned where it can locate buried water holes from memory. It then uses its trunk, tusks and legs to dig wells in dry river beds. The desert elephant digs for water even when there is surface water. Due to the limited **carrying capacity** of the desert, desert herds of elephants are generally smaller than savanna herds.

Discussion questions

1. Why do mother elephants repeatedly take their young to specific watering holes in the desert?
2. Why do you think that desert elephants would rather dig for water than drink surface water?
3. How have desert elephants adapted to suit the climate they live in? Compare with savanna elephants.

KEY TERMS

environment: the surroundings or habitat in which an organism lives

species: a group of living organisms made of up individuals that can produce fertile offspring when they reproduce

*carrying capacity: the number of a species which a region can support without environmental degradation

1.1 Introduction

Environmental management is not an easily defined term, as it has such a wide scope. Put simply, it is the management of the impact of human activity on the natural environment. Environmental management aims to ensure that ecosystem services and biodiversity are protected and maintained sustainably. Ecosystems are considered, along with ethical, economic and ecological variables.

Environmental management also attempts to identify the issues that arise between meeting the needs of people while protecting the needs of the environment.

The term environmental management can refer to a goal, philosophy, policy or vision. Environmental managers fill a wide range of positions. They include academics, policy-makers, company employees, non-governmental organisations (NGOs) and even individuals trying to influence how we interact with the planet. However, complex issues need more than a single individual's input. This is why different groups of people focus on concerns such as resource exploitation or pollution, bringing a whole range of specialist skills and knowledge to the table.

Environmental management requires the ability to think critically and to consider impacts over time, at both a local and a global scale.

To understand the environment, you need to understand Earth itself and how humans interact with the planet. It is important to consider a range of factors such as water, light, soils, and living organisms. Only then will you start to develop an understanding of the complex function of Earth and the life forms on it.

Continents and oceans

There are seven recognised **continents** on Earth: North America, South America, Europe, Asia, Africa, Oceania and Antarctica.

Water covers approximately 70% of Earth's crust, forming oceans and seas. There are five oceans (Figure 1.2):

1   the Pacific Ocean
2   the Atlantic Ocean
3   the Indian Ocean
4   the Arctic Ocean
5   the Southern Ocean.

There are many seas on the planet, too. For example, the Mediterranean Sea, the Black Sea (Figure 1.3), the Dead Sea and the Caribbean Sea. These are smaller bodies of water than the oceans, and are typically partially enclosed by land.

ACTIVITY 1.1

Fact finder: the oceans

1   Work with a partner. Select two oceans and find out some facts about them. Then **compare** the information about the two oceans. Look for similarities and differences, and present the information in a table.

    Examples of facts to look up might include: location, size, deepest point, temperature, salinity, widest point, interesting information.

2   Discuss your findings with the class to see how different oceans compare.

KEY TERM

continents: the main continuous expanses of land found on Earth (Europe, Asia, Africa, North and South America, Oceania, Antarctica)

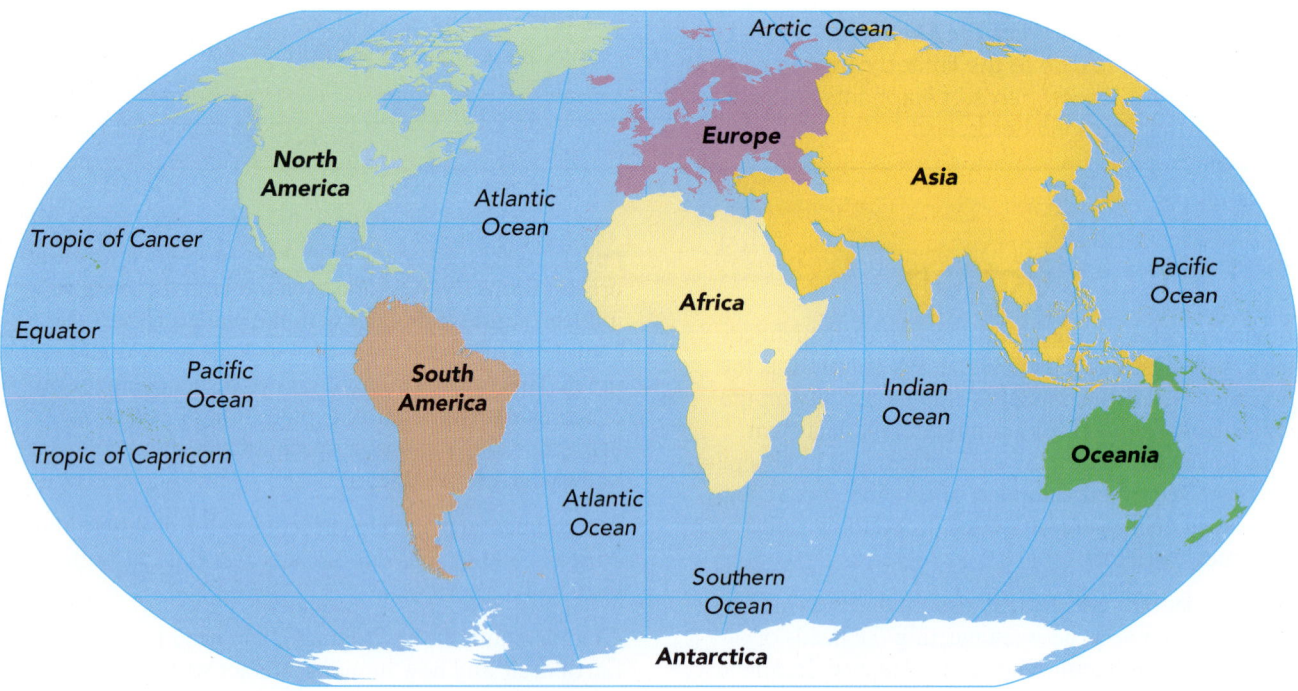

Figure 1.2: A map of the world showing the seven continents and five oceans.

22

1 Introduction to Environmental Management

Figure 1.3: The coastline of the Black Sea, near Cayyaka village, Türkiye.

There are a number of reasons why some countries are LICs and others are HICs. These include factors such as climate, resource availability, the frequency of natural disasters and social factors. We will explore these in Chapter 3.

Figure 1.4: Very dry desert climates like the Ennedi Mountain region in the Sahara desert make it difficult for any population to develop because of the lack of water.

> **COMMAND WORD**
>
> **compare:** identify/comment on similarities and/or differences

Classifying countries

The World Bank classifies countries according to their level of development, including economic, social, cultural and technological development. Countries fall into three main categories, defined in 2022 as:

1. **Low-income countries (LICs)** have a **gross national income (GNI)** per capita (per person) of US$1086 or less (e.g. Malawi and Nepal).
2. **Middle-income countries (MICs)** have a GNI per capita of more than US$1086 but less than US$13 205 (e.g. South Africa and Vietnam).
3. **High-income countries (HICs)** have a GNI per capita of more than US$13 205 (e.g. Switzerland and Germany).

LICs have weaker economies. As a result, portions of the population may live in extreme poverty, with lower levels of education and poorer standards of living.

HICs have stronger economies, with higher levels of education, higher standards of living and lower levels of extreme poverty.

> **KEY TERMS**
>
> **Low-income countries (LICs):** countries that have the weakest economies and are least developed. The category is determined by the low GNI per capita of US$1086 or less
>
> ***Gross national income (GNI):** the total amount of money earned by a nation's people and businesses. This is used to measure a nation's wealth
>
> **Middle-income countries (MICs):** countries that have started to develop, with growing industry and GNI per capita increasing (more than US$1086 but less than US$13 205)
>
> **High-income countries (HICs):** countries that have strong, well-developed economies and a good standard of living, where the GNI per capita is more than US$13 205

ACTIVITY 1.2

Fact finder: level of development

1. Research three countries, including one LIC and one HIC. Create a table using the example below to compare the differences and/or similarities between them.

   Include factors such as: GNI per capita, types of economic activity taking place, climate, political stability (war or peace), natural resources available, level of education, birth and death rates.

Factors	Country 1 *Country name	Country 2 *Country name	Country 3 *Country name
e.g. GNI per capita			
Climate			

2. Did any of the information surprise you?

Sustainability

The practice of **sustainability** prevents the depletion or degradation of Earth's natural resources, thus ensuring that the environment and the needs of future generations are protected. Sustainability is about making changes to look after the planet and to keep Earth's **ecosystems** healthy. These changes protect the environment, our natural resources (such as water, air, soil, oil, coal, natural gas, rocks and minerals) and society.

In the past, global resources have been used in an unsustainable way. For example, over-hunting has resulted in the almost total loss or disappearance of some species, such as the American Bison (Figure 1.5).

Using rivers and oceans to dispose of rubbish and waste has resulted in the **pollution** of both freshwater resources and coastal areas that provide much of the food that humans consume. Emissions from industry and power stations have polluted the environment, and in some cases, the pollution makes the air harmful to breathe. Another example of unsustainable practices is the rapid felling of large areas of forest, known as deforestation. Forests help to clean water and air and provide habitat the for many species. The replanting of many deforested areas is an attempt to reverse the damage (Figure 1.6).

Figure 1.5: The North American Bison is a species that used to roam the plains in large herds. Many of these herds were hunted to near-extinction.

Figure 1.6: Trees planted by the Australian Rainforest Foundation, an organisation committed to helping the rainforest recover from the unsustainable felling that has affected much of the forest cover.

Sustainably managing resources forms an important part of the sustainability model. This means managing our resources in a way that ensures their sources do not run out. For example, using coal and oil in an unsustainable way will result in the supply running out before they re-form, a process which takes approximately 300 million years. Instead, alternative forms of energy can be developed, such as wind and solar power, two sources of energy that are continually generated. Another example of sustainable resource management is forest management. Trees can be harvested and used as long as a replanting plan is put in place, and the rate

1 Introduction to Environmental Management

of tree harvesting is managed. This way, forests can supply wood while still being protected.

Sustainability is important for both humans and the planet. Without pollution-free air and water, or available food, both humans and ecosystems suffer. With continued over-consumption, humans may cause decreased **biodiversity**, mass species extinction and the decline of the current human population due to lack of food, clean water and clean air.

KEY TERMS

sustainability: the ability to meet the needs of the present without compromising the ability of future generations to meet their own needs

ecosystem: a biological community of organisms interacting with each other and the physical environment

pollution: the presence or introduction into the environment of a substance which is harmful or has poisonous effects, for example polluted water is harmful to drink

biodiversity: the number of different living organisms found within an ecosystem or region

ACTIVITY 1.3

Fact finder: local sustainability

Research an example of a resource that is not being managed sustainably near you. Write a brief summary of the situation (100–200 words) and include the following factors.

a   **Explain** why you do not think that this resource is being used sustainably.

b   **Outline** what could be done to solve or reduce the problem.

1.1 Questions

1   **Identify** the seven continents and five oceans.
2   **Define** the term 'sustainable'.
3   **Explain** why sustainability is important.
4   In approximately 200 words, **describe** the underlying causes for some countries being LICs while others are MICs or HICs.

COMMAND WORDS

explain: set out purposes or reasons / make the relationships between things evident / provide why and/or how and support with relevant evidence

outline: set out main points

identify: name/select/recognise

define: give precise meaning

describe: state the points of a topic / give characteristics and main features

give: produce an answer from a given source or recall/memory

1.2 The water cycle

The **water cycle** (Figure 1.7) is the continuous movement of water on, above and below the surface of Earth. Locally it is an **open system**, which means that the amount of water in a specific ecosystem can fluctuate as water can move both out of and into that ecosystem. However, globally, it is a **closed system**: there is a constant amount of water on the planet. Although the amount of water on Earth remains constant, the relative amount of water in its major reserves can vary over time. Major reserves are large stores of water and include ice stores, fresh water, sea water, ground water and atmospheric water.

KEY TERMS

water cycle: the process in which water moves from the sea, into the atmosphere, onto and into land and back into the sea

*open system: a system in which material can either be lost or gained

*closed system: a system in which material is neither being created or lost

25

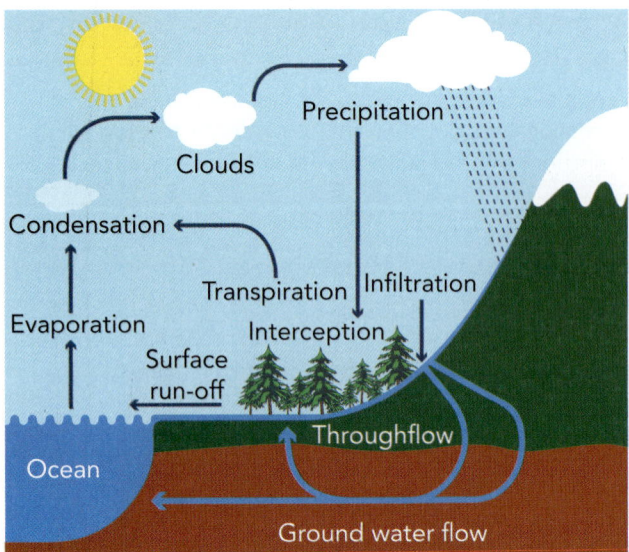

Figure 1.7: The water cycle.

Phases of the water cycle

Evaporation

During the **evaporation** phase of the water cycle, water changes from the liquid to the gaseous phase through the uptake of energy, cooling the environment. This process is driven by energy from the sun, which heats up surface water. As the water evaporates, any impurities such as salts, dust and pollution are left behind, to give purified water in the form of water vapour. This water returns to Earth later in the cycle as **precipitation** so the water stores are not reduced: the water just moves between phases of the water cycle. During this phase, fresh water reserves are replenished with the purified water.

Transpiration

Transpiration (Figure 1.8) is the process by which water vapour moves through and out of a plant (mainly through the leaves, but in some plants through the stem as well) before being released into the atmosphere.

Atmospheric conditions affect rates of evaporation and transpiration. For example, when **humidity** is high, transpiration slows down, as the moisture gradient between the water source (i.e. the plant) and the air decreases. Moisture moves from an area of high moisture to an area of low moisture more rapidly, so where this gradient is reduced the rate of evaporation slows down. On a windy day, evaporation, transpiration and evapotranspiration rates all increase as the wind is

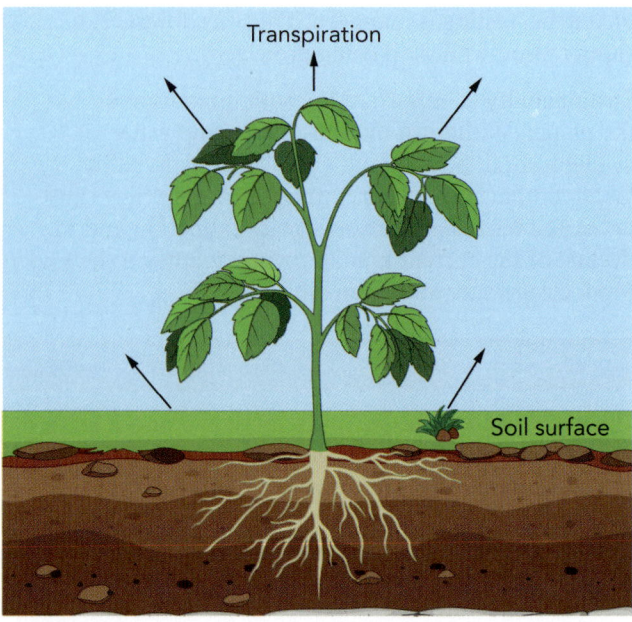

Figure 1.8: Transpiration is the process of water moving from the soil and up through the plant where it exits through pores in the leaves and/or stem.

constantly moving the moisture away from the surface of the water source (e.g. plant surface). As a result, the moisture gradient increases.

Condensation

During the **condensation** phase of the water cycle, water changes from gas to liquid through the release of energy, warming the environment. This stage occurs when warm moisture-laden air meets colder air (or rises to higher, colder altitudes). When condensation occurs, the air becomes saturated and clouds form.

> **KEY TERMS**
>
> **evaporation:** the process by which liquid turns to gas
>
> **precipitation:** water that falls to the ground as rain, snow, hail and sleet
>
> **transpiration:** water lost through the leaves of plants
>
> **humidity:** the percentage of water vapour in the air
>
> **condensation:** the process by which a gas changes into a liquid due to cooling

1 Introduction to Environmental Management

INVESTIGATIVE SKILLS 1.1

Testing for transpiration

This experiment shows that water is lost through plants during the process of transpiration. It also demonstrates how important plants are in the water cycle.

You will need:
- A small potted plant.
- A clear plastic bag, large enough to fit completely around the plant.
- A small watering can.
- Masking tape.

Getting started
- Think about how damage to the plant may impact your results. Take care when working with the plant to ensure you do not harm it.

Method
1. Place the plastic bag over the plant, and tape it securely around the base of the plant (Figure 1.9).
2. Water the plant, making sure the soil is damp.
3. Place the plant in the sun, or under a heat lamp.
4. Create a hypothesis linked to transpiration for the following question: What do you think will happen inside the plastic bag when the plant is put in the sun?

Observations
- What do you note happening over time?
- Why do you think this is happening?
- As a class, discuss what you see.

Questions
1. How might observations change with different plants, for example, a cactus and a plant that normally grows in the shade?
2. As a class, discuss whether the leaves inside the bag belong to an open or closed water system. Compare that to the whole of the potted plant – is it an open or closed water cycle?

Note: it is also possible to carry out this experiment on a tree or bush outside by taping a bag round the end of a branch and observing the change.

Tape seals plastic bag around the base of the plant stem.

Figure 1.9: A potted plant with a plastic bag placed over it and secured to the stem to capture any moisture lost through the leaves.

SELF ASSESSMENT

- Did you take a step by step approach to carrying out this investigation? Why is such an approach important?
- How could you change your approach to get a better result?
- Did another team get results that were easier to observe? Discuss their approach to see what you can learn from them.

Precipitation

Precipitation is when condensed water in clouds falls back to the surface of Earth as rain, snow, hail, sleet or fog. The amount of precipitation varies around the planet due to many factors. These include rates of incoming solar radiation, global wind patterns, ocean temperatures and the distribution of land and oceans on the planet.

Different areas around the planet get different quantities and types of precipitation. For example, areas of desert (cold or hot) get less than 250 mm of precipitation a year, while tropical rainforests can get in excess of 2000 mm per year. In contrast to these two areas, precipitation in the polar regions and at high altitudes is in the form of snow rather than rain.

ACTIVITY 1.4

Precipitation around the world

1   Different parts of the world have different amounts and types of precipitation annually.

   a   Investigate the annual rates of precipitation at the following locations and create a table to record your findings:
   - the village of Cherrapunji in India
   - the Atacama Desert
   - Antarctica
   - the Australian Outback
   - Panama City.

   b   Did any of the amounts of precipitation surprise you?

   c   Was the precipitation that you found for Antarctica higher or lower than you expected?

   d   Would you classify Antarctica as a desert?

Interception

Interception (Figure 1.10) describes the process of precipitation being blocked from reaching the ground by plant foliage/canopy, or even leaf litter. The water eventually evaporates back into the atmosphere without being absorbed by the soil.

Through-fall and stem-flow

Through-fall (Figure 1.10) describes the process of those raindrops that are intercepted by plants collecting on the leaves and falling through the foliage and onto the soil below. This differs from **stem-flow** which is when the water trickles along the branches and the trunk of the tree to the ground.

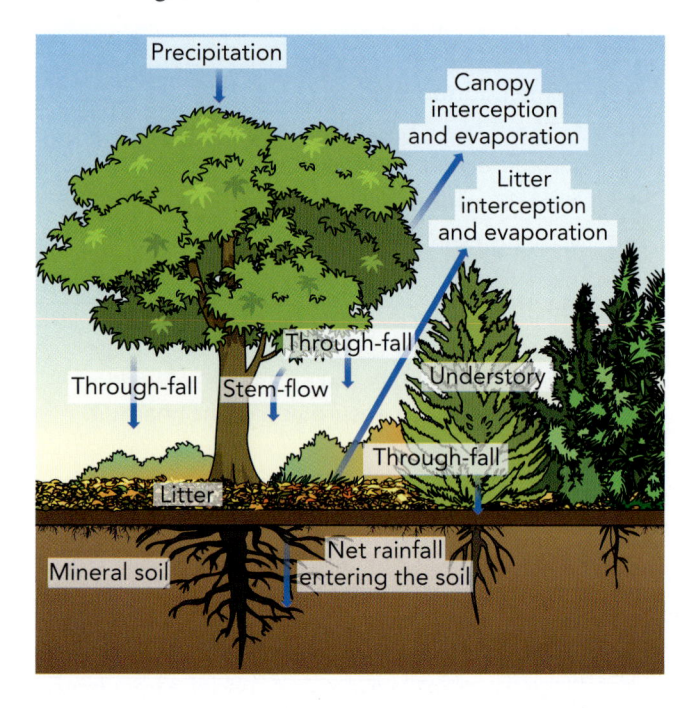

Figure 1.10: Typical pathways that precipitation can follow in a forest.

Runoff or surface water flow

Runoff, or surface water flow, is when the water from precipitation collects on the surface of the land and flows over the ground. Runoff occurs when precipitation exceeds the rate of infiltration or when the soil is saturated, compacted or when water cannot pass through it.

KEY TERMS

interception: the blocking of rainfall by vegetation, preventing it from reaching the ground

*through-fall: rain that falls through the leaves and branches of plants

*stem-flow: rainfall that reaches the ground in a forest by draining down the trunks of plants

runoff: the draining away of water as overland flow

Runoff can enter lakes and rivers replenishing surface fresh water stores. It can also infiltrate into the soil and recharge the groundwater stores and **aquifers**.

Infiltration

Infiltration is the process during which water is absorbed by the soil. The rate of infiltration is the maximum amount of water that can infiltrate into the soil in any given conditions. When soils are already saturated after a long period of rain, the rate of infiltration decreases and surface runoff increases. This is because the pores in the soils are already filled with water. The existing soil moisture plays a significant role in flooding during periods of prolonged rainfall as it reduces the amount of water than can be absorbed by the soil.

Where soil is compact or has a high clay content (low permeability), infiltration rates are low. In sandy or well aerated soils, infiltration rates are high. This is because water is able to move into the soil relatively easily (high permeability).

Through-flow

Through-flow refers to water that is flowing through the soil. This water moves laterally through the soil and can re-enter surface water as base flow (water that enters from ground water into the bed of a river) or as surface springs. Through-flow normally occurs when the soil is saturated.

> **TIP**
>
> Do not mix up *through-flow*, which is water that is flowing through the soil, and *through-fall*, which is water that falls through the leaves and branches of plant.

Groundwater

Groundwater is water below Earth's surface, usually within a few hundred metres. However, it can be found up to 4 km deep. **Groundwater flow** is the water moving though the ground (usually slower and deeper than through-flow). It may eventually recharge surface water, or may move from one aquifer to another. For example, an aquifer where water is being **abstracted** becomes depleted, so the water from a neighbouring aquifer can move laterally from the one aquifer to the other.

> **ACTIVITY 1.5**
>
> **The water cycle**
>
> **Sketch** a simplified diagram of the water cycle and label it with the correct terms. Include definitions for each term that you use.
>
> Make sure to use clear lines and labels when completing a scientific drawing.

1.2 Questions

1. Define the terms 'infiltration', 'impermeable rock' and 'through-fall'.
2. Briefly describe the water cycle, including the correct terminology.
3. Compare and **contrast** open and closed water systems.
4. Describe how high antecedent soil moisture (saturated soil) can make a flood event worse.

> **COMMAND WORDS**
>
> **sketch:** make a simple freehand drawing showing the key features, taking care over proportions
>
> **contrast:** identify/commment on differences

> **KEY TERMS**
>
> **aquifer:** an underground layer of permeable rock in which water is stored in the rock pores
>
> **infiltration:** the movement of water into the soil from the surface
>
> **through-flow:** the lateral transfer of water downslope through the soil
>
> **groundwater:** the water found underground in cracks and spaces in the soil, sand and rocks
>
> **groundwater flow:** water which flows under the ground until it reaches the surface, often through boreholes or wells
>
> ***abstraction:** the process of taking water from a ground water source

1.3 The structure and composition of the atmosphere

The **atmosphere** is the layer of gases, vapour and dust particles which envelop Earth. The lower layer of the atmosphere varies in thickness from approximately 6 km at the poles to approximately 18 km at the equator (Figure 1.11). The atmosphere is similar to the skin around a fruit because it protects the planet.

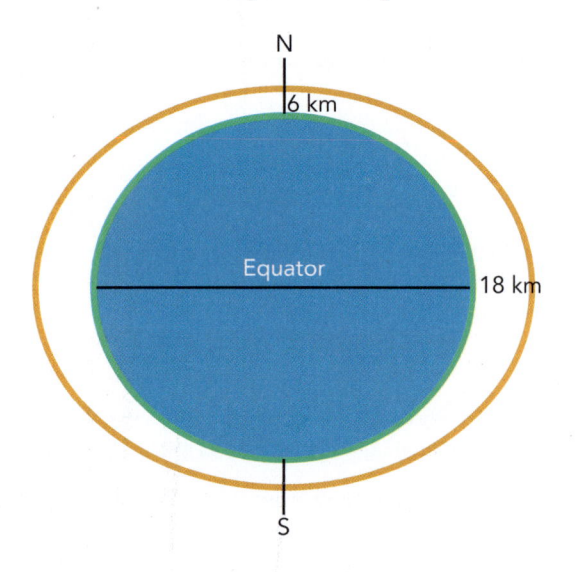

Figure 1.11: The atmosphere around the planet.

Three major components make up Earth's atmosphere: nitrogen (78%), oxygen (21%) and argon (0.93%). The remainder is carbon dioxide (0.03%) and small amounts of other gases. Water vapour is found in the lowest layer of the atmosphere and makes up an average of 1% at sea level and an overall average of 0.4%.

Water vapour and carbon dioxide (CO_2) are **variable gases**. This is because they vary either spatially or over time. Despite making up a small percentage of the gases in the atmosphere, they are important because of their ability to absorb heat. Currently the amount of CO_2 is increasing due to industrial activities that involve the burning of fossil fuels, which are high in carbon. An increase in these gases results in more heat being absorbed and held in the atmosphere.

The molecules that make up the atmosphere are found in greater density near Earth's surface due to the pull of gravity. As one moves away from Earth's surface, the air thins rapidly and the concentration of gases decreases, although the proportions of the gases stay the same.

> **KEY TERMS**
>
> **atmosphere:** the envelope of gases, vapour and dust, that surround Earth
>
> ***variable gases:** the concentration of the gas can differ either spatially or over time

The structure of the atmosphere

The atmosphere contains four distinct layers (Figure 1.12). These are divided vertically based on temperature. The layers vary in thickness depending on where you are on the planet. Higher amounts of solar radiation at the equator result in the atmosphere being thicker there than it is at the poles. The four layers also vary in temperature, as shown in Figure 1.12.

> **ACTIVITY 1.6**
>
> **Atmospheric gases**
>
> Draw a pie chart representing the proportions of the main gases in the atmosphere.
>
> Include the following gases: nitrogen (78%), oxygen (21%), argon (0.93%), carbon dioxide (0.03%) and water vapour (0.4%).

1 Introduction to Environmental Management

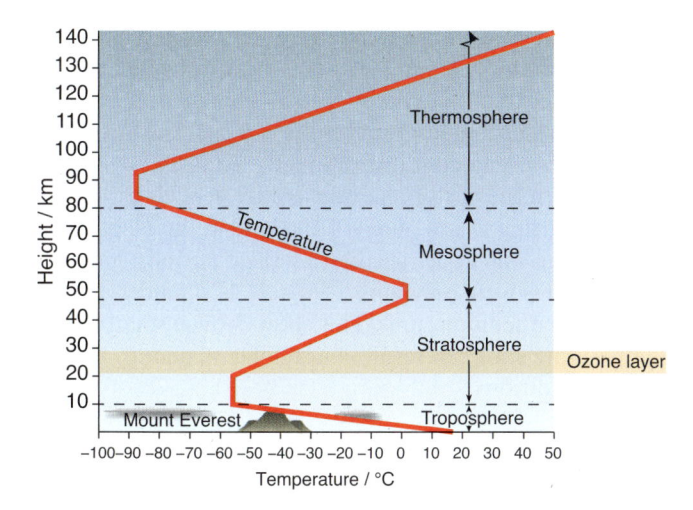

Figure 1.12: The structure of Earth's atmosphere, showing the change in temperature with altitude.

The ozone layer is an important characteristic of the stratosphere as it protects Earth from most of the incoming UV radiation, which is harmful to life. Plants and phytoplankton are unable to survive high levels of UV radiation. Without plants and **phytoplankton**, life on Earth would be very different as the food source that nearly all organisms rely on would not exist. The ozone layer also protects humans and animals, reducing the risk of skin cancers, **cataracts** and damage to immune systems.

> **KEY TERMS**
>
> **troposphere:** the lowest zone of the atmosphere that extends from Earth surface to a height of approximately 10 km
>
> **stratosphere:** the zone of the atmosphere above the troposphere where the ozone layer is located
>
> **ozone layer:** a layer of the stratosphere rich in ozone (O_3) molecules, which absorb much of the incoming UV radiation
>
> **mesosphere:** the zone of the atmosphere above the stratosphere
>
> ***phytoplankton:** algae found in the upper parts of the ocean, the algae photosynthesise, capturing energy from the sun to live and grow
>
> **cataracts:** a cloudiness of the lens of the eye which makes vision blurry

The troposphere

The **troposphere** is the layer of the atmosphere found closest to Earth's surface. It is often referred to as the 'weather zone' because this is where clouds and storms form.

The air in the troposphere is very well mixed. It is heated from the ground by longwave radiation so it is warmest near Earth's surface and cools with increased altitude.

The temperature in the troposphere drops by approximately 6.5 °C with an increase in altitude of 1 km, and drops to approximately −50 °C at the top (6–18 km from Earth surface) of the troposphere.

The stratosphere

The **stratosphere** is above the troposphere. In this zone the temperature increases from −50 °C (6–20 km above Earth's surface) to approximately −10 °C. This is due to the **ozone layer** which is approximately 15–40 km above Earth surface.

The ozone molecules absorb ultraviolet (UV) radiation (shortwave radiation) which heats up the air around them, resulting in an increase in temperature in this layer. Bacterial life survives in this zone and some birds are known to fly this high, making it part of the biosphere. Above the stratosphere is the **mesosphere**.

The mesosphere

The mesosphere is characterised by strong east to west winds. It is the zone in which most meteors burn up as a result of colliding with gas molecules. In this layer, the temperature decreases as altitude increases. It is still not a well understood part of the atmosphere.

The thermosphere

The thermosphere is the fourth layer of the atmosphere. In this zone, temperature increases with altitude, reaching up to 2000 °C. The air has very few gas molecules at a low density, so a small change in energy can cause a large change in temperature.

> CAMBRIDGE INTERNATIONAL AS LEVEL ENVIRONMENTAL MANAGEMENT: COURSEBOOK

ACTIVITY 1.7

Layers in the atmosphere

In groups, prepare a poster of the atmosphere showing how temperatures change with altitude in each layer. Include the names of each layer, the location of the ozone layer and describe the conditions in each layer.

The greenhouse effect

The **natural greenhouse effect** (Figure 1.13) is a natural process that causes the warming of Earth's surface and surrounding atmosphere.

When the sun's energy reaches Earth's atmosphere, some of the incoming shortwave UV solar radiation (insolation) is reflected back into space. The rest is absorbed by Earth's surface and radiated as longwave radiation (infrared radiation). Some of this infrared radiation is trapped by **greenhouse gases** in the atmosphere, heating them up. Greenhouse gases include water vapour, carbon dioxide (CO_2), methane (CH_4), nitrous oxides (NO_x), ozone (O_3) and some artificial chemicals such as **chlorofluorocarbons (CFCs)**.

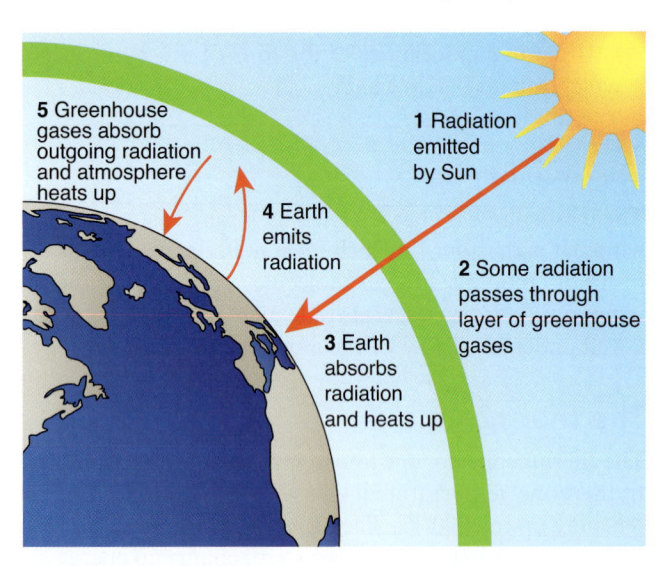

Figure 1.13: The natural greenhouse effect is caused by the ability of the gases and dust in the atmosphere to retain some of the heat entering Earth's atmosphere from the sun.

The natural greenhouse effect is required for life on Earth. Without it, the planet would be approximately 33 °C colder than it is today. A small change in Earth's surface temperature results in a significant change in weather conditions, which affects life on Earth. In the past, there were colder periods called ice ages, during which much of the land in the Northern Hemisphere was under ice. During these periods, global temperatures were only approximately 4 °C colder than current temperatures. This shows how a small temperature shift can result in a significant change in global conditions.

TIP

The greenhouse effect is a natural phenomenon and should not be confused with the **enhanced greenhouse effect** which is driven by human activity. The enhanced greenhouse effect is covered in Chapter 8.

KEY TERMS

natural greenhouse effect: the warming of the atmosphere by gases, found naturally in the atmosphere, trapping the heat from the sun

greenhouse gases: gases in the atmosphere that absorb infrared radiation

chlorofluorocarbons (CFCs): nontoxic, nonflammable chemicals containing carbon, chlorine, and fluorine, that are used in the manufacture of aerosol sprays, foams and packing materials, solvents, and refrigerants

enhanced greenhouse effect: an increase in the warming of the atmosphere, over and above the natural greenhouse effect, through gases produced by human activities. These gases increase the amount of infrared radiation being retained in the atmosphere, trapping heat from the sun

1.3 Questions

1. **State** three examples of greenhouse gases.
2. Define the terms 'variable gases' and 'natural greenhouse effect'.
3. Discuss the changes in temperature with altitude in the troposphere, stratosphere and mesosphere.
4. Explain the natural greenhouse effect.

> **COMMAND WORD**
>
> **state:** express in clear terms

1.4 Ecosystems

The **biosphere** is the life-supporting zone where the air (atmosphere), the water (**hydrosphere**) and land (**lithosphere**) meet. Life in the biosphere varies according to the availability of water and amounts of insolation. As a result, different **biomes** and ecosystems exist within the biosphere, and what type they are depends on where they are located on the planet.

A biome is a large scale ecosystem that is defined by the climate and dominant vegetation in an area, for example a tropical rainforest. In contrast, an ecosystem has a smaller geographic area than a biome, for example a small wetland, or a specific type of forest.

An ecosystem is where the **biotic** components, such as plants, animals and other organisms like bacteria, and the **abiotic** components, such as the soil or climate, interact to create a specific combination of life forms. The abiotic or physical environment that affects the development of an ecosystem includes elements such as temperature, humidity, water, oxygen, salinity, available light and pH.

Within an ecosystem each factor depends either directly or indirectly on every other factor. For example, trees in the Amazon rely on high rainfall levels. If the amount of rain changes, so does the number of trees present. This can affect other living organisms and may eventually result in soil changing, as the nutrients in the ecosystem are affected.

The scale of an ecosystem can vary. Ecosystems can be very small. For example, a pond is an ecosystem (Figure 1.14), as is a small wetland in a forest. The wetland is not isolated from the forest, and it interacts with the greater ecosystem. However, wetland differs from its surrounding landscape in that it has standing water and therefore specific plants and animals survive in it. An oasis in a desert is another example of an easy-to-identify ecosystem. The oasis is significantly different from its surrounding landscape due to the availability of surface water in a normally arid biome.

When considering the structure of an ecosystem, the smallest element is the individual (a single frog or plant). Each individual is a member of a **species**. A species is a group of organisms whose individuals can breed to produce offspring. A group of individuals of the same species living in an ecosystem is known as a **population**. A mobile population or individual may move between ecosystems. This movement can cover short or long distances. For example, a population of swallows will spend summer (April–September) in the northern hemisphere and then fly south to spend summer in the southern hemisphere (October–March). They are still considered a population.

> **KEY TERMS**
>
> ***biosphere:** the zones of Earth where living organisms can survive
>
> ***hydrosphere:** all the water on Earth's surface, including lakes, seas, ground water and atmospheric water such as clouds
>
> **lithosphere:** the rigid outer layer of Earth
>
> **biomes:** large-scale ecosystems that are typically defined by climate and the dominant vegetation type – for example, tropical rainforests or hot deserts
>
> **biotic:** living organisms (e.g. plants)
>
> **abiotic:** climate, soil type, slope angle and non-living things or things without life are all abiotic factors that influence the structure of an ecosystem
>
> **species:** an organism whose individuals breed to produce offspring
>
> **population:** a group of organisms of the same species living within an ecosystem

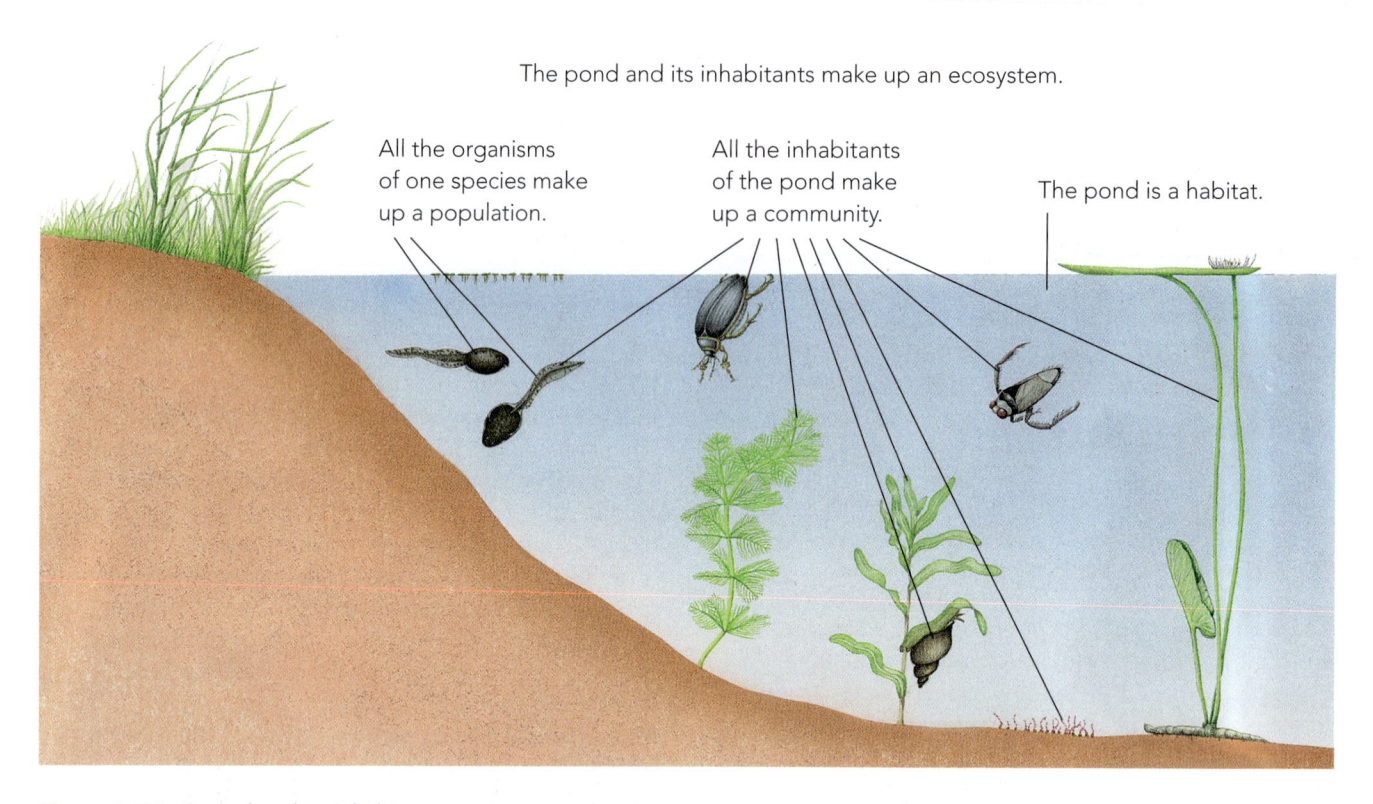

CAMBRIDGE INTERNATIONAL AS LEVEL ENVIRONMENTAL MANAGEMENT: COURSEBOOK

The pond and its inhabitants make up an ecosystem.

All the organisms of one species make up a population.

All the inhabitants of the pond make up a community.

The pond is a habitat.

Figure 1.14: A pond and its inhabitants are an example of an ecosystem.

Figure 1.15: Part of a community of wildlife gathers around a watering hole in Tanzania, Africa.

The next level in the structure of an ecosystem is the **community** (Figure 1.16). This is made up of the different populations living together, for example, grasses, trees, insects, mammals and birds. They interact with the abiotic components to form the ecosystem.

Ecosystem: all the species and the environment

Community: all the species

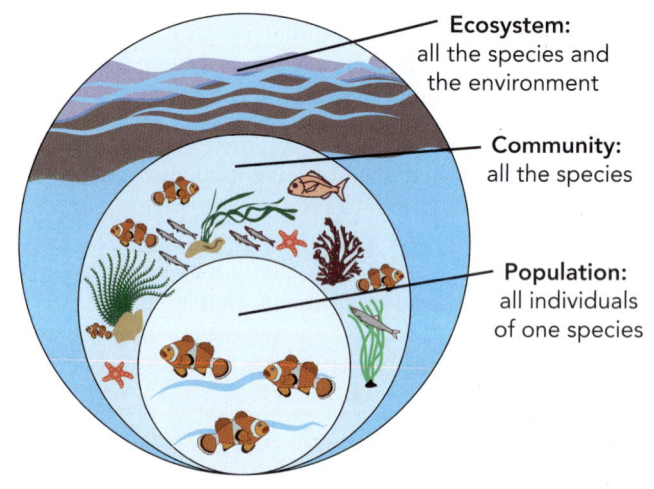

Population: all individuals of one species

Figure 1.16: The relationships between populations, communities and ecosystems.

> **KEY TERM**
>
> **community:** the different populations that live together in an ecosystem

34

1 Introduction to Environmental Management

In an ecosystem, the community lives in a **habitat** which is the place that an organism lives. A habitat meets all the environmental conditions that an organism needs and has adapted to in order to survive. For an animal, that means gathering food, accessing water, finding a mate and reproducing. For a plant, that means the correct soil, amount of water, light and space necessary for successful reproduction.

The main components of a habitat are water, shelter, food and space. Every living organism has a habitat that it has adapted to. This will vary from organism to organism depending on their survival needs. For example, a leopard requires not only the right amount of food, water and shelter, but also enough territory (space) for it to survive. **Competition** with other leopards or species may cause this habitat to become unsustainable. For example, if a leopard, lion and hyena are all hunting the same prey in the same area, there may not be sufficient food available for all of them. If the competition is too high, an animal has to find a new territory. In addition to this, if humans start using land or building in the area, the territory shrinks and the habitat becomes unsuitable for the leopard. The particular habitat needs of a species are defined as the **ecological niche**. The ecological niche is the role that a species fills within an ecosystem. For example, a lion fills the niche of being an apex predator (a predator at the top of the food chain with no natural predator of its own), preying on other large herbivores like antelope, while the African black-footed cat (Figure 1.17) is the primary predator of small mammals and birds. These two cats do not fill the same niche, so they can exist in the same ecosystem without any competition.

> **KEY TERMS**
>
> **habitat:** the habitat is the place that an organism makes its home. It meets all the environmental conditions that an organism needs for survival
>
> **competition:** the relationships between organisms that need the same resource in the same space
>
> ***ecological niche:** the role and position that a species fills in an ecosystem, including the conditions and feeding needs necessary for the survival of the species

Biotic and abiotic factors of ecosystems

Biotic factors

Biotic factors in the biosphere are living organisms (plants, animals, insects, micro-organisms) and their interactions which include competition, predation, grazing and decomposition.

Examples of some of these interactions are:

- Green plants provide oxygen to living organisms through photosynthesis.
- Animals breathe out carbon dioxide which plants use to generate food and energy through photosynthesis.
- Plants provide food and habitat for animals. However, plants in turn rely on organisms in the soil that carry out decomposition. This process releases the nutrients from decaying material that plants need to grow.
- Certain organisms are involved in plant reproduction – e.g. bees pollinating flowers and helping to fertilise one plant from another.

Figure 1.17: An African black-footed cat.

- Competition within a species (**intra-specific** competition), for example males competing to mate with females. For example, a male deer will compete with other male deer for the right to mate with the females in a herd.
- Competition between species (**inter-specific** competition), where different species compete for the same resources (for example, lions and hyenas preying on the same antelope species).
- Predation, where one organism feeds on another one, such as a lion feeding on an antelope.
- Grazing, where herbivores feed on grasses and other plants. The plants act as a food source, while grazing impacts the plant species present. When too many herbivores feed on the plants, overgrazing occurs and this results in a change to the types or amount of plants present in an area.

Figure 1.18: Living organisms, such as the plants found in this forest in the Seychelles, are examples of biotic factors that play a role in ecosystem form and function.

Trophic levels and food chains

Biotic components are made up of the producers and consumers. They fill a specific position in a **food chain**.

There are usually five main **trophic levels** within an **ecological pyramid** (Figure 1.19), the first of which is made up of the **primary producers**. Examples of primary producers are plants, algae and phytoplankton. They all use light energy (solar radiation) to carry out photosynthesis and produce their own food source.

The other four trophic levels are made up of **consumers** who cannot produce their own food source, and must consume other organisms in order to obtain nutrients. The second trophic level is made up of **herbivores** (primary consumers). Herbivores only eat primary producers (plants) to obtain their energy.

The third and fourth trophic levels are made up of **omnivores** and **carnivores**. Consumers in the third level (omnivores) are known as the secondary consumers as they eat organisms from the first two trophic levels, that is plants and animals. Those in the fourth level (carnivores) are known as tertiary consumers as they consume organisms from the trophic level below them.

Trophic level five consists of the apex predators (carnivores, an example of which is seen in Figure 1.20). These animals have no natural predators and form the top layer of an ecological pyramid.

> ### KEY TERMS
>
> **intra-specific:** between individuals of the same species
>
> **inter-specific:** between individuals of different species
>
> **food chain:** the feeding sequence of organisms indicating the flow of energy as one species is consumed by the next, from the primary producer through to the apex predator
>
> **trophic level:** a group of organisms within an ecosystem which fill the same level within a food chain
>
> **ecological pyramid:** a graphic representation of the relationship between organisms at different tropic levels in an ecosystem
>
> **primary producer:** the organism within a food chain that produces its own food source through photosynthesis
>
> **consumer:** an organism that cannot produce its own food, and must eat other organisms in order to obtain nutrients
>
> ***herbivore:** an organism that only eats plants, also known as a primary consumer
>
> ***omnivore:** an organism that eats both meat and plants, also known as a secondary consumer
>
> ***carnivore:** an organism that only eats meat, also known as a tertiary consumer

1 Introduction to Environmental Management

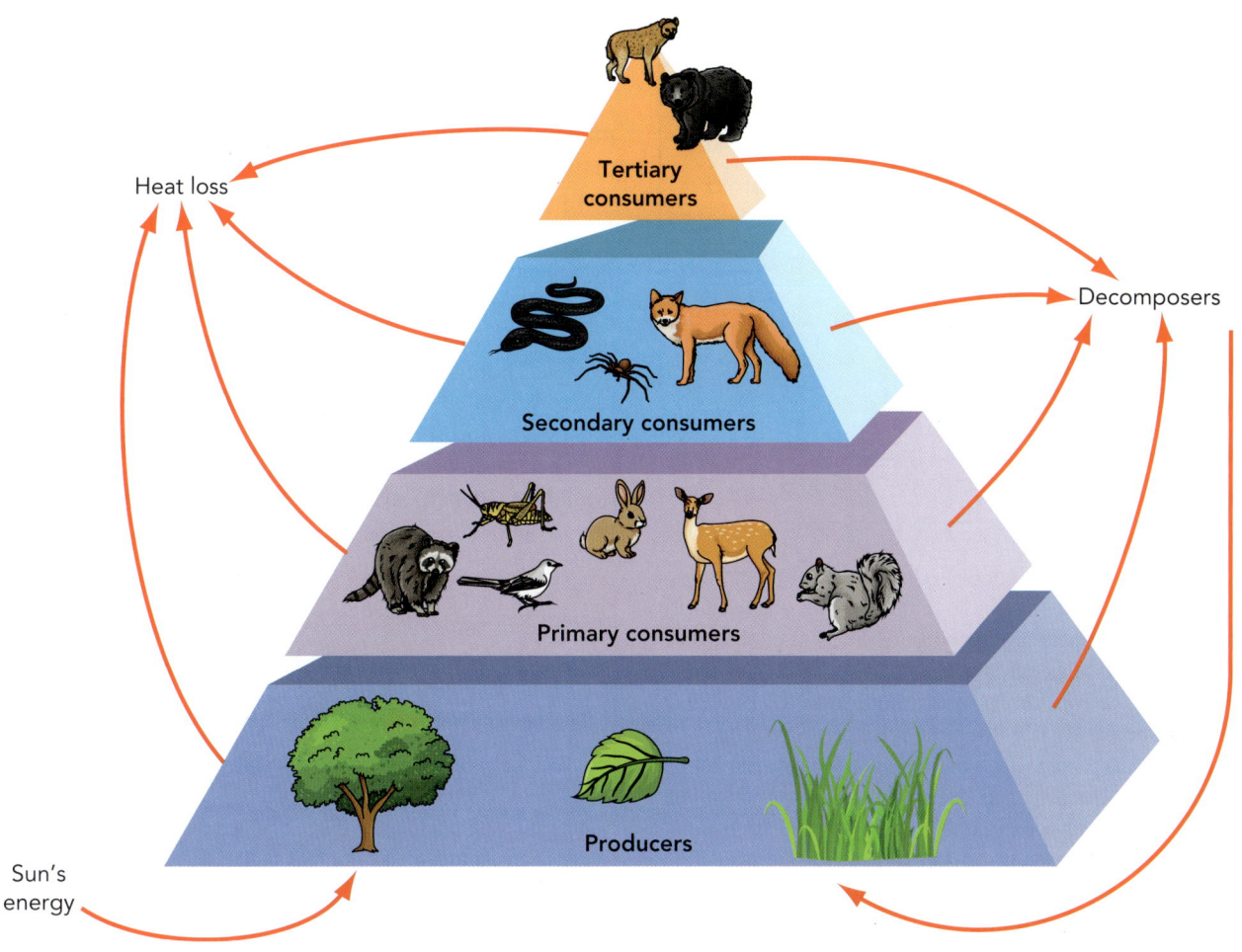

Figure 1.19: An ecological pyramid showing how the ascent through the trophic levels corresponds with the decrease of nutrients available. Only 10% of the energy is passed up to the next level, thus reducing the amount of energy available as one ascends the pyramid.

Decomposers (fungi and bacteria) do not belong to a specific trophic level. However, they play an important role in the consumption of dead plant and animal material, converting them into nutrients and energy that are released back into the soil and used by plants for growth. They are effectively the 'recyclers' of an ecosystem (Figure 1.19).

KEY TERM

decomposer: an organism that breaks down organic material

Figure 1.20: The lion is an example of an apex predator with no natural predators. It takes the position at the top of the ecological pyramid.

CAMBRIDGE INTERNATIONAL AS LEVEL ENVIRONMENTAL MANAGEMENT: COURSEBOOK

> ### ACTIVITY 1.8
>
> **Organisms in different trophic levels**
>
> 1. Find examples of organisms at each of the following levels within an ecological pyramid:
>   - primary producer
>   - primary consumer
>   - secondary consumer
>   - tertiary consumer
>   - decomposer.
> 2. Create a table to record your findings.

The **limiting factor** in a **food web** or ecological pyramid is the amount of energy available to each trophic level. Only about 10% of the energy created/consumed at each level is converted into **biomass**. The rest is lost through waste products, respiration, reproduction and growth (Figure 1.19). This means that by the time the apex predator consumes its prey, very little of the energy captured by the primary producer is available for its use. In the fourth trophic level only approximately 0.001% of the energy captured by the primary producer in that food chain is available for use by the consumer. The length of the food chain is therefore limited by the amount of energy transferred between each level of the food chain.

A food chain (Figure 1.21) is the feeding sequence of organisms indicating the flow of energy as one species is consumed by the next, from the primary producer through to the apex predator. The arrows in a food chain point in the direction of energy flow.

A food web (Figure 1.22) in an ecosystem is where many different food chains cross-link to form a web. All the plants and animals in an ecosystem form part of this complex food web.

The more complex the food web, the more stable the ecosystem. If there is a large number of primary producers and prey, the disappearance of one species means that their consumers can still survive, as they have an alternative food source. The smaller and less complicated the food web, the more fragile the ecosystem, and the more likely that any changing factor within the ecosystem (pollution, water availability, disappearance of a species) could result in its collapse.

> ### KEY TERMS
>
> **limiting factor:** anything that may slow population growth, or constrain population size. The term limiting factor can also be used in other contexts to refer to any factor that can slow or reduce the chance of an event occurring
>
> **food web:** the connection of all the individual food chains within a community
>
> **biomass:** the total quantity or weight of organic material in an ecosystem (1), or, plant material used as an energy source (2)

Figure 1.21: An example of a food chain which shows the flow of energy from the primary producers (plants), to the primary consumer (a grasshopper), to the secondary consumer (a flycatcher).

1 Introduction to Environmental Management

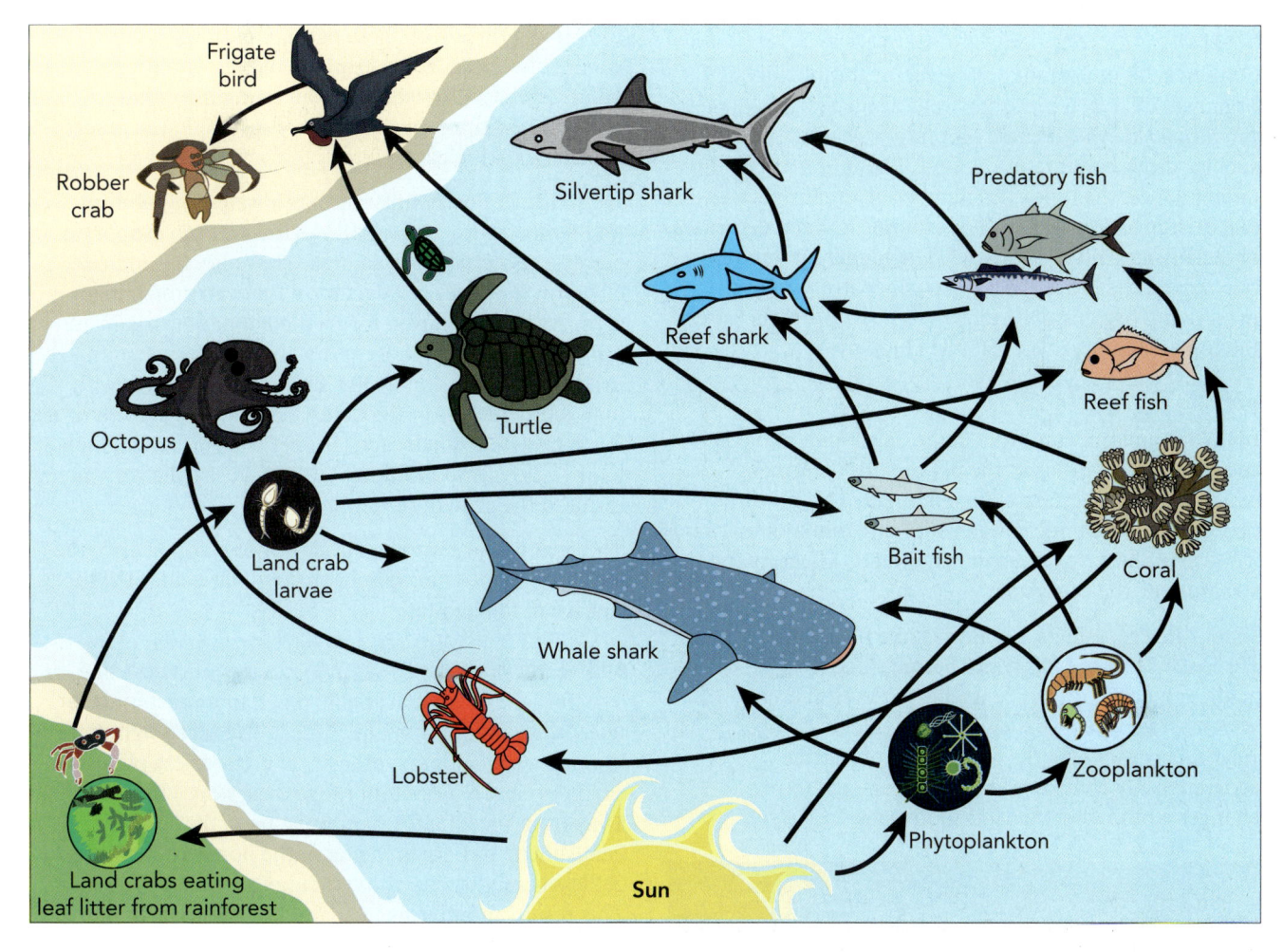

Figure 1.22: An example of a food web which shows the complexity of energy pathways with different food sources for different organisms.

ACTIVITY 1.9

Creating a food chain or food web

1. Working in pairs, and using an ecosystem of your choice, research and then draw a food chain. Give examples of species at each trophic level of the food chain, and show the direction of the flow of energy.
2. Now draw a simple food web for the ecosystem you have chosen. Include a maximum of three organisms at each trophic level.

TIP

When asked to draw a food web or food chain, you just need to include the names of the organisms. You do not need to draw an image of the organisms themselves.

Abiotic factors

The abiotic factors of the ecosystem are all of the non-living elements that play a role in determining which species live in the area. Examples include the temperature, humidity, water, oxygen, salinity, light and pH.

Oxygen

Oxygen is an important abiotic factor for most living organisms. Oxygen is used by cells as an energy source, providing them with energy for important functions such as growth and reproduction. Therefore, oxygen has an impact on the types and number of organisms that can exist in an ecosystem. For example, oxygen decreases in density as altitude increases. The higher up a mountain one moves, the lower the oxygen concentration, until most living organisms struggle to survive, as is the case at the top of Mount Everest in the Himalayas.

pH

pH plays an important role in the development of an ecosystem. Variations in the pH of soils determine which plant species are able to flourish. This, in turn, determines which microorganisms and animal life can live in the surrounding area, directly affecting the structure of the ecosystem.

In marine ecosystems, organisms like phytoplankton are sensitive to pH. As water becomes more acidic, phytoplankton are unable to develop their skeletal structure and therefore struggle to survive. Phytoplankton form the foundation of marine ecosystems. A change in their numbers due to pH changes would affect marine food webs.

Figure 1.23: A well-balanced soil pH allows for good plant growth and encourages the development of a healthy ecosystem.

Temperature, water and humidity

Temperature, humidity and water availability all play an important role in the formation of ecosystems. Some species have adapted to living in harsher environments (hot and cold deserts or deeper marine zones). However, even in these zones, life has upper or lower limits of both water availability and temperature that can be tolerated. Outside of these parameters, organisms cannot survive. A basic example of this would be a polar bear, which is suited to surviving in extreme cold temperatures, but would not survive in the desert, another extreme ecosystem.

Humidity is closely linked to both water availability and temperature, as heat allows for water to evaporate, increasing humidity in an area. Humidity has a direct effect on the types of vegetation that can grown in an area. As the vegetation forms the habitat of organisms, humidity has an impact on the type of ecosystem that can form. For example, low humidity and water with high temperatures in the desert result in the growth of a few hardy plants like cacti. In the Amazon forest, high temperatures and humidity with good supplies of water result in the growth of dense forest.

The greater the extreme (either hot or cold, or dry), the fewer the number of species that will have adapted to survive in the habitat.

Light

The amount of light available plays an important role in ecosystem development. Areas of reduced light have lower levels of photosynthesis. Therefore, less plant life can flourish. In mountain ranges, the slopes that face away from the equator (south facing in the southern hemisphere and north facing in the northern hemisphere) get less sunlight than those facing the equator. This results in differing availability of both heat and light, leading to contrasting ecosystem development on different mountain slopes.

In marine ecosystems, the top layer of the ocean receives light, allowing for photosynthetic organisms to flourish. Below approximately 200 m, no light penetrates. This impacts the ability of photosynthetic organisms to live in the deep ocean and therefore the types of organisms found in those ecosystems. For example, plants are not found in the deepest parts of the ocean, and animals found there often look very different to the fish we are most accustomed to. They often eat the carcasses of marine creatures which have sunk.

Salinity

Salinity is a measure of the amount of salt in water or soils. Changes in salinity have a direct and indirect effect on an ecosystem. Salinity can influence the pH, levels of oxygen and nutrient balance in plants, affecting primary productivity. It can also interfere with the ability of plants to absorb nitrogen, reducing plant growth and stopping plant reproduction.

Salinity also affects aquatic organisms because it affects how fast water evaporates, how much oxygen is available in the water and how many nutrients aquatic plants can absorb. These factors all affect the growth of aquatic organisms like zooplankton and therefore marine food webs.

Figure 1.24: Mountain slopes in Italy are exposed varying levels of light, which influences the ecosystems found on different slopes.

Limiting factors

Limiting factors are resources that restrict the development, growth, size and distribution of a species within an ecosystem. They can be either biotic or abiotic. Examples of limiting factors are fresh water in a desert and light in aquatic ecosystems. Competition between species, food availability, disease, predation or lack of space are also limited if the population numbers are impacted.

The population size of a species that can be supported by an ecosystem is called the carrying capacity. When demand exceeds the availability of a resource (such as a type of food), then the species has exceeded its carrying capacity. An example is grazing for antelope in the savanna. When an area becomes overgrazed the numbers decrease due to starvation or migration (movement of animals from one region to another). Exceeding the carrying capacity is a limiting factor.

CASE STUDY

Adapting to the climate of a hot desert

Organisms are able to adapt to different conditions. In some cases adaptations are unusual, with some species being found in regions where the abiotic factors are considered too harsh for them to survive. Biodiversity, is low in these regions due to the harsh conditions. It is therefore only highly specialised species that survive.

The Namib desert beetle

The Namib desert beetle (Figure 1.25) is native to the Namib Desert along the west coast of southern Africa. This region receives as little as 1.4 cm of rain each year, making it an extremely dry desert. However, the desert beetle has developed a way of keeping hydrated. It collects water on its bumpy back in the early mornings when the sea-fog rolls inland.

To drink water, the beetle stands on a small ridge of sand with its long, thin back legs helping to angle its body upwards at 45-degrees. The beetle spreads out its hardened wings and catches the fine fog droplets on them. Water droplets stick to the **hydrophilic** bumps on the wings, and then roll down towards the beetle's mouth so it can drink. In this way the Namib desert beetle is able to live in an environment that otherwise would not have sufficient water for its survival.

KEY TERM

*hydrophilic: a water-loving material which attracts water, and is capable of holding onto it

> CAMBRIDGE INTERNATIONAL AS LEVEL ENVIRONMENTAL MANAGEMENT: COURSEBOOK

CONTINUED

Figure 1.25: The Namib desert beetle lives in one of the world's driest climates with only 40 mm of rainfall per year. It has adapted to survive by collecting water from early morning fog, also known as fog basking.

Desert frogs

Some amphibians normally associated with wet climates have adapted to live in deserts, or climates that are exposed to long periods of drought. They do this by employing various strategies. Some have modified appendages for burrowing. They burrow as much as 1.5 m below the surface and away from the heat. Here they can spend months and even years at a time, coming out to reproduce in temporary pools that form after rains.

The African Pixie frog (Figure 1.26) is the second largest frog in the world and is known as a bullfrog due to its huge size and deep voice. This frog has adapted to survive in a hot, dry climate, as the region it lives in is subject to cyclical long term droughts.

Yet this frog is able to survive underground during drought conditions for up to 7 years. How does it do this?

Pixie frogs live in the southern African savanna near fresh water sources such as lakes or ponds. However, during periods of drought they survive in burrows 15–20 cm underground. Previously it was thought that the pixie frog died during the long periods of drought typical of its habitat. However, when the ground starts to dry out, the frog secretes a layer of mucus around its body which forms a specialised **mucus sac**. The frog survives in the sack in a state of **hibernation**, awaiting the return of the rains. Once the rain returns the water dissolves the mucus sac and the frog wakes up.

Questions

1. In groups of two or three, research an organism, like the pixie frog or Namib desert beetle, that has used an unusual way to adapt to an ecosystem where you would **not** expect it to survive. Consider biotic and abiotic factors, food chains and energy flows in your discussion.

2. Present your findings to the class. Discuss how the organisms that have been investigated have evolved different adaptation mechanisms to survive.

Figure 1.26: A male Pixie frog showing its distinctive yellow underbelly.

KEY TERMS

*mucus: a slimy, sticky substance that coats, protects and moistens the surface it covers

*hibernation: a period of time when a plant or animal remains in a dormant or inactive state resembling sleep

Photosynthesis and respiration

All living organisms need a continuous supply of energy to stay alive. Most organisms achieve this through the processes of **photosynthesis** by primary producers (Figure 1.27) and the process of **aerobic respiration**. During photosynthesis, organisms capture energy from the sun and convert it into stored energy. During aerobic respiration, this captured energy is made available for use.

There is a relationship between aerobic respiration and photosynthesis, where one uses the reactants of the other. Photosynthesis uses energy from the sun (sunlight), water and carbon dioxide to produce glucose and oxygen. During aerobic respiration, chemical reactions in the cell break down glucose molecules and release CO_2, energy and water. The equations for photosynthesis and respiration are as follows:

> Photosynthesis:
>
> $CO_2 + H_2O \rightarrow C_6H_{12}O_6 + O_2$
>
> carbon dioxide + water → glucose + oxygen
>
> Respiration:
>
> $C_6H_{12}O_6 + O_2 \rightarrow CO_2 + H_2O$
>
> glucose + oxygen → carbon dioxide + water
>
> *Note: the two equations are each other's opposite.*

Most living organisms depend on a cycle of photosynthesis and aerobic respiration for survival. The oxygen which plants give off during photosynthesis is inhaled by animals and transported from the lungs to the cells via the blood to allow for aerobic respiration. The CO_2 produced in this process is released from the body via respiration and absorbed by plants for use in photosynthesis. The cycles in producers and consumers are therefore closely linked.

Photosynthesis

Producers (e.g. plants, phytoplankton) capture sunlight to produce energy rich carbohydrates such as glucose through the process of photosynthesis. They do this through the use of **chlorophyll**-containing organelles called chloroplasts. Chlorophyll is a green pigment in the leaves of all green plants, which absorbs light to provide energy for photosynthesis to occur.

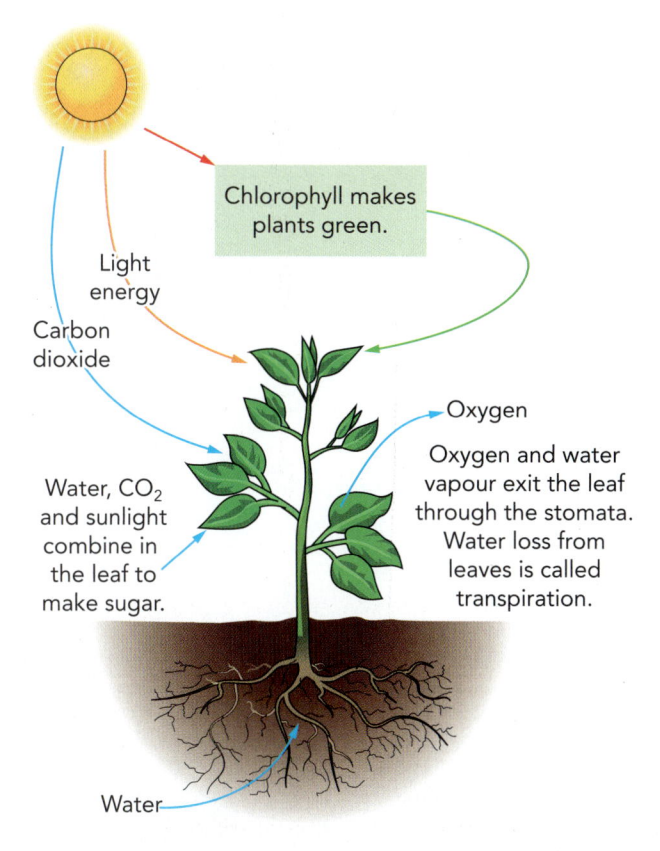

Figure 1.27: In photosynthesis, the plant uses incoming light energy, carbon dioxide and water to produce energy in the form of glucose. The plant gives off oxygen as a by-product.

Photosynthesis is the foundation of the energy flow within most ecosystems on the planet. It is important to note that plants undergo both photosynthesis and aerobic respiration.

KEY TERMS

photosynthesis: the process by which plants synthesise glucose using carbon dioxide, water and energy from sunlight

aerobic respiration: the chemical reactions in cells that break down glucose molecules and release energy, carbon dioxide and water

chlorophyll: green pigment in the leaves of all green plants, which is responsible for the absorption of light to provide energy for photosynthesis

The availability of water, the concentration of CO_2 and the availability of light are all limiting factors in the potential rate of photosynthesis in a primary producer (Figure 1.29). The rate of photosynthesis increases with the availability of light; if temperature or available carbon dioxide are limited then the rate of photosynthesis becomes constant or is reduced. If any of these three factors are absent, there is no photosynthesis.

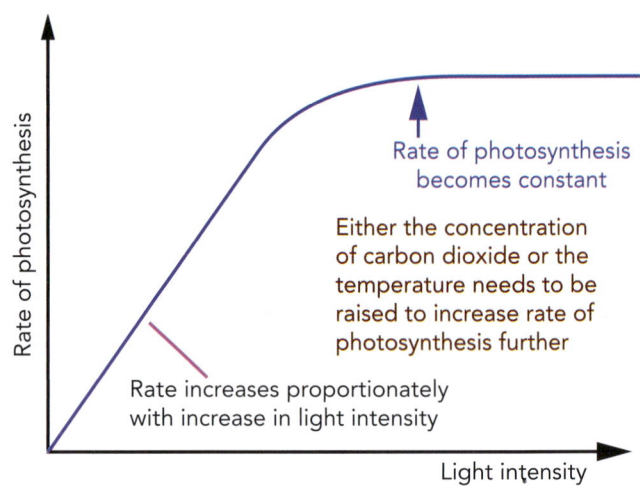

Figure 1.29: The rate of photosynthesis versus light availability.

Figure 1.28: All three of these primary producers, the Algave plant, the Coconut palm and the spear mint plant use chlorophyll to carry out photosynthesis.

INVESTIGATIVE SKILLS 1.2

Photosynthesis experiment

Organisms that photosynthesise capture their energy from the sun, forming glucose and giving off oxygen. This experiment tests for the rate of photosynthesis in a plant. In pairs or groups of three, carry out the following experiment.

You will need:

- fresh spinach leaves (keep them in water to maintain freshness)
- a hole puncher or a hard metal or bamboo straw
- sodium bicarbonate (baking soda)
- rainwater or tap water that has been left to stand for 24 hours
- dishwashing liquid
- 2 plastic syringes (10 ml or larger, no needle required)
- 4 clear beakers
- glass markers

1 Introduction to Environmental Management

CONTINUED

- 1 measuring cylinder
- 1 measuring spoon (1/8 teaspoon) or scales (to weigh 0.6 g)
- Light source (sunlight or artificial)
- Stopwatch.

Safety

- Are any harmful chemicals being used in this experiment? Check the components you are going to be working with and make sure you can handle them safely.
- Work carefully with all glassware to prevent breakages and cuts.

Getting started

- Before you start ensure that all glassware is clearly labelled.
- Label your syringes A and B.
- Prepare any solutions that you will be using and store them in labelled containers.
- Make sure your lamp is set up and ready for use once you have completed your preparations.
- **Predict** what will happen with the circles of spinach leaf when they are placed in water containing bicarbonate of soda and exposed to light. **Justify** your answer. Revisit your prediction when you have completed the experiment. Were you correct?

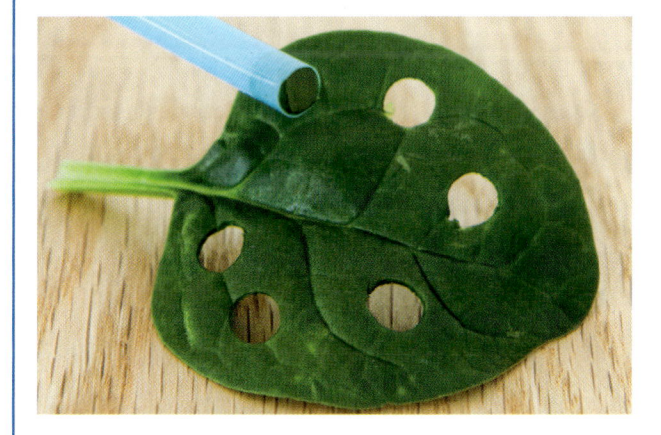

Figure 1.30: An image showing how to cut circles out of the spinach leaf with the straw.

Method

1. Label the beakers: (a) water, (b) bicarbonate solution, (c) detergent solution, (d) bicarbonate and detergent solution.
2. In a beaker mix 0.6 g (1/8 tsp) Sodium bicarbonate into 300 ml of unchlorinated water (rainwater or tap water that has stood for 24 hours prior to use).
3. In the second beaker add 200 ml of water and 1 drop of liquid dishwashing detergent to create a solution. Mix carefully and do not create bubbles.
4. Pour half (150 ml) of the bicarbonate solution into the third beaker and add 1 drop of the dishwashing solution. Again, make sure there are no bubbles.
5. Using the hole punch, create 20–30 discs from the spinach leaves. Avoid the veins and edges of the leaves: punch from the flat areas.
6. Label the syringes (a) water and (b) bicarb/dish liquid solution. Remove the plungers from the syringes and place half of your discs into each syringe. Replace the plungers and slowly depress them to expel as much of the air as possible. Take care not to crush the spinach discs.
7. Using the syringes, pull up approximately 3 ml of plain water into syringe (a) and the bicarbonate/detergent mixture into syringe (b). Then gently tap the syringes to suspend the leaves in the solution.
8. Placing your fingers over the end of each syringe, pull back the plunger to form a vacuum. Maintain the vacuum for about 10 seconds while gently swirling the solution. Repeat this step until you see the all the discs start sinking to the bottom of the syringe. If they do not, start again with fresh discs and a new solution.
9. Pour the discs from syringe (a) into the beaker with water and from syringe (b) into the beaker with the bicarbonate/dish washing liquid solution. Make sure that the discs sink to the bottom and do not stick to the sides.

CONTINUED

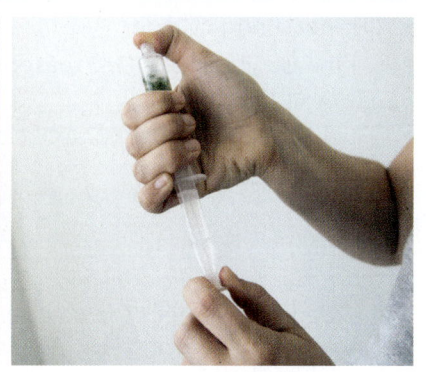

Figures 1.31–1.33: Preparing the syringe with the spinach discs and either the water or the bicarbonate solution.

Figures 1.34–1.36: Decanting the spinach discs into a cup and exposing them to a light source as per steps 9 and 10.

10 Expose the beakers to a light source (sunlight or artificial light), then start the timer.

11 Watch the leaf discs, noting any tiny bubbles forming around the discs.

12 Copy the table below into your exercise book. In the table, record the number of discs that are floating each minute, until all the disks in one beaker are floating.

Time / Minutes	1	2	3	4	5	6
Number of discs in water						
Number of discs in bicarb						

13 When all the discs have floated, put the beaker in a dark cabinet for 15 minutes and see what happens.

Investigative skill questions

a **Consider** why bicarbonate of soda was used in this experiment.

b Which experiment is the control and why should you run both at the same time?

c What happened to the spinach discs as they gave off oxygen during photosynthesis?

d What happens to the spinach discs when you turn the light source off?

e Which two compounds were necessary to allow for photosynthesis?

f What unknowns are there in the experiment?

> **CONTINUED**
>
> **g** Record your results in a graph using different symbols for the two experiments. Make sure to label your axes and the symbols.
>
> **Extension activity**
>
> Try this experiment with different types of leaves or in different conditions and see if the leaves behave differently. Always avoid plant veins when punching the circles.
>
> Consider the variables in order to make the results comparable.
>
> Discuss the differences that you observe with your class to understand the potential causes.

> **COMMAND WORDS**
>
> **consider:** review and respond to given information
>
> **predict:** suggest what may happen based on available information
>
> **justify:** support a case with evidence/argument

The carbon cycle

The **carbon cycle** is a series of processes by which carbon is reused in nature. In this cycle, carbon travels from the atmosphere into organisms and then into Earth before being released back into the atmosphere. Most carbon is stored in rocks and sediments, while smaller amounts are stored in the oceans, atmosphere and living organisms. In the carbon cycle, carbon flows between the various carbon stores. Some of these exchanges are rapid while others are slow. Any alteration in carbon cycle flows can result in one store taking up more carbon while less is stored in another. This principle is central to understanding how changes to the balance of carbon in the atmosphere cause changes to the global climate.

The carbon cycle, like the water cycle, should be thought of as a system. There are inputs, stores, flows and outputs that transfer carbon from one carbon store to another, changing carbon balances within the carbon cycle.

Carbon forms the foundation of life on Earth. Living organisms are made of carbon, we consume carbon and our lives are built on carbon economies. Carbon is stored on Earth in the rocks, atmosphere, oceans, soils, plants, animals and fossil fuels.

The carbon cycle

The carbon cycle (Figure 1.37) refers to the movement of carbon through living organisms on Earth during their lifetime, and within the biosphere. Carbon atoms continuously move from the atmosphere to the Earth

> **KEY TERM**
>
> **carbon cycle:** the flow of carbon between various carbon stores

and then back to the atmosphere. Most of the carbon on Earth is stored in rocks, sediments and fossils, while the rest is located in the atmosphere, oceans and living organisms. Fossilization of organic matter results in the formation of carbon energy sources such as coal and oil. The timescales on which the different processes of the carbon cycle take place vary from days to decades between living organisms and the atmosphere and hundreds of millions of years during the fossilisation process, in which carbon moves between the Earth's crust and the atmosphere.

Carbon is a versatile molecule that is able to form a large variety of complex organic molecules; even DNA is built around a carbon chain. During photosynthesis, energy is captured and stored in the carbon chains that form glucose. When carbon chains are broken down in glucose during the process of aerobic respiration, stored energy is released. Therefore, carbon is a good source of fuel for living organisms.

Organisms that photosynthesise (plants and phytoplankton) and the process of photosynthesis itself form a vital part of the carbon cycle. The organisms take in CO_2 from the atmosphere and store the carbon in their cells as glucose. Therefore, photosynthesis influences different carbon stores. The carbon energy stored by plants can be released in a number of ways:

- The plant breaks it down and uses the energy for respiration, reproduction and growth.
- The plant gets eaten by a primary consumer, which then incorporates the carbon into its own system, or releases it as CO_2 through respiration.
- The plant decomposes and carbon is reabsorbed back into the atmosphere.
- The plant is burned, and the carbon is released back into the atmosphere as CO_2, leaving a residue of ash.
- When the plants become fossilised and form oil and coal, the **combustion** of these fossil fuels results in the release of stored carbon.

In all of these processes, CO_2 is released back into the atmosphere, making it available for absorption in the carbon cycles.

The carbon cycle (Figure 1.37) has been closely linked to the plant growing seasons, with CO_2 in the atmosphere

> **KEY TERM**
>
> **combustion:** the burning of an item, e.g. the burning of fossil fuels to use their energy

increasing during slow growth periods (winter) and decreasing during fast growth periods (summer). Photosynthesising organisms therefore play a vital role in the carbon cycle.

The carbon cycle forms an important aspect of life on Earth as carbon is a greenhouse gas and can trap the heat from the sun.

Carbon stores and carbon sinks

Carbon is stored in long and short-term carbon stores known as **carbon stores** and sinks. A **carbon sink** is a natural or artificial store that absorbs carbon through the physical and biological mechanism of the carbon cycle. When carbon is released from a carbon sink, that sink becomes a **carbon source**. For example, when

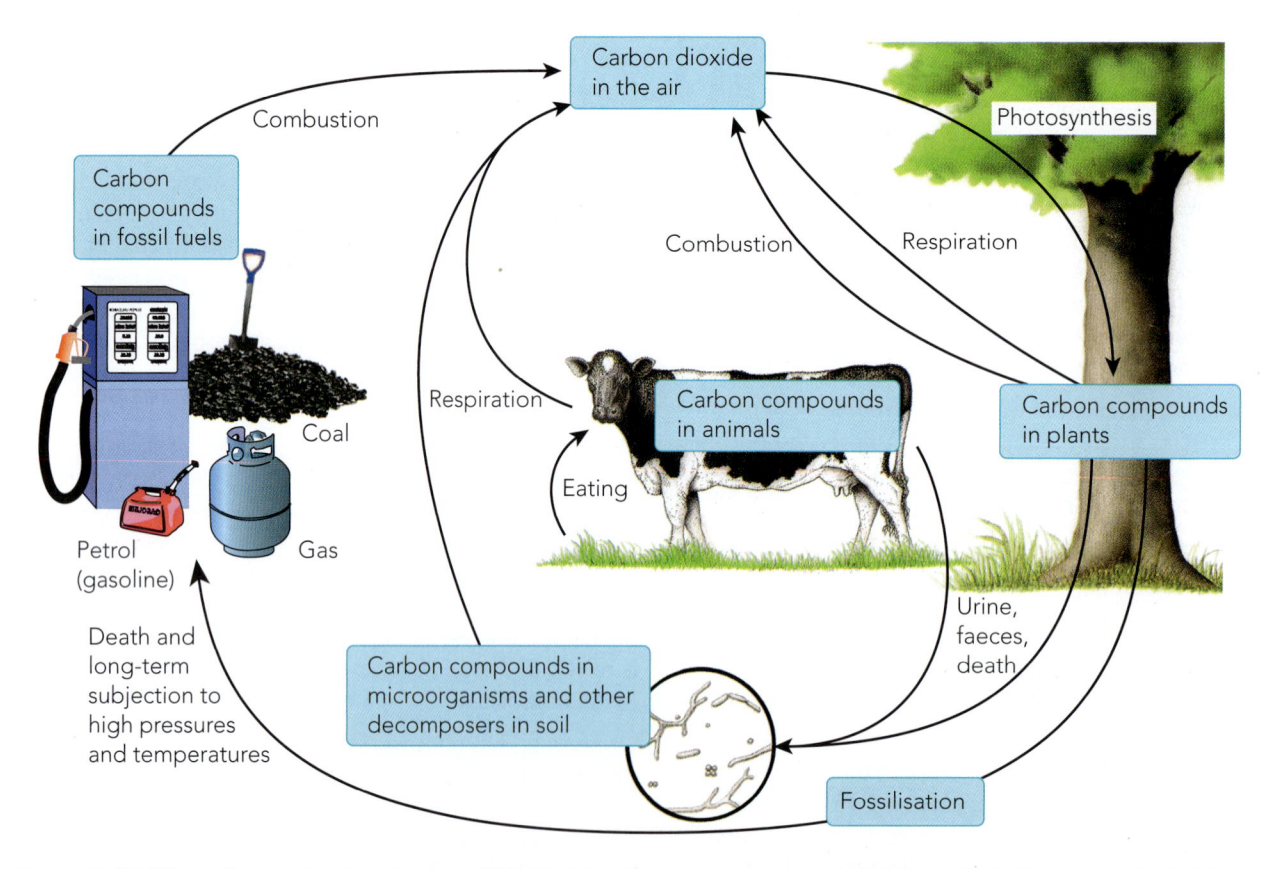

Figure 1.37: The carbon cycle, showing how CO_2 can move from one store to another in a relatively short period of time (weeks, days and even hours).

1 Introduction to Environmental Management

humans mine coal and burn it, the fossil fuel becomes an atmospheric carbon source, producing CO_2. The mangrove trees shown in Figure 1.38 show how vegetation can act as a carbon sink and then become a carbon source when cut down.

The following are the main carbon stores:

- Biosphere: as organic molecules in living and dead organisms.
- Atmosphere: mainly as CO2 but also as methane CH4.
- Cryosphere: in Arctic and tundra regions where the soils are frozen and contain plant materials.
- Pedosphere: as organic matter in soils (remains of dead plants and animals).
- Lithosphere: as fossil fuels and sedimentary rock deposits such as limestone.
- Hydrosphere: in the oceans as dissolved atmospheric CO2.
- Hydrosphere: in the oceans as calcium carbonate shells of marine organisms.

The largest store is in the lithosphere where most of the carbon is stored as fossil fuels, sedimentary rocks, marine sediments and organic carbon.

1 Photosynthesis   2 Carbon storage   3 Carbon release

Figure 1.38: How mangrove trees absorb and store CO_2 during photosynthesis (step 1) as part of the carbon cycle. They then become a carbon source (step 2), releasing it back to the atmosphere when disturbance occurs (step 3). Protecting mangroves and forests is an important step in maximising carbon stores.

When considering carbon storage, it is important to understand that the biological processes that result in the storage of carbon are sensitive to disturbances (Figure 1.38). Long-term carbon sinks require millions of years to form (coal, oil, gas). If rapid change occurs, these stores are released back into the atmosphere, changing the carbon balance within the natural system. This can cause change, with carbon building up in stores (layers of the atmosphere) that previously did not have high levels of carbon, and becoming depleted in sinks (oil, coal and gas stores).

> ### KEY TERMS
>
> **carbon stores:** carbon is stored in carbon sinks, as organic material in organisms, soil, fossil fuels and the oceans
>
> ***carbon sink:** this is anything that absorbs more carbon from the atmosphere than it releases
>
> ***carbon source:** a carbon store that releases more carbon that it stores
>
> **fossilisation:** the process through which organic material is replaced with mineral substances in the remains of an organism. It is a physical, chemical and biological process the preserves the plant and animal remains over time
>
> ***cryosphere:** parts of Earth's surface that are made up of ice, including ice, snow, glaciers, ice sheets and areas of permafrost
>
> ***pedosphere:** the outermost layer of Earth that is made up of soils

1.4 Questions

1. **Define** the terms biome, ecosystem and habitat, and explain how their meanings differ.
2. **Explain** how a change in the rate of photosynthesis can impact the amount of carbon dioxide in the atmosphere.
3. Briefly explain the importance of photosynthesis in the carbon cycle.
4. Select a local ecosystem and write 150–200 words about it. **Summarise** the key biotic, abiotic and limiting factors, and discuss how they affect biodiversity. Sketch a food chain for species found in the ecosystem.

> ### COMMAND WORD
>
> **summarise:** select and present main points without detail

49

EXTENDED CASE STUDY

Masai Mara: the Great Migration

Each year over two million zebra, wildebeest, eland and gazelles migrate from the southern Serengeti in Tanzania to feed on the grasses of the Masai Mara in Kenya (Figure 1.39). This migration is not a single event, with animals moving between two points. Instead, the Great Migration refers to the constant year-long movement of these great herds. The migration forms an 800 km long, clockwise circular pattern through the Serengeti and Masai Mara.

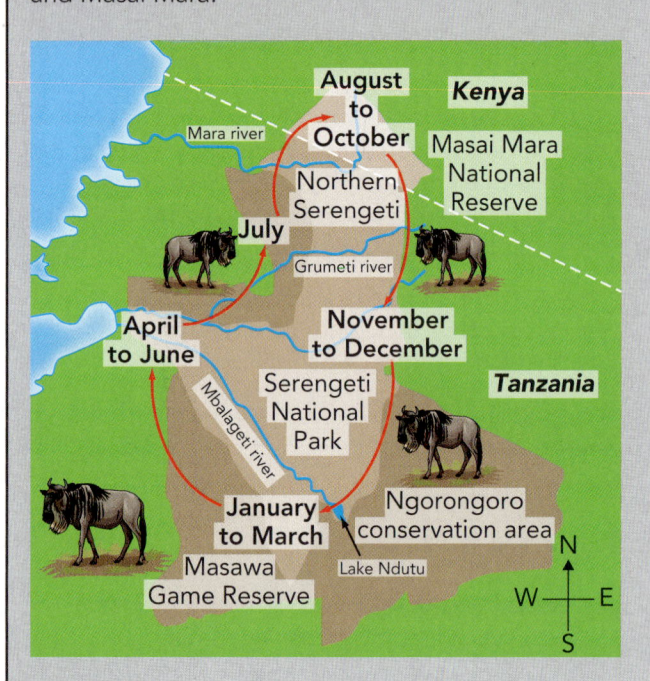

Figure 1.39: The migratory routes of the great herds.

The animals have to cross a number of rivers in this migration, including the Mara (Figure 1.41) and the Grumeti Rivers. Despite the dangers of the crossing, which include some of the largest crocodiles in the world and swift water currents, the animals cross the rivers in search of food on the other side. Of the estimated 1.2 million wildebeest that cross the river each year, an average of 6250 die in the crossing.

The Mara river which empties into Lake Victoria is a key water source for the wildlife of the greater Serengeti–Mara Ecosystem. The individuals from migrating herds that die provide food for the crocodiles and fish of the river. Both the abiotic factor (the river) and the biotic factor (the animals that die in the river) support life. In fact, the two systems function as one, relying on each other for ecological balance.

Figure 1.40: The Great Migration: Large herds of animals made up of wildebeest, zebra, eland and gazelles follow a continuous path of migration around the Serengeti and the Masai Mara every year.

Figure 1.41: Wildebeest crossing the Mara river in Kenya. The wildebeest cross in large numbers, reducing the chances of any one individual being attacked by crocodiles. However, the mass of bodies increases the risk of drowning or being trampled.

1 Introduction to Environmental Management

CONTINUED

Abiotic and biotic causes of the Great Migration

The main abiotic cause of the Great Migration is thought to be the climate (Figures 1.42a and 1.42b), as the animals are following the rains. If one looks at the graphs for both the southern Serengeti (the southern-most point of the migration – Figure 1.42a) and the Masai Mara (the northern most point of migration, Figure 1.42b), there is no obvious difference in temperature or rainfall. Both regions have similar constant maximum temperatures throughout the year. The Masai Mara maximum temperature **ranges** between 25 °C and 29 °C and the Southern Serengeti ranges between 26 °C and 29 °C. At first glance, the rainfall patterns are also similar with the dry seasons falling in the middle of the year. However, there is a difference between the dry and wet seasons. The dry season in the Serengeti is much more pronounced than it is in the Masai Mara. It rains more regularly in the Masai Mara, which is located closer to the equator.

In January and February, the herds spread out across the Southern Serengeti, as there is still sufficient precipitation to support plant growth. It is in the south that the females give birth. From July through to October the animals move away from the extremely dry southern areas and move north following the rain and the areas where vegetation growth is still healthy. When rain comes to the region in October, the herds start to head south again.

The main biotic factor of this migration, which is the available vegetation, is closely linked to the climate. Plants require rainfall to grow. During the dry season the vegetation dies back and during the wet season it recovers. Although the climate is the cause, vegetation is equally important when discussing the reasons behind the Great Migration, as the great herds follow the food.

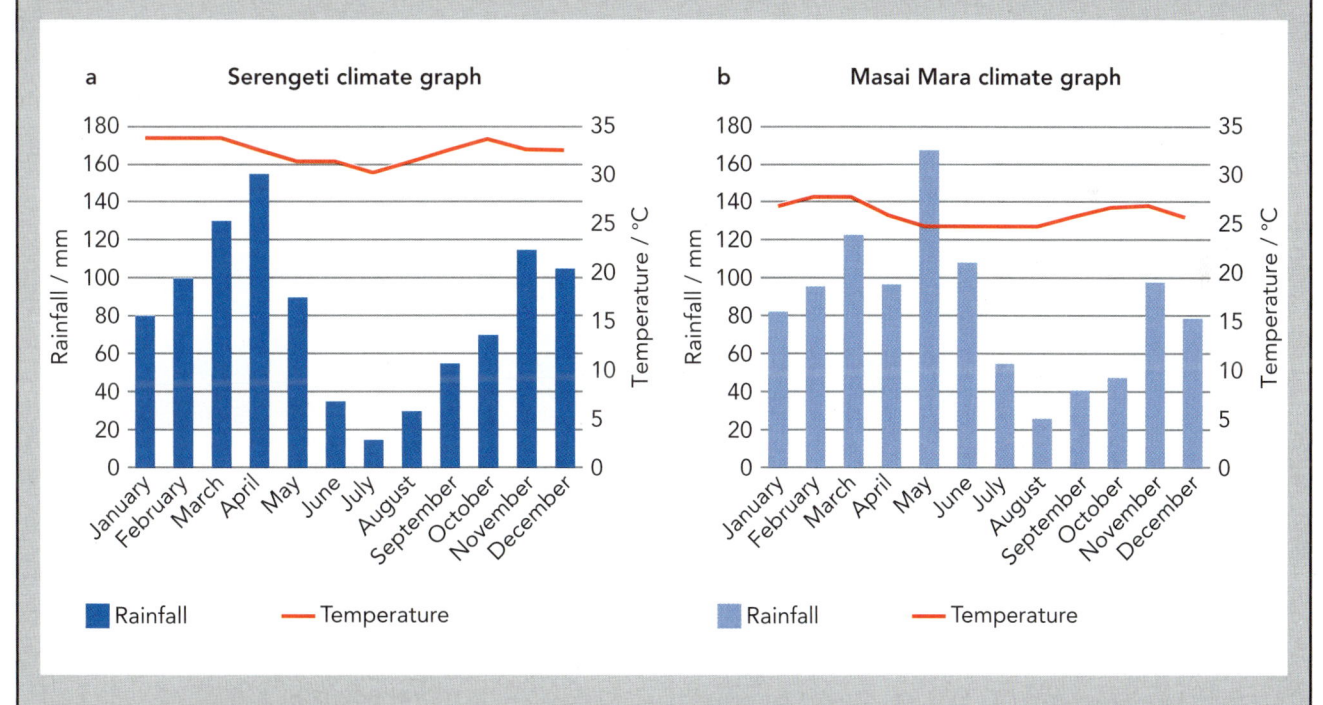

Figures 1.42: The climate graph for **a** the Southern Serengeti, which is in the south of the migratory route, and **b** the Masai Mara which is in the north of the migratory route, showing the maximum monthly temperatures and the average monthly rainfall.

CONTINUED

In following the food, the great herds add another biotic factor that needs to be considered: the pressure which grazing puts on the available food resources. In each area that the great herds move through, they graze. As grazing reduces the amount of available food, the herds need to move on to find a more abundant food source. This action of migrating protects the food supply in the long term, because the area that has been grazed recovers when the animals move on. When they return back to the area a year later the vegetation has returned and continues to support the herds year on year.

How the animals know where to find the food and when to move is still not understood, but some theories have been developed about their behaviour. Most of the evidence suggests that the weather patterns and the cycle of rainy and dry seasons have the greatest influence on the animals' movement. However, rainfall and precipitation within the region are unpredictable, so it is not possible to predict with any certainty exactly where the animals will be any time in the year, nor how long they will remain in an area.

Questions

1. **Calculate** the percentage of wildebeest that are estimated to lose their lives in the Mara river crossing each year.
2. Estimate the annual rainfall for the Masai Mara and for the Southern Serengeti.
3. **Describe** the resources that the great herds supply to the vegetation as they move through the area.
4. **Explain** why females give birth in October/November when they are in the south.
5. Describe the impact that fencing off areas of the Serengeti would have on the wildlife and the vegetation. Explain your reasoning.

Extended case study project

Carry out research to investigate:

1. How could tourism in the Serengeti/Masai Mara Reserves become a threat to the ecosystem and indigenous people?
2. Propose some sustainable solutions that could be employed to improve the tourism and practices of tour operators in the reserves.
3. Come to a conclusion as to how successful these sustainable practices could be if implemented.

Consider things like: the number of tourists, the vehicles being used, construction of lodges, waste and pollution from tourists, noise disturbances, the type of tourism, artificial barriers like fences, the access of the Masai people to the areas, soil erosion, habitat loss.

Produce a report of 600–800 words with your findings. The report should include the following headings:

Introduction, Tourism as a threat, Sustainable tourism opportunities, Conclusion.

KEY TERM

range: the difference between the upper and lower limits on a particular scale (e.g. temperature)

COMMAND WORDS

calculate: work out from given facts, figures or information

describe: state the points of a topic / give characteristics and main features

explain: set out purposes or reasons / make the relationships between things evident / provide why and/or how and support with relevant evidence

1 Introduction to Environmental Management

SUMMARY

There are seven continents, five oceans and many seas found on Earth.
LIC, MIC and HIC are classifications of the world's countries by their levels of economic development.
Environmental sustainability is the protection of the environment and its natural resources to protect the needs of future generations and nature.
The water cycle is the flow of water between land, waterbodies and the atmosphere.
The atmosphere is made up of gases, water vapour and dust particles. It is made up of four distinct layers, defined by their temperature characteristics.
The greenhouse effect is a natural process that causes warming of the atmosphere.
Ecosystems vary in size and biodiversity, and are made up of both biotic and abiotic components.
Organisms in an ecosystem provide resources for each other. The resulting energy flows are divided into trophic levels.
Photosynthesis and aerobic respiration are the processes by which the sun's energy, water and CO_2 are converted into stored energy in the form of glucose and oxygen, and then released for use during respiration.
The carbon cycle is the continuous movement of carbon atoms from the atmosphere into living organisms and stores within the Earth, then back into the atmosphere.

REFLECTION

- Does creating diagrams of the water cycle, the structure of the atmosphere or food webs help you understand the different concepts?
- Discuss approaches to learning terminology with a partner. Find out if they have methods that help them to understand and recall new words.

PRACTICE QUESTIONS

1 The short article below discusses a company's approach to producing shoes. Read the article and then answer the questions below.

Made for reuse

Logan's Eco-Clothing aims to strengthen their commitment to helping the planet tackle plastic waste. Their latest clothing range is made from plastics recycled from the ocean, and is 100% recyclable. All used products can be returned to Logan's Eco-Clothing, where they will be broken down and reused to create new clothing. In this way raw materials can be used repeatedly, reducing the pressure on landfill and the oceans.

    a Define the term 'sustainable'. [2]
    b Explain how the actions of Logan's Eco-Clothing are considered sustainable. [3]

2 Figure 1 represents some of the flows and stores of the water cycle in a forest.

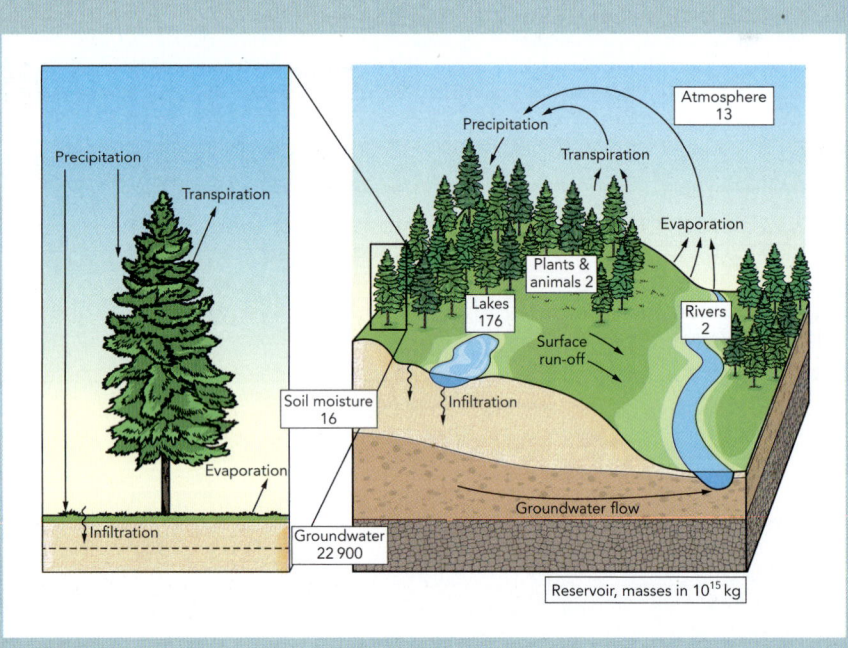

Figure 1

    a Define the terms *interception*, *infiltration* and *transpiration*. [3]
    b Use the figure to calculate the total amount of water being stored as surface water on the land. [2]
    c Describe one process, other than condensation, in which water changes state in the water cycle. [2]
    d Explain why the water cycle in a forest is considered an open system. [5]

CONTINUED

3 Figure 2 below shows the relationship between altitude and temperature in the atmosphere.

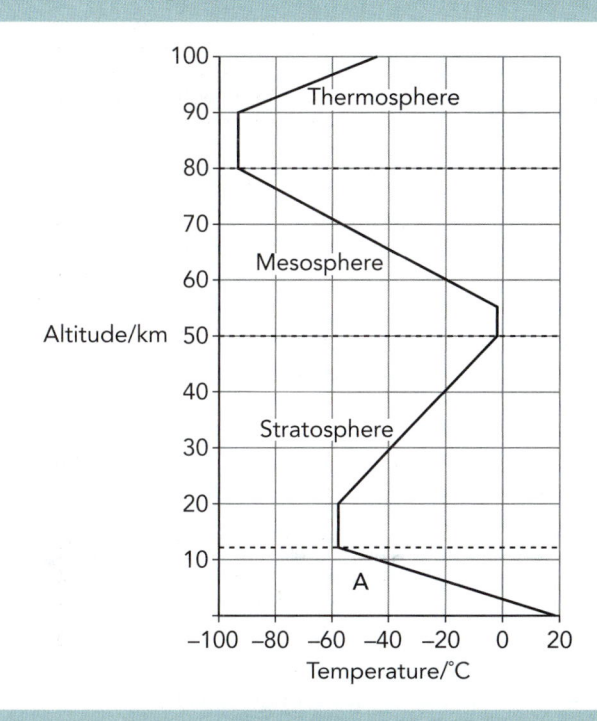

Figure 2

a Name the zone labelled A in the figure. [1]
b Using the figure, describe how temperature changes with altitude. [4]
c Identify in which atmospheric zone the ozone layer is located. [1]
d Explain why the ozone layer is important. [3]

CONTINUED

4 Figure 3 shows a marine food web.

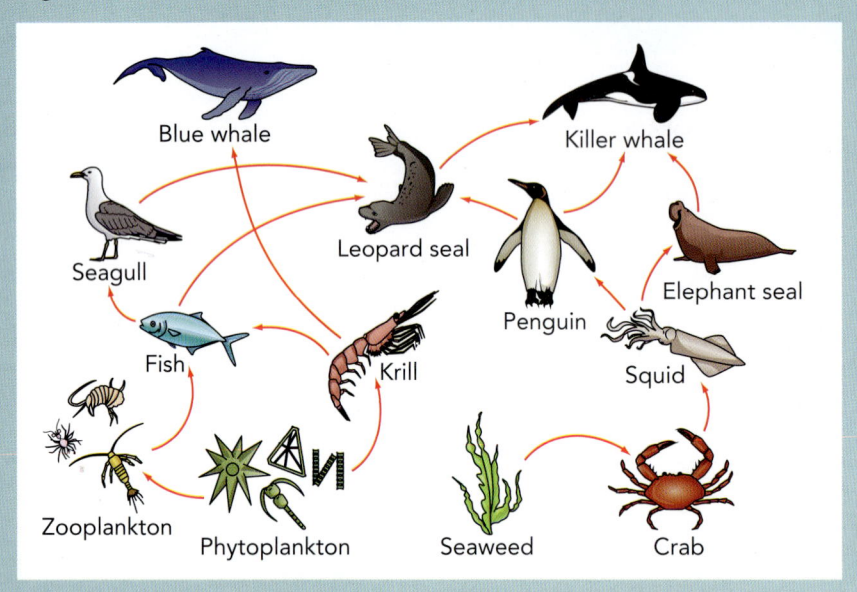

Figure 3

a Suggest the effects of the decrease in the number of elephant seals on the marine food web shown above. [4]

b Explain the importance of primary producers in this food web. [2]

c State how a decrease in the numbers of primary producers in this marine ecosystem would impact humans. [2]

5 The concentration of CO_2 in the atmosphere and light intensity are two factors which limit the rate of photosynthesis.

a Explain what is meant by *limiting factor* in relation to photosynthesis. [2]

Corals live in symbiosis with zooxanthellae (a photosynthetic algae), where the algae provide nutrients to the corals and the corals provide a protective environment for the zooxanthellae.

b Describe the effects of reduced quality and/or quantity of light on both the corals and zooxanthellae. [3]

1 Introduction to Environmental Management

SELF-EVALUATION CHECKLIST

After studying this chapter, think about how confident you are with the different topics.
This will help you to see any gaps in your knowledge and help you to learn more effectively.

I can	Needs more work	Getting there	Confident to move on	See Section
Identify and name the world's continents and major oceans.				1.1
Identify the income groups that the World Bank uses to classify countries.				1.1
Define the term sustainability.				1.1
Explain the need for the sustainable management of resources.				1.1
Describe the water cycle.				1.2
Draw and interpret diagrams of the water cycle.				1.2
State the major components and layers of the atmosphere.				1.3
Describe the location and role of the ozone layer.				1.3
Describe the natural greenhouse effect.				1.3
Define an ecosystem and the related terminology.				1.4
State both biotic and abiotic components of an ecosystem.				1.4
Explain biotic interactions using examples.				1.4
Define trophic levels within food chains and webs using the appropriate terminology.				1.4
Explain energy transfer and losses within food chains.				1.4
Define photosynthesis and respiration with the use of chemical equations.				1.4
Describe the carbon cycle and how it is related to photosynthesis and respiration.				1.4
Interpret and draw diagrams of the carbon cycle.				1.4
Explain what carbon stores and sinks are.				1.4

Chapter 2
Environmental Research and Data Collection

LEARNING INTENTIONS

In this chapter you will:

- identify and explain the scientific method
- design investigations, including the formulation of hypotheses and the interpretation of data, while understanding method limitations
- consider environmental research in the context of climate change
- explore strategies and methods for collecting environmental data
- find out about the use of technology in data collection
- use statistics to calculate population size, biodiversity, abundance, percentage cover and frequency.

2 Environmental Research and Data Collection

GETTING STARTED

1 Work in pairs to complete the following questions, and then discuss your answers with the class.

   a A researcher wants to record rainfall and temperature for a month. Design a table that could be used to collect the data.

   b How do systematic and random sampling differ?

   c Explain the meaning of the term 'climate change' and say what causes it.

   d Predict the findings of the following investigation: The amount of rainfall in a farming area is decreasing. An investigation is carried out to determine whether declining rainfall affects the growth of crops.

ENVIRONMENTAL MANAGEMENT IN CONTEXT

The role of the scientific method in environmental management

An **Environmental Impact Assessment (EIA)** is an evaluation of the potential environmental impacts of a proposed development project, such as a mine, a building or even a golf course. It takes into account the social, economic and environmental factors of the proposed development, and considers both positive and negative outcomes. It is a tool based on the scientific method and identifies possible areas of concern before the project starts.

Mineral sands mining, Mozambique

There is a proposal for a remote **titanium dioxide** beach mine in Mozambique (on the east coast of southern Africa). Due to a lack of research on the area, knowledge of its species, biodiversity, environmental fragility and social stability or interactions is very limited.

To decide if mining the area is environmentally acceptable, scientific studies need to be undertaken. These studies include identifying geology, vegetation types, plant and animal species present, levels of biodiversity, and social and cultural structures.

Areas of concern also need to be identified. Concerns could be natural or human. They might include risks to threatened habitats or species, or even threats to the project itself. Researchers who specialise in key areas (for example: botanists and zoologists) are then brought into the area to carry out various scientific studies, using the scientific method, to determine the species present. A scientific study attempts to identify as many species or areas of concern as possible, and then reports the findings. These studies use technology such as satellite images, on the ground sampling and interviews and discussions with local people.

The scientific method must be applied in a transparent and unbiased manner if the resulting data is to be reliable. Species identification must be correct and, where necessary, the international community will be asked to identify species that are rare or new to science.

KEY TERMS

Environmental Impact Assessment (EIA): the evaluation of the environmental consequences of a plan, policy, program, or project before a decision is made to move forward with it

titanium dioxide: a metal commonly found in plants and animals and the ninth most common element in Earth's crust. It is a white power that can be made into a bright white pigment. It is used in products such as paint, paper, plastic, ink, soap, food colouring and sunscreen

> **CONTINUED**
>
> **Discussion questions**
>
> 1 Why is it important to identify species in an area before starting a project like this mine?
> 2 How severe do you think the environmental impacts of such a project would need to be for it to be stopped?
> 3 How do you think that climate could negatively impact a project like this?
> 4 EIAs offer an opportunity for scientists to do research in unexplored parts of the world. Do you think this is beneficial? If so, why?
>
> **Figure 2.1:** Beach mining using a dredger in the sand and water, in Richard's Bay, South Africa. This is the same type of mining that would be used at the mineral sands mine in Mozambique.

2.1 The scientific method

The **scientific method** is a standardised process that allows you to test a **hypothesis**. Even when a hypothesis has been proven to be correct, it can be retested in the future. This could be in response to changes in technology or ways in which **data** is collected. Retesting can potentially change the initial outcome from a correct hypothesis to an incorrect hypothesis.

The basic steps of the scientific method are:

1 Make an **observation** that describes a problem or question.
2 Create a hypothesis.
3 Test the hypothesis using the data collected. Use well-defined methods to do this.
4 Draw conclusions using the data.

> **KEY TERMS**
>
> **scientific method:** a procedure that involves systematic observation, measurement and experiment to test hypotheses
>
> **hypothesis:** a precise, testable statement that a researcher makes, predicting the outcome of a study that is designed to answer a specific question
>
> **data:** a set of information, in the form of facts, numbers, measurements, or statistics, that can be used for analysis
>
> **observation:** to watch, view or note for scientific investigation

2 Environmental Research and Data Collection

Observations, questions and hypotheses

The first step in the scientific method is making observations. For example, you may observe that vegetation in an area is experiencing die back (areas of dead plant material). At this stage the observations are often **qualitative**, not **quantitative** (Figure 2.3).

> **KEY TERMS**
>
> **qualitative data:** data that is non-numerical, or descriptive. These data are collected through observations, interviews and focus groups
>
> **quantitative data:** data that is numerical, giving the quantity, range or amount of a variable. For example, monthly rainfall

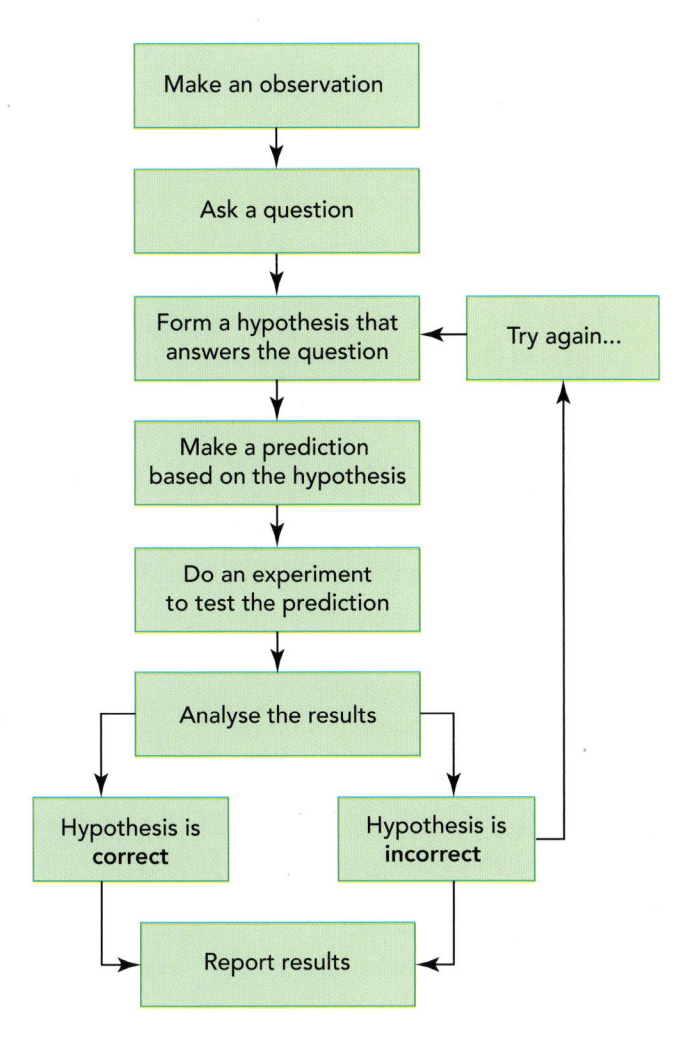

Figure 2.2: The scientific method starts with an observation and then moves through all the steps until a result is reported. The method includes a feedback loop to allow for retesting when a hypothesis is proven incorrect.

Basis for comparison	Qualitative data	Quantitative data
Definition	Qualitative data is information that can't be expressed as a number	Quantitative data is data that can be expressed as a number or can be quantified
Can data be counted?	No	Yes
Data type	Words, objects, pictures, observations, and symbols	Numbers and statistics

Figure 2.4: A comparison of qualitative and quantitative data, indicating the different types of data collected.

Qualitative data is information based on descriptive rather than numerical data. We know that vegetation is struggling to grow because we can visibly observe vegetation die back. Quantitative or numerical data can help investigate why the plants are dying off.

A hypothesis is formed to try and explain an observation. A hypothesis is one possible answer to the question 'why'? In the instance described above, the question may be 'Why are the plants in the area dying?' The hypothesis could be: 'Declining annual rainfall levels are causing plants to die off' or 'a toxic spill has resulted in soil contamination, causing the death of the vegetation.'

Figure 2.3: A scientist taking water readings to collect data for a research study.

61

ACTIVITY 2.1

Qualitative and quantitative data

1. Decide whether each of the examples in the table represents quantitative or qualitative data.

Data collected on the height of all the students in a class	The descriptive field work diary entries of a scientist
People's opinion regarding the best tasting dessert	Average daily temperature records for a town over a month
A question in a questionnaire collecting yes and no answers	The measurement in square metres for each of the classrooms in a school
A case study in a coursebook	The description of a polar bear
Photographs of wildlife	500 people attended a seminar

2. Create a new table. Rewrite the examples in two columns, one for qualitative data and one for quantitative data.

ACTIVITY 2.2

Setting a research question

In small groups, think of a research question for each of the following environmental observations:

1. The river Trent runs through a small town. The town has some industry, residential areas and very few parks. The vegetation on the river banks has not been protected, and many areas have been covered with concrete. There are points along the river where cars can drive through the water without the need for a bridge.

2. Two sites are being compared. One is open farmland that has extensive livestock grazing on it. The second site is in a natural grassland area that is protected from livestock or damage by human activities.

CONTINUED

Use the following guidelines to help you through this process:

- Keep the question short.
- Start with words like: What/Why/How/Is.

  For example:

  What is the most popular car colour in the town?

  Why do the birds migrate south in October?

  Is the increase in tourists damaging the beach?

- The question must only address one problem: it should not have multiple parts.

  For example:

  Is sunlight damaging the car that is not parked undercover?
  This is a short question with a single investigation.

  Is the sunlight damaging the car that is not parked undercover, or is the wrong chemical being used to clean the car?
  This is a long question and makes up two different studies.

- What sort of observations (for example, the amount of vegetation cover, number of species present in an area, water quality or level of pollution) would prompt you to investigate an environmental problem?

- Think about the type of data that you could collect to help you answer your question (measurements of vegetation cover, or number of plant species present).

At this stage a **prediction** can be made. This is a statement of the expected results. For example, 'Rainfall has been declining gradually over the preceding years, reducing available ecological water'.

KEY TERM

prediction: a statement of the expected results of an experiment if the hypothesis is true

A good hypothesis should be a statement, not a question. It should be a prediction, include cause and effect and it should be short. A hypothesis is an explanation of an observation that can be tested through experimentation. It is important to note that a hypothesis can be proven wrong. In these cases, the hypothesis is not changed to fit the results. Instead, the data is used to show that the hypothesis was incorrect.

Testing the hypothesis and managing variables

Next, you need to design an experiment which tests your hypothesis without **bias**. If the data does not answer the hypothesis, you will need to design an new experiment to find an answer. Alternatively, if you find a hypothesis to be untrue, you can propose and test a second hypothesis to find a cause for the observation.

To analyse the data using **statistics**, your experiment needs to produce quantitative data that can be analysed and either prove or disprove the hypothesis. In order to achieve this, the **variables** in the experiment need to be correctly managed.

> ### ACTIVITY 2.3
>
> #### Creating a hypothesis
>
> In small groups, use one of the questions that you created in Activity 2.2 and write a hypothesis for one of the environmental observations you read about in the activity:
>
> 1. The river Trent runs through a small town. The town has some industry, residential areas and very few parks. The vegetation on the river banks has not been protected, and many areas have been covered with concrete. There are points along the river where cars can drive through the water without the need for a bridge.
>
> 2. Two sites are being compared. One is open farmland that has extensive livestock grazing on it. The second site is in a natural grassland area that is protected from livestock or damage by human activities.
>
> Use the following guidelines to help you:
>
> - Keep the hypothesis short.
> - The hypothesis must be a statement.

> ### CONTINUED
>
> - The hypothesis should state what you think you are going to find.
> - The hypothesis must not be a question.
>
> Write out your hypothesis and present it to the class. Consider the differences between the hypotheses that each of you have written.

> ### KEY TERMS
>
> **bias:** when a scientist knowingly or unknowingly incorporates systematic errors into sampling or testing by selecting or encouraging one outcome over another
>
> ***statistics:** the practice of collecting, analysing and interpreting numerical data in large quantities. This includes ways of reviewing and drawing conclusions from the data. Statistics are a way to see patterns in numerical data or to determine whether data shows a difference between two treatments
>
> **variable:** a factor that can change in quality, quantity or size regarding the category of data that is being measured (e.g. rainfall)

Variables

A variable is any factor that can be manipulated, controlled or measured in an experiment. In other words, variables are conditions in the experiment that can change or be changed. Experiments contain different types of variables which are divided into **independent variables** and **dependent variables** (Figure 2.5).

> ### KEY TERMS
>
> **independent variable:** a variable that stands alone and is not changed by other variables. It is the variable being changed in the experiment to test the hypothesis
>
> **dependent variable:** a variable that depends on other factors. It is the variable being measured in the experiment

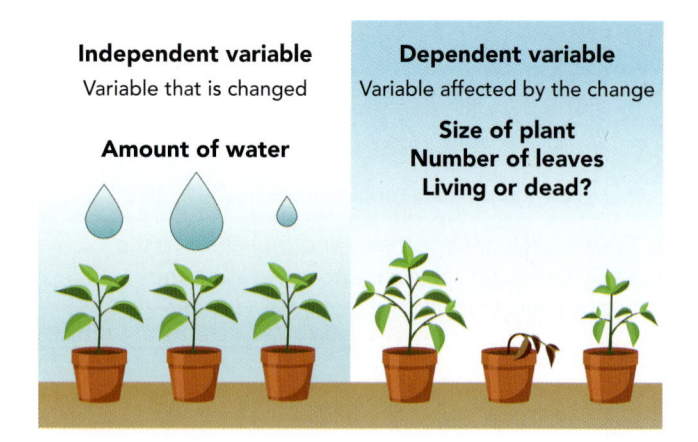

Figure 2.5: Dependent and independent variables. Water is the independent variable that is changed. The size of the plant, number of leaves and whether the plant lives or dies are the dependent variables.

The dependent variable is the one that is affected in an experiment. Its value depends on changes in the independent variable. The dependent variable is also the one that is observed or measured (e.g. the growth rate of the plant). As the independent variable is changed, the dependent variable is observed for change. Examples showing the difference between independent and dependent variables are given in Table 2.1.

In addition to these two variables, the **control group** (Figure 2.6) must also be included in an experiment. This group is treated in the same way as the experimental group, and all the variables should be exactly the same as the rest of the test groups, except for the independent variable. The control group is used to isolate the effect of the independent variable and determine the cause of the changes that you observe. Researchers change the independent variable in the treatment group but keep it constant in the control group. The results from the two groups can then be compared.

The independent variable is the one that is changed during the experiment. It is also the cause of any change in the experiment and its value is independent of other variables in a study (e.g. the amount of water or fertiliser being given to a plant).

The experiment	The independent variable	The dependent variable	Example of a control group
The effect of fertiliser on plant growth.	The amount of fertiliser being used.	The growth rate of the plant.	Plants that get no fertiliser, grown under normal soil conditions.
The effect of salt on water salinity	The amount of salt added to the water.	The salinity of the water.	Water that has no salt added to it.
Does the change in room temperature affect scores in a maths test?	The temperature in the room, which can be made cooler or warmer to test the effect.	The math test scores. Using a standardised test, you can determine if test scores vary if temperature is increased or decreased.	Maths scores achieved when temperature is set at a comfortable level.
Testing the impact of sunlight on plant growth.	The amount of sunlight the plant receives. This can be varied to test for the effect.	The growth rate or decline in the condition of the plant.	Plants getting a standard amount of daylight that allows for normal growth.
The effect of pH on the hatching of fish eggs.	The level of acidity or alkalinity of the water (the pH).	The hatching of the fish eggs.	Fish eggs kept at normal pH conditions for that species.
The effect of a sewage leak on aquatic organisms.	The sewage leak.	The aquatic organisms found downstream from the sewage leak.	**Samples** from a site upstream from the sewage leak that has not been affected.

Table 2.1: Examples of experiments showing independent and dependent variables.

2 Environmental Research and Data Collection

The control group gives the scientist the results that could be expected under 'normal conditions'. You may be testing the hypothesis that 'a plant's growth improves when liquid fertiliser is added to the irrigating water'. The control group would be plants that are given water with no added fertiliser, while the experimental group contains plants given water with liquid fertiliser. All other variables are kept constant for both groups.

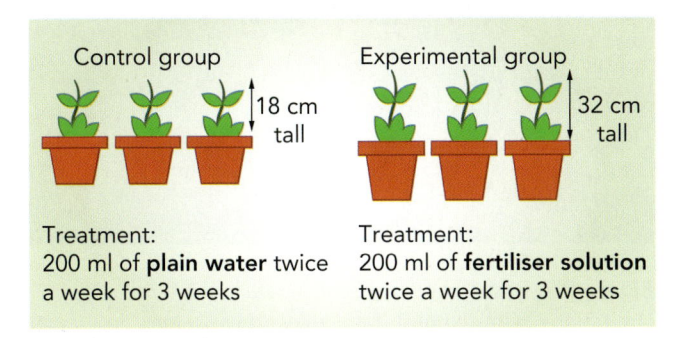

Figure 2.6: The control group and the experimental group in a plant experiment. The control group has the same variables as the experimental group, except for the fertiliser, which is the independent variable in this example.

When carrying out an experiment, a number of variables may affect your results. For example, when testing how plant growth responds to the amount of water, it is important to remember that plants can also be affected by sunlight, soil type, wind and pests. To test the effect of water alone in the experiment all the other potential variables need to be managed. These are referred to as **control variables** (Table 2.2).

Experiment	Variable being tested	Variables that are controlled
Testing soil quality on plant growth	Soil quality	Temperature, amount of light, amount of water, type and size of plant used
Testing the temperature of the room on test scores	Room temperature	Comfort of the learners, amount of light available, noise and other disturbances, length of time allowed

Table 2.2: Examples of control variables.

ACTIVITY 2.4

Identifying dependent and independent variables

1. Identify dependent and independent variables in examples **a** to **d**. Remember the variable that is affected is the dependent variable.

   First identify the two variables. Then determine the dependent and independent variables.

   a. The effectiveness of classical music on learners' reading comprehension.
   b. Learners who study for a higher number of hours get higher examination scores.
   c. Plants grow faster under fluorescent than natural light.
   d. Dogs that are taken for walks every day weigh less than similar dogs that are not walked as frequently.

2. Write out your answers. Use the same format as in Table 2.1.

KEY TERMS

control group: the group of test subjects left untreated or unexposed to the independent variable. The results from this group are then compared to the results of the test subjects

sample: a set of data (number of plants, number of species, plant distribution) taken from a larger population for measurement

control variable: any variable that is held constant in an experiment

A control variable is any variable that is kept constant in a study. The effect of the control variable is not part of the study and needs to be managed so that it does not influence the results. Examples of factors that may need to be controlled are:

- Temperature
- CO_2

65

- O_2
- pH
- light intensity.

If more than one variable is changed in an experiment when testing a specific independent variable, the data obtained will not be valid. This is because it will not be possible to determine which variable caused a change in the data.

Data interpretation

Data interpretation plays an important role in the successful completion of an experiment. During data interpretation and analysis, you can determine if your experiment's findings support or discredit the hypothesis. The interpretation of data gives meaning to the information that has been collected, and will help you identify how significant that information is. In some instances, the data may only partially support a hypothesis.

Limitations in the scientific method are its short comings, and the parts of an experiment that prevent scientists from producing **reliable data**. Experimental limitations can occur due to restrictions in research design, materials, methodology, available time and costs. Problems with the sample, sample size and issues with the instruments or techniques used to collect the information are all further examples of methodology limitations. Such limitations can be addressed by clearly identifying the potential problem and suggesting ways in which it could be addressed. Time and cost constraints are also common limitations. These need to be clearly identified and noted when carrying out research.

Limitations are an important consideration as a hypothesis must be testable to be proven or disproven. It must also be possible to repeat the experiment with similar results. Any issues identified by a scientist must be reported, allowing others to understand the potential limitations of the data being analysed.

Examples of limitations in scientific experiments include:

- Existing human knowledge: for example, 500 years ago, bloodletting was considered a sound scientific method of curing an illness. We now know that this is not an effective method of controlling disease.
- Human error (such as mixing the wrong concentration of a chemical that is being tested, or miscalculating the time required for the experiment) can result in inaccurate results.
- An experiment can have a narrow focus due to researcher bias. This means that the researcher may be expecting a specific answer and might therefore miss other information that could have given a different result.

Figure 2.7: Researchers collecting data in the field. Limitations in this type of data collection may include difficulty accessing remote locations for repeat data collection, or weather conditions.

A hypothesis that is consistently supported by good scientific data can result in a **scientific theory**. For example, carbon emissions into the atmosphere result in the increased ability of the atmosphere to retain heat. This is no longer a hypothesis; it is a theory. Theories can be used in **models** to predict what may happen in different situations. For example, if carbon emissions continue to rise, what are the expected outcomes?

Theories are not static. They can be modified and change as new data becomes available.

> **KEY TERMS**
>
> **limitations:** shortcomings in a study that can influence the information collected. These include research design, methodology, materials and time constraints
>
> **reliable data:** data that is reasonably complete and accurate, works towards answering the hypothesis in a clear and transparent manner and has not been inappropriately altered
>
> **scientific theory:** an explanation of an aspect of the natural world that has been tested repeatedly to verify it through the use of the scientific method
>
> **model:** a scientific model is the production of a physical, conceptual or mathematical representation of a real occurrence that is difficult to observe

ACTIVITY 2.5

Limitations in a scientific experiment

Copy and complete Table 2.3 by suggesting the limitations for each experiment. Examples are given for the first one.

Experiment carried out by a researcher:	Limitation
To test the effect of fertiliser on plant growth. The researcher tested the fertiliser on one plant.	• The sample size is too small. • There is no control to compare the experiment to.
To find out the impact of sunlight on plant growth. The researcher tested ten plants, and included a control group. The experiment was completed after two days.	
Sampling a population to find out their favourite food type. The researcher used a questionnaire, and asked people who looked friendly.	
To test the effect of a heat lamp at different temperatures on incubating eggs. The power was cut during the night for an unknown period of time.	

Table 2.3: Identifying limitations in scientific experiment.

ACTIVITY 2.6

Data analysis

A learner undertook an experiment to answer the question: How effective is human hair at absorbing oil spills?

The learner was testing the hypothesis that human hair is effective at absorbing oil spills.

The learner created ten small **containment booms** made out of stocking sleeves all the same size. Each containment boom was filled with the same weight of human hair (1 g). The learner then placed each boom in containers of equal size and an equal volume of liquid. However, booms 1–5 were placed in a mixture of water and cooking oil, while booms 6–10 were placed in water alone. The table below shows the results.

1. On graph paper, plot a bar chart showing the change in weight of the two sets of booms exposed to the different treatments (water and oil, and water alone).
2. Does the data support the hypothesis? Use information from Table 2.4 and the graph you have drawn to support your answer.
3. **Comment** on why you think the learner used cooking oil instead of a petrochemical product.
4. **Explain** why the learner calculated the averages.
5. The learner went on to dry out both sets of booms for a week. They then reweighed them. Explain why this was an important step in the experiment.

COMMAND WORD

comment: give an informed opinion

> CAMBRIDGE INTERNATIONAL AS LEVEL ENVIRONMENTAL MANAGEMENT: COURSEBOOK

CONTINUED

Booms 1–5 were placed in individual beakers with both water and oil			
Boom	Weight before experiment (g)	Weight after experiment (g)	Difference in weight (g)
1	1	28	27
2	1	22	21
3	1	22	21
4	1	10	9
5	1	24	23
Oil average	1	21.2	20.2
Booms 6–10 were placed in individual beakers with only water (control)			
Boom	Weight before experiment (g)	Weight after experiment (g)	Difference in weight (g)
6	1	2	1
7	1	2	1
8	1	4	3
9	1	2	1
10	1	2	1
Control average	1.00	2.40	1.40

Table 2.4: Results for an experiment testing if human hair is effective at absorbing aquatic oil spills.

KEY TERM

*containment boom: tubing that is normally filled with air to allow it to float on the surface of water and stop the movement of an oil spill. The boom acts as a barrier to prevent the spread of an oil spill on the surface of the water

2.1 Questions

1 **Define** the terms *bias* and *prediction*.
2 Compare the meaning of *dependent* and *independent* variable.
3 **Demonstrate** how you would form a hypothesis for an experiment.
4 **Describe** the main considerations when testing a hypothesis. If it helps, use a diagram to illustrate the process.
5 **Explain** what is required for a hypothesis to become a scientific theory.

COMMAND WORDS

define: give precise meaning

demonstrate: show how or give an example

describe: state the points of a topic / give characteristics and main features

explain: set out purposes or reasons / make the relationships between things evident / provide why and/or how and support with relevant evidence

2.2 Environmental management research in the context of climate change

Reliable environmental data collection and analysis is essential in understanding environmental investigations such as climate change. Reliable data helps policy makers to make well-informed decisions and to develop effective strategies for managing identified environmental problems.

There is a wide range of data available for nearly every area of environmental management. However, data is only reliable if it is collected in an **unbiased** manner. Where scientific data is collected with bias, the resulting data is unreliable as it supports a specific viewpoint. Such bias can be misleading and cause mistrust within both the scientific and wider community, undermining hypotheses that are being investigated. To achieve unbiased reliable data sets, the scientific method should be followed.

Some people believe that **climate change** exists and some do not. This results in political and social division. Within this discussion, there are also people who are confused by conflicting information. For the science of climate change to be taken seriously, and to convince those who do not believe, data must be collected in a transparent and unbiased manner.

Figure 2.8: Shrinking ice impacts polar bears' ability to move across it.

> **KEY TERMS**
>
> *****unbiased:** not affected or influenced by a person's beliefs or opinions
>
> **climate change:** detectable change in the global temperatures. It is also referred to as global warming

Development of climate change theory and data

Historical climate change data

Climate change is a significant and lasting change in the distribution of global weather patterns over time scales that range from decades to millions of years.

Historical global climate data, that is, global climate data collected by humans, started as early as 1880. Data from earlier than this is not considered to be accurate, as it was limited to specific regions.

Using this data is difficult. We need to consider how the data was collected in 1880. For example, what equipment was used, and the time of day or year the data was recorded, as even the calendar that was being used differed in some instances. Due to these factors, the information needs to be analysed, converted and processed before it can be meaningful in the climate change discussion. There is also a limited amount of historical global climate data available. As so many factors influence climates around the world, the greater the amount of data, the greater its reliability.

Development of scientific theory

Climate change has been scientifically recognised from the early 19th century, when greenhouse gases were first observed. At this time, ice ages and other natural historical changes to the global climate were also identified. Only late in the 19th century did scientists start to argue that human emissions from burning fossil fuels could affect the global climate.

Evidence of past periods of colder or hotter climates can be found in geological deposits, coastal landforms and ice sheets in the north and south poles. Scientists have found evidence to show a succession of climate changes (Figure 2.9). Data shows glaciers in areas that are now no longer covered in ice. This suggests a colder past.

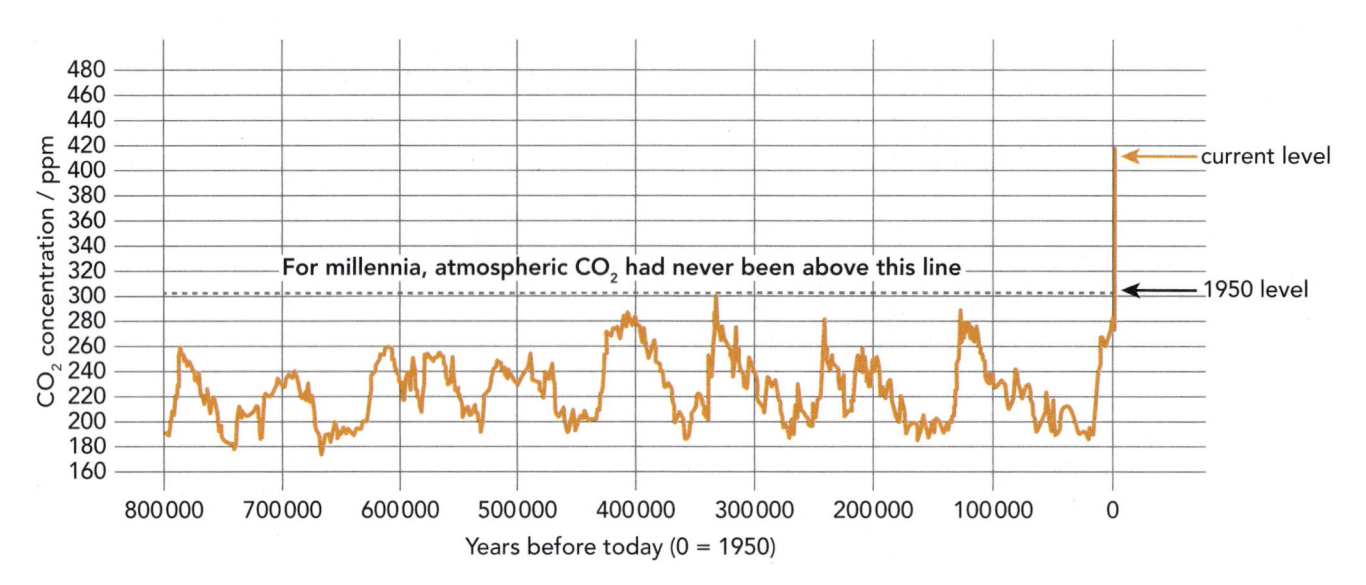

Figure 2.9: Estimated changes in the global amount of CO_2 over the last 450 000 years. This shows fluctuations in CO_2 in the past, as well as cyclical changes in CO_2 levels. The data also shows how current CO_2 levels exceed historical values.

Scientists have used data on CO_2 levels in the atmosphere to estimate historical global temperatures in the past. Figure 2.9 shows how CO_2 levels have changed.

When the idea of climate change was first proposed, it was dismissed. Over time, as more data has been collected, both climate change in the past and potential changes to the climate in the future have become established as an accepted scientific theory.

Advances in technology

As early as 1895, investigations into how infrared radiation was absorbed by different gases found that water vapour, methane (CH_4) and CO_2 absorb infrared. Over time, technology has resulted in the improvement of such experiments and, therefore, more reliable data. Technology has also made it possible to look further into the past to determine how climates behaved before human records began.

Gaining access to the polar regions and having the equipment to take samples of the ice there were two major developments in the investigation of climate change. Ice is one of the best available records for past climate estimates. It is formed as snow accumulates each year. The weight of the snow compresses the previous layers of snow, eventually forming deep ice sheets. When the snow falls, it traps air bubbles in the ice. Those bubbles preserve information about the atmosphere at the time it was formed. They then deposit climate information from that time onto the ice sheet, such as the gaseous composition of the air, or other atmospheric pollutants. Scientists access this ice by drilling out cores (Figures 2.10, 2.11) and analysing them.

Figure 2.10: Researchers drilling ice core samples for scientific research. These samples can be analysed to tell us about the climatic conditions at the time when the ice formed.

2 Environmental Research and Data Collection

Figure 2.11: An ice core showing a black band. This indicates a period of volcanic activity during the formation of the ice.

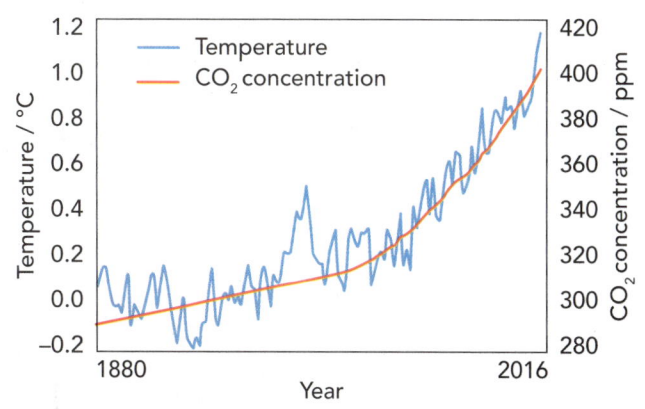

Figure 2.12: Increases in the temperatures and CO_2 levels in the atmosphere in the last 136 years. ppm refers to the concentration of CO_2 in parts per million. This is the number of carbon molecules for every million gas molecules in a sample.

Scientists use a range of technological developments to monitor climate change. For example, the same technology which makes it possible to analyse information in sediments and tree rings has also made it easier to determine changes to climates in the past. More recently, weather stations and precise weather equipment have resulted in reliable global data, while satellites can also track many different types of weather data, for example temperature, precipitation, ocean temperatures, greenhouse gas levels.

The ability to increase data reliability by using new technology helps to support climate change theory. The unreliability of data produced in the early days of climate change data collection has resulted in **sceptics** who do not believe in the phenomenon.

More recently, precise records of temperature, rainfall and greenhouse gases in the atmosphere have made it possible to prove that there are changes occurring, and that CO_2 levels are rising rapidly. Burning fossil fuels has been proven to release vast quantities of CO_2. This further supports the theory that humans are exacerbating recent climate change. Figure 2.12 shows how closely information taken from recent ice records links to data collected from the atmosphere.

KEY TERM

*****sceptic:** a person who doubts or does not believe in a concept or hypothesis

Scientific bias and misuse of climate data

Changes to global climate are complex. This means scientists face the challenge of proving without doubt or bias that carbon emissions are the only cause of the changes being experienced. Some people doubt the data that has been collected, and apply their own bias to interpret it. Some argue against the existence of climate change for reasons which range from misuse of data, to selective use of data, to lack of understanding of what the data suggests.

Bias and unreliable data have, in the past, led to the **false reporting** of scientific information regarding climate change. As a result, doubt has been cast over its causes. In addition, funding influences which research topics get addressed, and this leads to further bias. Groups who want to prove that climate change does or does not exist will fund research supporting their argument. This results in unbalanced data which has significant limitations in that it does not equally address all data. This is an example of **confirmation bias**, where a researcher seeks information that supports what they already think is true. Confirmation bias casts further doubt in the minds of the public and the media, who may not have a sound scientific understanding of climate change.

KEY TERMS

false reporting: the reporting of information that is false, fabricated or biased

*****confirmation bias:** when data that does not fit with the hypothesis is ignored. Data is then interpreted to support the hypothesis, even when some of it may not

Other arguments support climate change sceptics. Some argue that **volcanism**, **oceanic circulation**, changes to Earth's orbit around the sun and **solar variation** cause warming of the climate. The biased use of data related to these factors supports an argument against human causes for climate change. It also undermines the data that shows how human activities are causing climate change.

> ### KEY TERMS
>
> *volcanism: any process associated with surface discharge of molten rock, hot water or steam from inside Earth
>
> oceanic circulation: the large-scale movement of waters in the ocean basins through ocean currents and the oceanic conveyor belt system
>
> *solar variation: fluctuations in the amount of radiation output from the sun

Figure 2.13: A volcanic eruption, during which greenhouse gases are released back into the atmosphere.

Volcanoes
Volcanoes release carbon dioxide (CO_2), sulfur dioxide (SO_2) and dust particles into atmosphere. CO_2 emissions alter temperatures in atmosphere by trapping more heat. SO_2 and dust particles reduce levels of incoming solar radiation, potentially reducing atmsopheric temperatures. Volcanoes therefore can directly impact conditions in atmosphere.
Ocean currents
Ocean currents circulate heat around Earth, taking warmth from equator towards the poles and cooler water from poles to equator. Direction and strength of these currents can change as areas become warmer or cooler.
Insolation
Fluctuations in solar radiation and amount of energy reaching Earth can result in heating or cooling of atmosphere.
The orbit of Earth
The orbit of Earth around sun is thought to change over tens of thousands of years. If Earth receives more or less radiation as it is closer to or further from the sun, then the climate could be reacting to these changes.

Table 2.5: Natural causes of fluctuations in global climates.

The argument that volcanic activity, changes to ocean circulation, solar radiation fluctuations and Earth's orbit are responsible for the current rate of climate change is not supported by scientific data.

> ### ACTIVITY 2.7
>
> **The climate change debate**
>
> 1 Work in pairs. You and your partner should choose a different argument below to research. Make sure you each find at least three points to support your argument. You should then present the argument to your partner.
>
> a Climate change is a unrelated to human activity.
>
> b Climate change occurs primarily as a result of human activity.
>
> 2 Discuss your research findings with a new partner who chose the same statement as you. Did they include elements that that you had missed? Had they found anything you agreed/disagreed with?

CASE STUDY

Reporting on climate change: leading the reader astray

With climate change at the centre of many weather discussions and news reports, it is constantly in the media. As the frequency of floods, heatwaves and other extreme weather events increases, so does public awareness of the climate change discussion.

News articles that discuss the shrinking or expanding **sea ice** in northern winters show how the public can be misled. The title of one article on changing sea ice in Greenland read: 'For 25th year in a row, Greenland ice sheet shrinks'. (The United Nations News website (2022) *Climate change: For 25th year in a row, Greenland ice sheet shrinks*) This article discussed the long-term tracking of data on the Greenland ice sheet. It mentioned data that had been collected, and described the changes in the ice sheets around Greenland. It included information on how snowfall rates have changed and outlined the ice losses that were being observed. The article considered medium to short-term trends and what these changes might indicate. However, the title of the article suggested that climate change was causing long term reduction to the Greenland ice sheets. Figure 2.14 could be used to argue that the ice is receding in Greenland. However, unless images from previous years are compared to this image it is not possible to make that claim. Evidence must be presented to show the change.

In contrast, a year earlier, an article entitled 'The Media is lying about Greenland' was published (Real Clear Energy website (2021) *The Media Is Lying About Greenland and Climate Change*). This suggested that discussions around the loss of Greenland ice were alarmist, making the problem sound worse than it was, and that the media was promoting the climate change discussion. The article stated that in 2021, Greenland's **surface mass balance** of ice was higher than the 30-year average, in other words, that the amount of melting ice had decreased, and that more ice was forming. According to the writer of this article, the media was lying to the public about ice loss in Greenland. Climate change, the article attempted to persuade the reader, was a hoax.

Both articles contain valuable information. One discusses short to medium-term trends, while the other analyses recently recorded data. The two sets of data are important when discussing shifts in local climates. However, with climate change it is important to remember that one cannot look at a single incident and use it as evidence either for or against climate change. One needs to consider how the same incident fits into a long-term pattern. The greater the number of years and data sets, the more reliable the information. We cannot take one very hot year as sufficient evidence for climate change: we must consider the temperature trends for the location. The second article uses the fact that the public may not know about this concept to persuade readers against the existence of climate change. In other words, the writer is using the data to tell their preferred story.

Looking at Figure 2.15 and Figure 2.16, it is clear that the 30-year data suggests a decline in Greenland's ice coverage. However, if we look at the data for the year 2011–2012, we see a clear increase. Short-term data is of little value when compared to long-term trends. Misuse of data such as this can mislead the public. A report can tell two very different arguments, depending on the data used.

Figure 2.14: The ice sheet in Greenland showing areas where ice is not present.

KEY TERMS

sea ice: the ice that floats on the surface of the oceans and seas

*surface mass balance: the balance between the build-up of and loss of glacial surface

CAMBRIDGE INTERNATIONAL AS LEVEL ENVIRONMENTAL MANAGEMENT: COURSEBOOK

CONTINUED

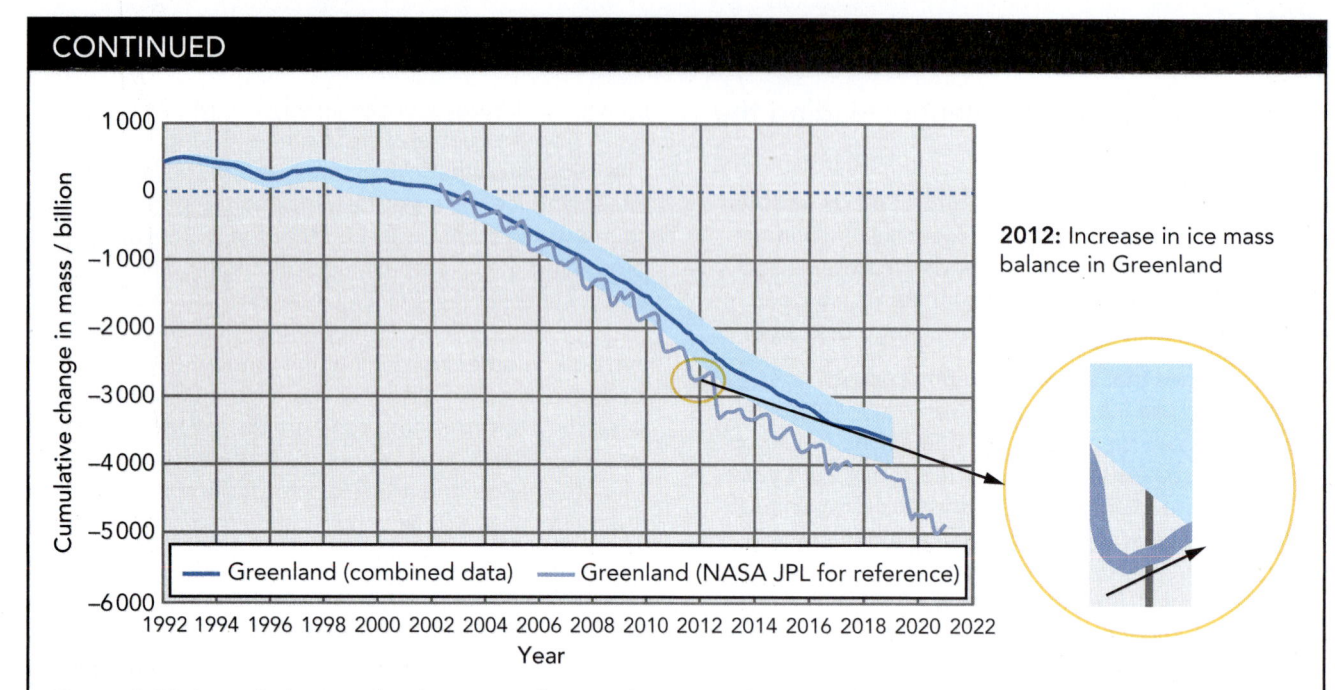

Figure 2.15: A graph showing the decrease in the cumulative mass balance of ice in Greenland.

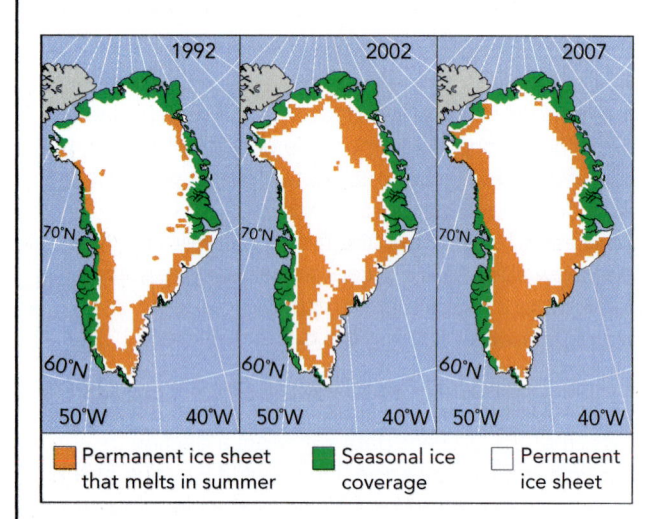

Figure 2.16: The change in Greenland ice coverage between 1992 and 2007.

a Consider the value of the data included within the article. How could it be used effectively within the climate change discussion?

b Did the heading of the article sensationalise the content in a bid to give it more relevance to the climate change debate than it perhaps deserved?

c Do you think the article is an example of responsible reporting?

d Do you think that some of the information in the article was 'fake news'?

e How do you think this article will influence public opinion?

Key search terms

News, shrinking/expanding sea ice, Antarctic ice, Arctic ice, climate change, climate sceptics, climate alarmist, wild fires, heat wave, fake news

Case study questions

1 Using some of the key search terms below, find a news article online. Read it and decide if it supports the climate change discussion or denies climate change. Is data used to support the discussion and influence the reader's viewpoint?

2 Write a summary on how the article could influence the climate change discussion. Use the points a–e above to guide you.

2 Environmental Research and Data Collection

CONTINUED

3 **Analyse** the different articles you have reviewed. Consider questions like:
   a Were some articles more informative than others?
   b How many of the articles promoted climate change scepticism? Did any seem alarmist?
   c Which article do you think was the most informative and why?

COMMAND WORD

analyse: examine in detail to show meaning, identify elements and the relationship between them

Climate models

Climate models (Figure 2.17) are generated from mathematical equations that use thousands of data points to predict climate change. These models remain highly uncertain despite huge amounts of scientific data. This is because some processes in the climate system occur on such a small scale, or are so complex that data is not available to reproduce them in the models. The uncertainty in the models can be used to support climate change scepticism if a model makes a prediction that is wrong.

Factors that the models need to account for include:
- temperature fluctuations
- wind patterns
- ocean currents
- land surface characteristics
- rates of ice melt.

Melting permafrost is increasing the rate of methane being released into the atmosphere. The complexity of this problem makes modelling very difficult.

The uncertainty of the models has led to false reporting of scientific conclusions, further fuelling arguments against the scientific claim that humans are responsible for the current climate change crisis.

KEY TERM

climate model: a computer simulation of Earth's climate system using mathematical equations. It seeks to simulate the outcomes of changes to factors that influence Earth's climate

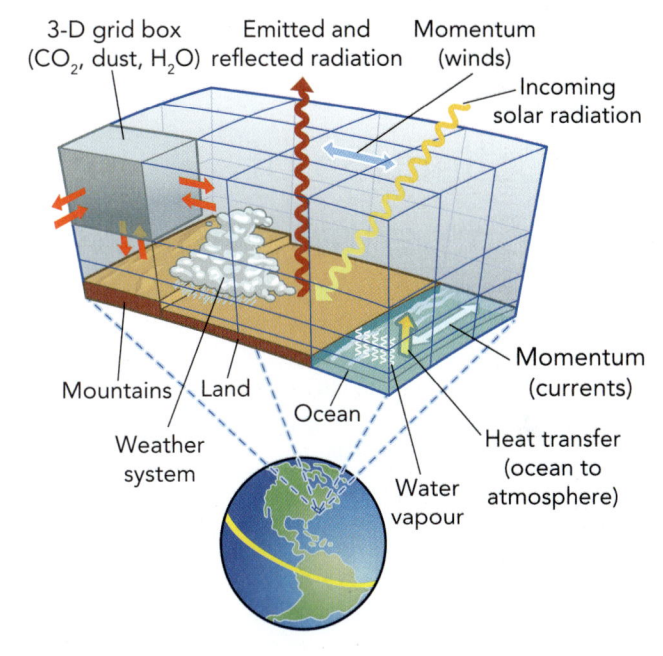

Figure 2.17: The concept of climate modelling, showing the complexity of data that interacts and needs to be considered when trying to predict changes to global climates.

2.2 Questions

1 **Define** the term climate change.
2 List the three main arguments concerning climate change.
3 **Give** two pieces of scientific evidence to show that the climate has been colder or hotter in the geological past.
4 **Explain** why reliable data is so important in scientific research.
5 Explain why ice cores are considered one of the best available records of past climate estimates.
6 Briefly describe the benefits and limitations of climate models.

> CAMBRIDGE INTERNATIONAL AS LEVEL ENVIRONMENTAL MANAGEMENT: COURSEBOOK

> **COMMAND WORDS**
>
> **define:** give precise meaning
>
> **give:** produce an answer from a given source or recall/memory
>
> **explain:** set out purposes or reasons / make the relationships between things evident / provide why and/or how and support with relevant evidence

2.3 Collection of environmental data

A sample is a set of data (number of plants, number of species, plant distribution) that is taken from a larger population for measurement. The sample needs to represent the variable being investigated. This ensures that the results can be used to generalise the findings to the population or area as a whole.

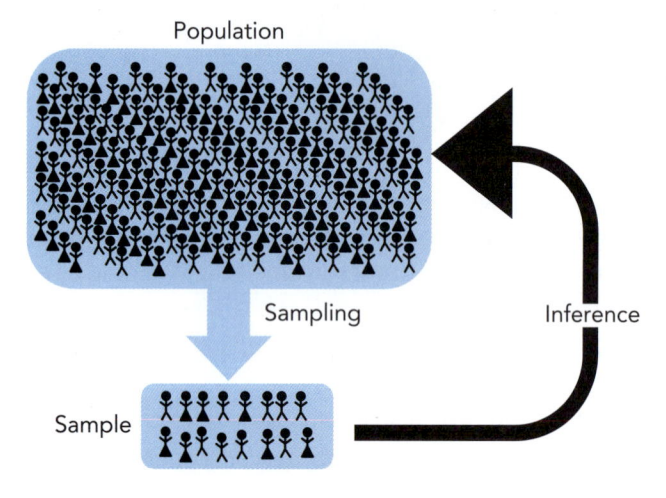

Figure 2.18: A diagram representing a population and indicating how a sample is taken from it.

Random and systematic sampling

There are various sampling strategies (Figure 2.19) you can employ when carrying out field work to collect **primary data**. Primary data may be used by another researcher to support or dispute their findings. It then becomes **secondary data**.

The strategies used for collecting data include random and systematic sampling.

Random sampling is when sample points are selected using random numbers. These can be generated using a computer system, or by creating a grid pattern of the study area. Each grid point is allocated a number. These numbers are then selected randomly, and those are the points sampled. This helps to minimise researcher bias. Random sampling is useful when the population size or the size of the individual samples is relatively small, and all individuals have an equal chance of being sampled (see Section 2.1).

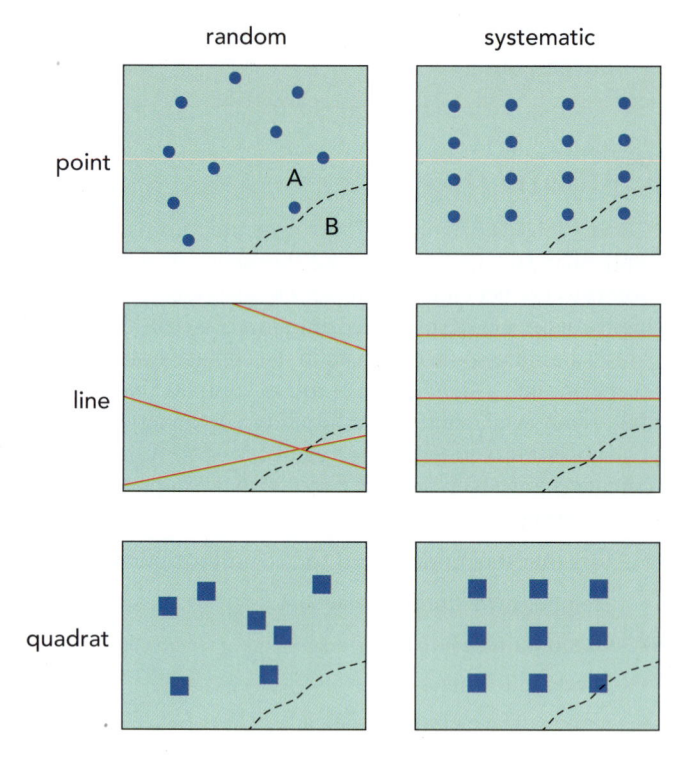

Figure 2.19: Different sampling methods.

Systematic sampling is when a regular pattern is used to identify sample points, e.g., asking every third house on the left-hand side of the street to answer a questionnaire or sampling every fifth square metre on a field grid pattern. Systematic sampling is more effective when data does not show patterns and there is a low risk of data manipulation by the researcher.

As sample sizes become larger and a researcher needs to create multiple samples, the sampling process can become expensive and time consuming.

KEY TERMS

*primary data: information that is collected by the researcher (e.g. rainfall which is collected daily and recorded)

*secondary data: data that is collected by somebody else in a separate investigation (e.g. climate data from the local airport)

random sampling: samples based on drawing names/numbers out of a hat or using a computer program to give a random list

systematic sampling: choosing a sample based on regular intervals rather than random selection

ACTIVITY 2.8

Sampling strategies

Read the following three research project examples. Decide whether random or systematic sampling should be used. Write an explanation supporting your choice.

1. A research team is seeking opinions about household cleaning products in a town of 300 000 people.
2. A factory has 70 assembly line employees that the supervisor must carry out 'spot observations' on without bias.
3. A 5 km² mountainside side has a grassland ecosystem. The research team wants to estimate biodiversity across the area.

Suitability of sampling strategies and sample size

Different objectives require different sampling strategies. For example, if the objective is to measure the total release of toxins into a river, a 24-hour sample can be taken. However, if accidental releases are being monitored then continuous, ongoing individual samples need to be taken. Therefore, when planning an environmental study, it is important to carefully consider the strategies and methods that are being employed.

Researchers need to consider the size of the sample needed, how easy the sampling location is to access, existing knowledge of the environment being studied and previous similar studies that can guide the strategy chosen.

When it comes to sample size, the minimum size that will give meaningful results is 100, while the maximum sample size is 10% of the population, as long as this number does not exceed 1000. For example, if a survey is carried out on five towns with a total of 7200 households, it will not be possible to survey every household, due to cost and time constraints. In this case, the researcher will typically select a representative population from those 7200 homes. The sample size will depend on the level of accuracy the researcher is seeking. In this example, 10% would require a survey of 720 homes. However, imagine if the town had 20 000 households. Ten percent of 20 000 is 2000, which exceeds 1000. In this case, the maximum sample size would be 1000.

Where the population size is less than 100, then all of the individuals or sample points need to be included in the survey.

The minimum number is chosen when:

- budget and time are limited
- a rough estimate is required
- data is not going to be divided into different groups for analysis
- similar answers/outcomes are expected.

The maximum number of samples is chosen when:

- there is sufficient time and budget.

Whether you are using your own methods or ones previously used by others, you should look for bias, precision (how precisely data is collected), the efficiency of the process and if methods used are appropriate for the study.

2.3 Questions

1. **Define** the terms *sample* and *primary data*.
2. **State** how random and systematic sampling differ.
3. **Identify** the factors to consider when determining the minimum or maximum sample size needed.
4. **Calculate** the sample size for the following two examples:
   a. A researcher wants to find out the percentage of trees in a forest that have been infected with a fungus using random sampling. The area of forest being checked has 2750 trees. How many trees should the researcher include in the survey?
   b. A researcher is investigating the number of households that recycle their waste in three different towns:

- Town 1: has 98 households
- Town 2: has 10 700 households
- Town 3: has 2 970

How many houses should be sampled in each of the towns in order to get fully representative data?

> **COMMAND WORDS**
>
> **define:** give precise meaning
>
> **identify:** name/select/recognise
>
> **calculate:** work out from given facts, figures or information

2.4 Data collection techniques and data analysis

If you want to carry out data analysis at the end of the experiment, you need a sample large enough to allow you to draw sound conclusions from it. It is therefore important to decide how many samples you are going to take to get reliable results. This varies depending on the field work being carried out. For in-lab experiments, repeating a test five times gives sufficient baseline data to carry out statistical analysis of the information. The more data you collect, the more reliable your results will be. However, note that too much data can also be problematic as it is difficult to process and apply in a meaningful way. See Table 2.6 for methodologies.

Sampling techniques

Various sampling techniques are available for field work. It is important to select the correct one for the task at hand. Types of sampling in field work include point sampling, line sampling and area or quadrat sampling.

With all sampling techniques, there are benefits and limitations. Some of the limitations include sampling bias; lack of skill identifying species, resulting in misidentification or under-/over-counting of species; harm to individuals that are sampled; the length of time needed to obtain statistically sound samples; and the cost of working for long periods of time in the field. Many of the techniques are not difficult to implement, so the ease of use and low cost of producing the equipment is a benefit in the field. In addition, these methods result in quantitative data that can be used to generate statistics and graphs.

Point sampling

Point sampling is one of the most common approaches to estimating plant cover of a site. It is carried out by selecting a number of points within an area (either randomly or systematically) and determining how many of those points hit a specific plant species. In this way, total cover of the species can be calculated as a percentage of the number of times the points land on the plant species being investigated (see Figure 2.19).

Line sampling

In this approach to sampling, data is collected along a straight line and, for example, every plant that is touching the line is noted down. This is a time consuming approach, and so a combination of point and line sampling can be used to sample specific points along the line.

In transect/line sampling, the transect is laid out and **quadrats** are randomly or systematically placed along the transect (see Figures 2.19, 2.20).

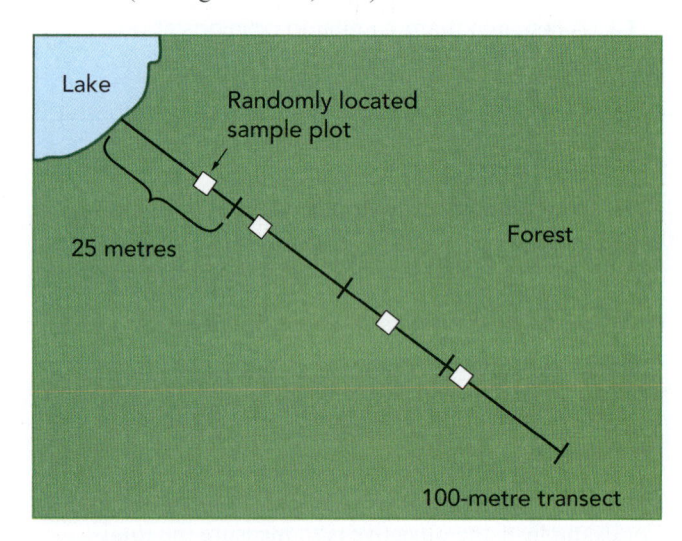

Figure 2.20: Line or transect sampling.

Area or quadrat sampling

Data is collected within a 4-sided quadrat of a pre-determined size. This technique can be combined with line sampling, as Figure 2.21 shows.

Quadrats are used to sample **sedentary** species, or species that move very slowly, such as snails. When sampling,

2 Environmental Research and Data Collection

the species is either present or absent. Data can then be used to determine the **frequency** of each species throughout a whole sample area.

An open frame quadrat is the simplest way to collect data for a species list. A quadrat is a square frame, usually made of wire or wood of specific dimensions (for example, 0.25 m × 0.25 m). The frame is placed on the ground so the plants found inside it can be studied.

> **KEY TERMS**
>
> **quadrat:** typically one square metre samples, selected for assessing the local distribution of plants or animals
>
> ***sedentary:** organisms that do not move, such as plants or rocky shore species like the barnacle
>
> **frequency:** how often a specific species (e.g. plant) occurs in a sample

Quadrat sampling can be used to determine the following:

- Number of individual plants.
- Number of individuals of a specific species.
- Number of different species.
- Estimated percentage cover (the % area of the quadrat covered by a specific species).

The information can either be expressed as percentage cover or as a local frequency. The quadrat can also be used randomly or systematically on a grid.

Limitations:

- Estimation can result in bias, miscalculation and, ultimately, unreliable data.

Benefits:

- Easy, quick and inexpensive.

A gridded quadrat is a quadrat that has been divided into smaller squares (a grid).

The number of grid squares that a plant is present in is counted and the percentage of cover is calculated. This method is more reliable than the open quadrat. However, the researcher needs to have sufficient knowledge to to identify different plant species. Where species are very similar, this can become difficult.

Figure 2.21: Quadrats can be used to sample an area or along a transect.

A point quadrat

A point quadrat (Figure 2.22) is a T-shaped frame used to carry out point sampling. The bar of the T has ten holes in it. To sample vegetation, a long pin is stuck through each hole. The plants that the pin 'hits' are identified and counted. To collect data on an abundance of a plant species, count the first hit made by the pin on each different plant (i.e. if a plant is large enough to be hit by two or three pins, it must only be counted once, but if there two plants of the same species hit by separate pins then it counts as two).

Limitations:

- If repeat sampling is required, the points are not easy to relocate

Benefits:

- Relatively quick to carry out.

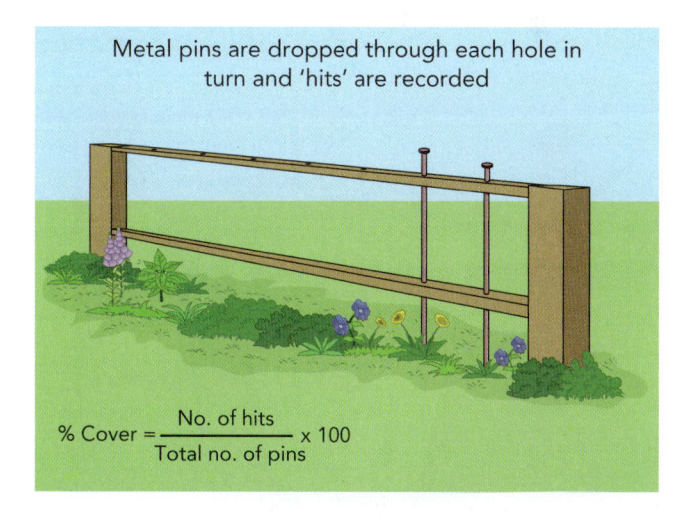

$$\% \text{ Cover} = \frac{\text{No. of hits}}{\text{Total no. of pins}} \times 100$$

Figure 2.22: An example of a point quadrat frame. The pins are dropped through the frame and the plant that the pin lands on is the one that data is collected for.

Pitfall traps

A pitfall trap (Figure 2.23) is a trap used to sample populations of insects that are active on the ground. This trap usually consists of a cup or beaker that is buried so that the lip of the container is level with the surface of the ground. The insects fall into the trap and can be identified when the trap is checked.

Some pitfall traps have a hole in the base to allow water to drain out in the event that it rains during the night, while other traps have antifreeze added to the cups. The antifreeze kills insects, preventing predatory insects from eating prey insects, and preventing the sample insects from escaping the trap.

Limitations:

- Predatory insects may eat other insects in the trap, so often alcohol or another substance is put into the trap to kill the insects.
- Identifying the insects requires training and skill.
- It takes time to visit the sites, the amount of sites included in a sample site must be limited.
- If left for too long, the trap can fill with water so insects are killed or washed away.
- The traps can only catch small insects, and flying insects could escape.
- Urban areas do not lend themselves to this type of sampling, as their surfaces tend to be hard, and containers need to be sunk into the ground.

Benefits:

- Traps are easy and inexpensive to set up.
- Traps can be left for long periods of time to increase the chance of catching samples.

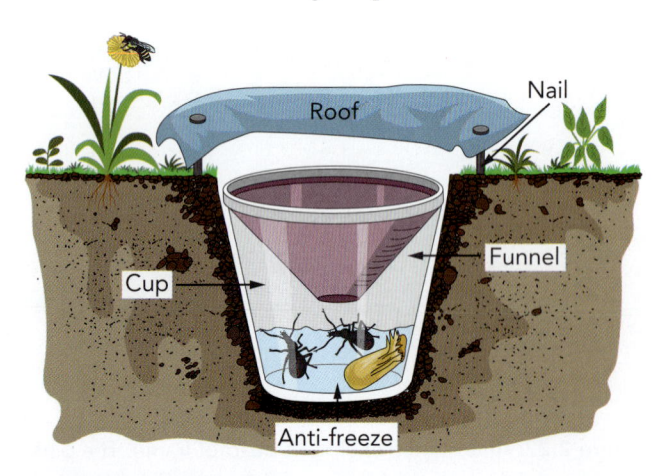

Figure 2.23: An example of a pitfall trap.

Sweep nets

Sweep nets (Figure 2.24) are nets that are mounted on a pole. They are used to collect insects and other invertebrates from long grass. The net can either be raised above the plants and moved in a figure of eight motion, or kept in the vegetation and swung in a pendulum motion. Once you have completed the sample, keep the net closed so that the insects cannot escape.

Limitations:

- Heavy vegetation does not lend itself to this type of sampling.
- Not all insects are equally 'catchable' so the sample may not represent all species present.
- Time consuming and could damage plants.

Benefits

- This method is inexpensive and easy to use, and does not require a lot of time or skill to complete. Volunteers can assist in the collection of samples.

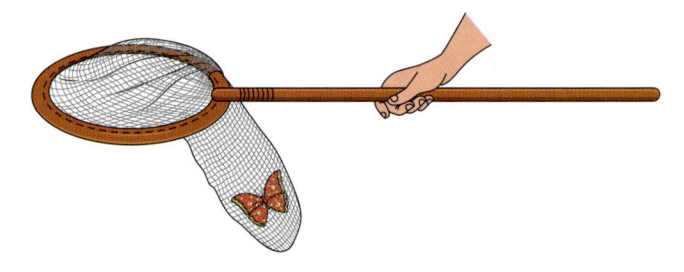

Figure 2.24: An example of a sweep net used to catch flying insects.

Beating trays/nets

A beating tray consists of light-coloured cloth (Figure 2.25) that is stretched across a square or rectangular frame. The frame is then held underneath a tree branch or a shrub and the foliage is shaken. The insects that fall from the branches land on the cloth and can be examined and identified.

Limitations:

- Flying insects may fly off as soon as they fall onto the tray.
- Plants can be damaged.
- Time of day can influence the number and type of species captured.

Benefits:

- The method is relatively easy to use.
- A representative sample can be obtained in a short period of time.

Figure 2.25: A beating tray being used to collect insects from bushes and trees.

Kick sampling

Kick sampling is widely used to sample invertebrates living on the bed of a river or stream. The sweep net is placed downstream of the point of sampling, with the water flowing into it. The researcher kicks the substrate of the riverbed for a measured period of time in each sample and the dislodged organisms are collected in the net for identification later (Figure 2.26).

Limitations:

- Small species may be missed if the net size is too large.
- Species that are stuck to rocks will be missed.
- This method is also difficult to carry out in very silty or shallow waters.

Benefits:

- It is quick to carry out and cost effective.

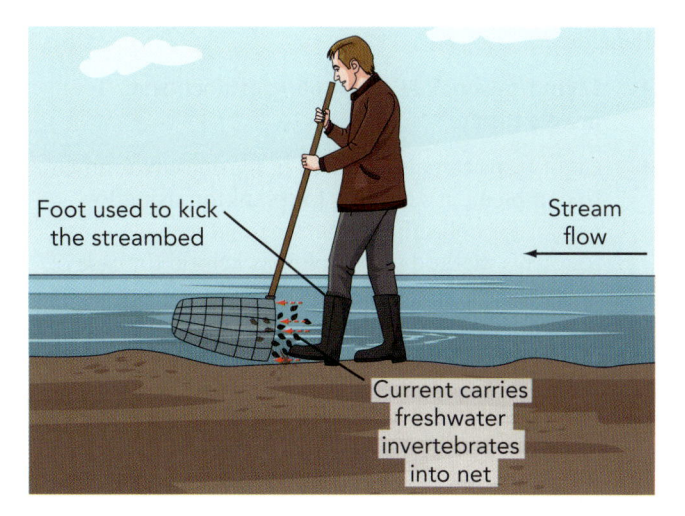

Figure 2.26: Kick sampling is carried out in shallow waters, by either a single person or working in pairs.

Light traps

Light traps (Figure 2.27) attract certain insects. The source of light can include fluorescent lamps and UV lights. These traps are widely used to collect data on nocturnal moths.

Limitations:

- Night-time temperatures and humidity may limit the abundance and types of species caught.
- Some insects are attracted by light at long range but repelled at short range, so their species and numbers may not be caught in the sample. To solve this problem, some light traps have a large base to catch insects.

Figure 2.27: Light trap for insects.

Benefits:

- Light traps make the sampling of nocturnal insects easier.
- Light traps can also be used in marine environments to complement sampling for plankton using tow nets or for collecting organisms that may avoid tow. This method of sampling is helpful for ease of identification and collection.

Capture–mark–recapture

The capture–mark–recapture technique is used to estimate the size of a population where it is not practical to catch all the individuals within it.

This is a live trap technique, where the individuals are trapped and then marked harmlessly. The investigator then records information about them (gender, weight, length, colouring, species and how they were marked,) before releasing them back into the population where they were trapped. When the area is resampled, those individuals that have previously been caught are easy to identify, and their recapture is recorded along with all the individuals captured for the first time. For larger animals and even insects, marking using tags may be used to identify the sampled organisms later (Figure 2.28).

Limitations:

- It is possible that the trapping or marking of an individual lowers its chance of survival.
- The marks may not be easy to see the next time the individual is trapped.
- An individual trapped once may have a lower chance of being trapped again through learned behaviour.
- Live trapping is used, so traps need to be checked regularly to ensure individuals come to no harm.

Benefits:

- Harm to the individuals can be kept to a minimum.
- This type of sampling can be carried out in remote locations and the exact sample point can be revisited at later dates.

Figure 2.28: A penguin which has been tagged and released to track the movement of the individual. When an area is resampled, it is easy to identify this individual.

Water turbidity

Turbidity is the cloudiness or haziness of water due to sediments (see Figure 2.29). Higher levels of turbidity have many negative impacts on aquatic life. Suspended sediments that cause turbidity can contain pollutants. They also block aquatic plants' light and smother organisms. Turbidity is measured by the amount of light scattered by the material in the water when light is shone on the sample. The higher the level of scattered light, the higher the turbidity of the water.

2 Environmental Research and Data Collection

> **KEY TERM**
>
> **turbidity:** the cloudiness or haziness of water; the lower the visibility the higher the turbidity

A Secchi disc, which is a 20 cm disc with alternating black and white quadrants (Figure 2.30), is a simple way to measure turbidity. The disc is lowered into a body of water until it is no longer visible, and the depth is measured. Secchi depth values that are low indicate low visibility and high turbidity. Secchi depth values that are high indicate high visibility and low turbidity.

Limitations:

- Turbidity is measured using simple visual assessment. However, this is susceptible to researcher bias. Equipment is available to give precise readings to avoid this problem.

Benefits:

- It is easy to make quick visual comparisons between sites.
- It is easy to make comparisons at a site over time.

The methods for each technique are detailed in Table 2.6.

Low turbidity ⟶ High turbidity

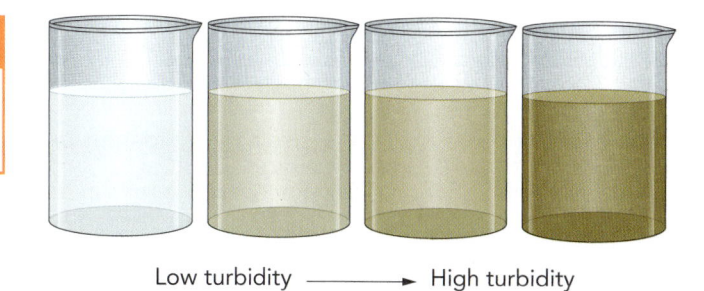

Figure 2.29: Image showing different levels of turbidity in water samples. As the sediments increase in the water, so does the level of turbidity.

Figure 2.30: Scientists in Antarctica using a Secchi disc to measure turbidity.

Technique	Method
Random and Systematic Sampling	Random sampling 1 Divide the area to be sampled into a grid using measuring tapes. 2 Using computer-generated numbers, or by pulling numbers out of a hat, use random numbers to generate coordinates within the grid on to sample. 3 Sample at identified point *using appropriate sampling equipment* 4 Collect relevant data, identifying any species or factors that need to be recorded 5 Record results in a table. 6 Calculate the average (mean) of the data collected. Systematic sampling 1 Extend a measuring tape from one side of the habitat to the other. 2 Sample at 0m on the tape *using appropriate sampling equipment*

Technique	Method
	3 Collect relevant data
	4 Use a key to identify each species (if relevant)
	5 Record the results in a table.
	6 Move the sampling equipment along the measuring tape.
	7 Repeat step 3–5 at regular intervals along the measuring tape.
	8 Continue until the full length of the measuring tape has been sampled.
	9 Calculate the average (mean) of the data collected.
	For both random and systematic sampling, ensure you standardise your sampling at each point to make data comparable. For example, the length of time you spend sampling a point, the size of the area sampled, or the number of samples taken.
Quadrat	Use the random sampling procedure with a quadrat:
	1 Place a quadrat at each coordinate identified in the random sampling process.
	2 Count the numbers of each species in each quadrat.
	Belt transect method
	The belt transect method is used when there is a gradual change from one side of a habitat to another, such as the change in light between the outer edges of a forest to the centre.
	1 Using the systematic sampling procedure, place the quadrat at 0m and then count the numbers of each species.
Point quadrat	After using either the random or systematic sampling process described above:
	1 Place a point quadrat at each coordinate identified in the random or systematic sampling process
	2 Push the pins of the point quadrat down through the holes and record the different plants that each of the pins touch ('hit')
	3 When using the point quadrat to measure abundance, only count the first hit made by a pin on an individual plant. Do not count the plant more than once.
Pitfall traps	After using either the random or systematic sampling described above:
	1 Choose a location for your trap, on flat ground, near vegetation
	2 Use a spade to dig a small hole
	3 Place the clean trap cup into the hole. Fill in any empty space around the cup with soil. Make sure that the top of the pot is level with the ground
	4 If you want avoid rainwater building up in the trap, make a hole in the cup
	5 Leave your trap overnight. If you are leaving it during the day, check it at least every few hours

2 Environmental Research and Data Collection

Technique	Method
	6 Empty the trap into a tray see what insects you have caught. Use ID guides to help you identify the insects
	7 Record your findings. Make a note of what you caught, the date and location.
	8 You could sketch the insects you have caught, or take photographs
	9 Carefully release the insects, returning them to a safe, sheltered place in the area that you found them.
Sweep nets	After using either the random or systematic sampling described above:
	1 Choose an area you want to sample
	2 Keep the net path parallel to the ground
	3 Tilt the net opening so the lower edge of the rim is slightly ahead of the upper rim
	4 Swing the net from side to side in half circle motion. Sweep one stroke per step as you walk through the sample area, or along the sample line
	5 In short vegetation, swing the net as deeply as possible
	6 In taller vegetation, sweep only deep enough to keep the upper edge of the sweep net opening even with the top of the plants
	7 Empty the trap into a tray see what insects you have caught. Use ID guides to help you identify the insects
	8 Record your findings. Make a note of what you caught, the date and location.
	9 You could sketch the insects you have caught, or take photographs
Kick sampling	After using either random or systematic sampling:
	1 Find an area suitable for shallow running water with a gravel or muddy bottom
	2 Hold a fine-mesh net in downstream of where you are standing
	3 Use one foot to kick the bottom of the stream. Organisms dislodged from the bed of the river will be washed into the net
	4 Always take the first sample at the lowest point upstream, then work back upstream.
Light traps	After using either random of systematic sampling:
	1 Install the light trap near or within the field where you want to trap flying insects
	2 When using electric bulb, make sure that the bulb and wiring are not in contact with water to avoid electrocution
	3 Place the shallow basin with soapy water underneath the light to capture insects that fall
	4 Use the light trap from early evening - 6:00 PM - until 10:00 PM
	5 Collect the trapped insects daily. Use ID guides to help you identify the insects
	6 You could draw the insects you have caught, or take photographs.

Technique	Method
Turbidity	This technique takes place in a body of water, on a boat.
	After using systematic or random sampling to identify your sample points:
	1 Sample between 10am and 3pm on a bright, calm day, because the angle of the sun and waves or ripples can affect readings
	2 Never wear sunglasses while making the reading because this can affect visibility of the Secchi Disc
	3 Lower the Secchi Disc into the water on the shady side of a boat, or out of the shade your body is casting
	4 Slowly lower the Secchi Disc into the water until it is no longer visible
	5 Record the depth
	6 Slowly raise the disc until it just becomes visible. Record the depth
	7 Repeat this at least 3 times in order to take average the depths from steps 1 and 2 to get the Secchi depth
	8 Repeat sampling at the same points at a later date to be able to determine if changes in turbidity are occurring.

Table 2.6: Methods for using equipment in the field.

INVESTIGATIVE SKILLS 2.1

Calculating population density using fieldwork techniques

Carrying out field work is an essential part of environmental management. Data makes up the information that is used to make decisions regarding the health of an ecosystem. For example, data can help to gauge threats to an ecosystem or to decide if ecosystem rehabilitation is required. It is important to know how to undertake fieldwork.

In this investigation you will calculate the estimated number of coloured paper circles in a sample area, using the quadrat technique.

You will need:

- Coloured paper. Use either small squares or circles punched out of the paper.
- Tape or chalk to create a grid (or quadrats if you have them available).
- The numbers 1 to 32 written down on individual pieces of paper (small squares).

Safety

- When working outdoors, always be aware of your surroundings. Note any dangerous wildlife in the area (e.g. snakes, insects, lions, bears, water or cliff hazards).

2 Environmental Research and Data Collection

CONTINUED

- Work in teams. Always know where the rest of your team is.
- Do not wander away from your group.

Getting started

- Review the use of a quadrat before you start. The grids you are going to create represent quadrats.
- Consider the meaning of the terms **population density** and **population size** Make sure you understand the difference between the two Do you understand what information you are looking for?
- Work through the tip box on calculating the sample area and make sure you understand it.

The investigation

This investigation has been designed to take place in a classroom. However, it replicates an investigation into the density and population size for flowers or plants growing in a given area of a field.

A learner wants to know how many circles of colour paper are there on a 2 m² desk. They divide the desk into 32 squares and carry out random sampling to estimate the number of circles of colour paper x. The learner then calculates the estimated population size of colour paper x using the formula:

Population density = $\frac{\text{Number of individuals counted}}{\text{sample area, square metres}}$

The learner uses the population density which was calculated using the first formula to calculate the population size. This time they use the following formula:

Population size = density × total sample area

Once complete, the learner has an estimate of the number of coloured circles in the total sample area.

Method

Working in pairs or small groups prepare:

1. Using coloured paper, punch a small handful of circles (or cut small shapes). The number of circles does not have to be known.
2. Measure out an area on a desk/floor of 1 m × 2 m (total area = 2 m²).
3. Create a grid of squares using tape or chalk.

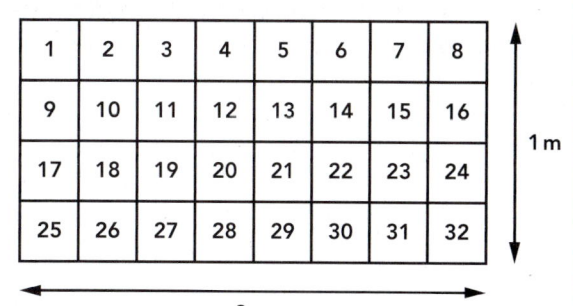

1	2	3	4	5	6	7	8
9	10	11	12	13	14	15	16
17	18	19	20	21	22	23	24
25	26	27	28	29	30	31	32

1 m × 2 m

Figure 2.31: The layout of the gridded quadrat required for the investigation.

4. Sprinkle the paper shapes over the sample area.
5. Place individual pieces of paper numbered 1–32 into a bag.
6. Take one of the pieces of paper out of the bag at random.
7. Count the number of paper shapes in the corresponding squares which are on the desk. For example, if you take out the piece of paper with 22 on it, count the number of shapes in square 22 on the desk.
8. Repeat this for four (0.25 m²) of the squares, recording the number of coloured circles in each square.

Questions

a. Why were only four squares sampled?
b. Use the data and the formulae above to calculate the estimated population density and size of the paper shapes.
c. Count the number of circles in four different quadrats and multiply each of them by 32 to get an approximate population size.
d. How much do the results for the four quadrats differ from each other. How much did these results differ from the result obtained when the formula was used?
e. Which answer do you think is more reliable, the individual square or the one from the formula? **Justify** your answer.
f. Are your results directly comparable to another table in the class? Explain your answer.

> CAMBRIDGE INTERNATIONAL AS LEVEL ENVIRONMENTAL MANAGEMENT: COURSEBOOK

> **KEY TERMS**
>
> **population density:** the number of individuals of a species living in a specific unit of area (e.g. square metre or mile)
>
> **population size:** the number of individuals in a population

> **COMMAND WORD**
>
> **justify:** support a case with evidence/argument

> **TIP**
>
> **Calculating the sample area**
>
> The sample area is not the 2 m². It is the area that coloured paper was counted in.
>
> For example, if you sampled four of the 32 squares then:
>
> One square is 25 cm x 25 cm = 0.0625 m².
>
> There are four of those, so it is
>
> 0.0625 m² x 4 = 0.025 m².
>
> The sample area in this example is 0.025 m².

Questionnaires and interviews

A **questionnaire** is a research tool in which a series of questions is created for respondents to complete. The survey is used to collect information on social issues, such as how the COVID-19 pandemic affected the food security of a population.

Most questions in a questionnaire are **closed questions**. These are questions which need a yes or no answer or where the respondent can choose their answer from a range of tickable boxes. This allows statistical analysis of the quantitative data. **Open questions** seek a longer answer from the respondent and can also be included to collect qualitative data.

In research, a **pilot survey** is a small-scale preliminary study that is conducted to test the proposed investigation before it is used in the full study. The pilot study determines if the proposed research method is reliable.

It can produce both unbiased data and a procedure that is easy to replicate. The pilot study helps the researcher decide if changes need to be made to the investigation before time and money are invested in it.

Limitations:

- Completed questionnaires and interviews may contain unanswered questions or dishonest answers.
- The person filling in the questionnaire may interpret or understand the question in a different way from the one intended.
- It is hard to convey feeling or emotions through these types of surveys.

Benefits

- They are inexpensive to carry out.
- They are practical and achieve fast results.
- Their information is also comparable and easy to analyse.

An **interview** is the collection of data in a direct, in-depth conversation (either in person or on the phone) held between the researcher and the respondent. It is carried out for the purpose of research and the two parties interact on a one-on-one basis. It allows for a full exchange of information.

> **KEY TERMS**
>
> **questionnaire:** a form with a series of questions for respondents to complete which is designed to seek data for an investigation
>
> ***closed question:** yes or no answers, or answers where the respondent can select an answer from tickable boxes
>
> ***open question:** questions that allow the respondent to give a free-form answer with opinions and detailed information
>
> **pilot survey:** a survey carried out prior to a full-scale study. Designed to identify areas of concern or areas for improvement before the full study is carried out
>
> **interview:** when people meet face to face, or via the phone, with one person asking questions and another answering them

The interviewer clarifies ideas and questions to the respondent and the respondent can ask questions and answer in a way that is fully comprehensive. Interviews can be carried out in person or over the telephone. An interview consists of open questions.

Limitations:

- People are required to conduct the interviews, and this can be costly.
- The quality of data collected by the interviewer can also vary, depending on their ability. Unlike questionnaires, the sample size is also limited.

Benefits:

- Face-to-face interviews allow for more accurate screening.
- They also offer an opportunity to capture both verbal and non-verbal cues (facial expressions and body language).
- The interviewer can keep the discussion on track.

Figure 2.32: An interview being carried out.

ACTIVITY 2.9

Creating a questionnaire

You have been asked to take part in an investigation on the subject of 'The Great White Shark and Ecotourism'.

As a class, create a survey of approximately ten questions to answer the following question:

Do local surfers think that chumming (attracting great white sharks with bait for the benefit of cage divers) increases the chance of a shark attack on surfers?

1. Identify your research aims and the goal of the questionnaire.
2. Define your target respondents (make the survey anonymous as it is not about the individual but about the overall data collected). Your questions should seek out the right individuals and exclude the wrong ones.
3. When setting the questions try to find answers to these questions –
   - Are respondents aware of a shark attack problem?
   - What is their opinion on humans feeding sharks to attract them?
   - Ask yourself: will my survey questions generate the information I am looking for?
4. Run a pilot study by giving your questionnaire to students in another group. This will check your questions for clarity.
5. Revisit your questionnaire to determine how it could be improved. Summarise is no more than 3 sentences how you would improve it.

As a class, look at the questions you have set. Discuss the types of graph you could create to show your results.

Create a final written or typed copy of your questionnaire, showing both your questions and layout.

Extended activity: Set three interview questions you could ask a surfer that would reveal more information for your investigation.

89

Data analysis using statistical tools

Figure 2.33: Data collected from around the planet is analysed using a wide variety of statistical tools.

Once data has been collected, you can process your results in a number of ways. Statistical analysis of data is vital to make sense of the information. It can be used to determine patterns in data, or it can be used to show differences between data sets.

A wide variety of statistical tools can be employed.

Lincoln index

The Lincoln Index is used to calculate estimated population size. It is carried out using the capture–mark–recapture technique and the following equation:

$$N = \frac{n_1 \times n_2}{m_2}$$

N = estimate of population size

n_1 = number of individuals caught in first sample

n_2 = number of individuals (both marked and unmarked) captured in second sample

m_2 = number of marked individuals recaptured in second sample.

Typically, an area is sampled for organisms such as insects, using either random or systematic sampling. Every organism captured in each grid is marked and then released. Later, sampling is repeated in the same location, and the numbers of marked and unmarked individuals are counted. This data is then used in the Lincoln index.

WORKED EXAMPLE 2.1

USING THE LINCOLN INDEX

Question

After random sampling, 200 ducks were captured, tagged and released. Four weeks later, 180 untagged and 20 tagged ducks were sampled at the same location.

Step 1 $n_1 = 200$

Step 2 $n_2 = 180 + 20 = 200$

Step 3 $m_2 = 20$

Step 4 $N = 200 \times 200 / 20$

Answer $N = 2000$

This indicates that at the location where sampling was undertaken there is an estimated duck population of 2000.

Use the following data and the Lincoln index to calculate population size. Write out the steps you have taken to get to the answer.

A capture–mark–recapture method was used to determine the number of fish eagles on an island. The following data was recorded:

1. The number of fish eagles initially caught, marked and released = 22

2. The total number of fish eagles caught in the second sample = 14

3. The number of marked fish eagles in the second sample = 2

2 Environmental Research and Data Collection

Figure 2.34: An African fish eagle catching a fish from a river.

Note that the Lincoln index makes some assumptions:

1. That there is no emigration or immigration by the population. In other words, the population being studied is a closed one.
2. The time between samples being taken is small compared to the life span of the organism being sampled.
3. The marked organisms have mixed with the rest of the population in the time between samples being taken.

Simpson's index

Simpson's index is used to calculate the estimated biodiversity in an ecosystem. It takes both the number of species present and the abundance of each species into account. Simpson's index can be used with data that shows types of species, genera or families. It is carried out using quadrat and transect sampling techniques and the following equation:

$$D = 1 - \left(\Sigma\left(\frac{n}{N}\right)^2\right)$$

D = diversity

Σ = sum of (total)

n = the number of individuals of each type present

N = total number of individuals of all types present in the sample.

Diversity ranges from 0 to 1 with 1 representing very high diversity and 0 representing zero diversity. The closer the result is to 1, the higher the diversity.

Figure 2.35: A school of fish in the Sucuri River, Brazil.

WORKED EXAMPLE 2.2

USING SIMPSON'S INDEX

Question

The diversity of ground cover vegetation in a given area is tested using sampling with random quadrats. The number of plant species in each quadrat, as well as the number of individuals of each species, is recorded.

Species	Number (n)
a	5
b	8
c	2
d	7
e	3
Total	N = **25**

> **CONTINUED**
>
> Watch what happens when you put these figures into the Simpson's index:
>
> **Step 1**
>
> $D = 1 - \left(\Sigma \left(\frac{n}{N} \right)^2 \right)$
>
> **Step 2**
>
> $D = 1 - \left(\left(\frac{5}{25} \right)^2 + \left(\frac{8}{25} \right)^2 + \left(\frac{2}{25} \right)^2 + \left(\frac{7}{25} \right)^2 + \left(\frac{3}{25} \right)^2 \right)$
>
> **Step 3**
>
> $D = 1 - ((0.2)^2 + (0.32)^2 + (0.08)^2 + (0.28)^2 + (0.12)^2)$
>
> **Step 4**
>
> $D = 1 - (0.04 + 0.1 + 0.006 + 0.08 + 0.01)$
>
> **Step 5**
>
> $D = 1 - 0.236$
>
> Answer: $D = 0.76$
>
> Here the 0.76 represents probability. This means that if you picked any two individuals out of the sample at random there would be an 76% chance of the individuals being different. This means there is a high biodiversity.
>
> Now calculate biodiversity in a river using the following data and Simpson's index. Write out the steps you have taken to get to your answer.
>
> | Species | Number (n) |
> |---|---|
> | Trout | 83 |
> | Bass | 65 |
> | Carp | 110 |
> | Catfish | 9 |
> | Total (N) | N = |

Percentage cover and frequency using quadrat data

Percentage frequency reflects the probability of a species being found in a single quadrat in a sample area.

$$\text{frequency} = \frac{\text{number of quadrats in which the species is found}}{\text{total number of quadrats}}$$

% frequency =

$$\frac{\text{number of quadrats in which the species is found}}{\text{total number of quadrats}} \times 100$$

Percentage cover can be calculated using the same method. Percentage cover is the area covered by a species, rather than the number of individuals present. For example, with plants like mosses or grasses, it is difficult to count individual plants. Instead, the researcher estimates the percentage area covered by the plant in each quadrat. That information is then used to calculate average percentage cover across all the sampled quadrats.

Figure 2.36: A catfish.

> **KEY TERM**
>
> **percentage cover:** a measure of how much space an organism is taking up as a proportion of a specified area

WORKED EXAMPLE 2.3

USING THE PERCENTAGE FREQUENCY EQUATION

An investigation was undertaken to determine the difference in vegetation cover between ground cover where chickens are present and ground cover where they are not. Systematic sampling using transects was used in each area. In the area with chickens, squash flowers were found in six of the 30 quadrats, while in the area free of chickens it was present in 20 of the 30 quadrats.

% frequency in area with chickens = $\frac{6}{30} \times 100 = 20\%$

% frequency in area without chickens = $\frac{20}{30} \times 100$
$= 66\%$

The number of quadrat samples taken affects how reliable the results are. This means the number of samples must be large enough to be accurate, but not so large as to be unrealistic to complete. As the data is quantitative there is less risk of researcher bias.

Now calculate the frequency in the following example. The crown-of-thorn starfish feeds on coral. It is a predator that is native to the coral reefs. However, the overfishing of the main predator of the star fish, the giant triton (a large sea snail), for its meat and shell, has resulted in a significant increase in star fish numbers. This increase is having a negative impact on the corals. As a result, the ecosystem is out of balance.

Use the following data to calculate the frequency of dead coral when crown-of-thorns starfish are present.

Using random sampling and quadrats, an investigation was undertaken to determine the number of dead coral in areas where crown-of-thorns starfish outbreaks are present.

In areas with a starfish outbreak, corals were found to be dead in 27 of the 35 quadrats. In areas where no starfish were present, corals were dead in two of the 35 quadrats.

1. Calculate the frequency of dead coral. Show how you arrived at your answer.
2. Discuss the conclusion/s you could draw from these results.

Figure 2.37: An area with many chickens showing how the chickens have removed all the plant life compared with an area with good ground cover and no chickens.

Figure 2.38: An example of a crown-of-thorn starfish that feeds on coral in the Indian Ocean.

CAMBRIDGE INTERNATIONAL AS LEVEL ENVIRONMENTAL MANAGEMENT: COURSEBOOK

SELF ASSESSMENT

- Have you worked through the examples? Do you understand how to use each of the statistical formulae?
- Were you able to work through the statistics activities successfully, using the examples?
- Do you understand what the results of your calculations mean?

Estimated abundance using the ACFOR scale

Estimated **abundance** can be calculated using quadrat data and the qualitative ACFOR scale. This provides faster results than percentage cover, but it is more subjective. The categories are:

- **A**bundant
- **C**ommon
- **F**requent
- **O**ccasional
- **R**are
- none.

Using this scale, plants identified in each quadrat are allocated an abundance letter: A, C, F, O or R. If a plant is not present, it would be recorded as N. Any plant that covers most of the quadrat would be an A. The benefit of this technique is that it is easy to use. However, limitations include researcher bias (plants with flowers are often overestimated, while inconspicuous plants are missed or underestimated). Estimates also leave a margin for error, and the researcher needs to have strong botanical knowledge to identify similar but different plant species.

To minimise researcher bias, use a gridded quadrat (10 × 10 squares each) and count the number of squares that are at least half-occupied.

KEY TERM

abundance: calculating abundance means counting the number of a specific organism present. Abundance can be low, with few individuals present. Where abundance is high many of the identified organisms are present

2.4 Questions

1 **Define** the terms sedentary and pilot study.
2 **Explain** the difference between an open quadrat and a gridded quadrat.
3 When would you use the Lincoln index as a statistical tool?
4 What can the data collected using quadrats be used to determine?
5 **Explain** how either pitfall traps or kick sampling are used in the field for sampling a mobile species.
6 When is the use of the capture–mark–recapture technique most appropriate?
7 Briefly **explain** how a questionnaire and an interview differ.

COMMAND WORDS

define: give precise meaning

explain: set out purposes or reasons / make the relationships between things evident / provide why and/or how and support with relevant evidence

Planning an environmental investigation

An environmental investigation is a process that follows the scientific method. It is used to identify environmental risks or existing problems. Environmental investigations can range from undercover investigations (to determine if environmental harm is being caused through illegal activities, for example), to using data in order to ensure that a proposed development causes minimal environmental harm.

An environmental investigation first needs to have a question or an identified problem. From this, the researcher develops their hypothesis. Once the hypothesis is set, a plan needs to be designed for the collection of data in order to prove or disprove the hypothesis.

There is normally a principle underpinning the hypothesis. This is the reasoning behind the hypothesis that has been proposed. For example, your hypothesis might be: Rainfall is decreasing over time in the town of Garrick. The principles underpinning this hypothesis could be discussions concerning climate change records for the area and changes to the vegetation (a decrease in vegetation, wildfires occurring more frequently, desertification taking place etc.). This hypothesis could be proven incorrect. For example, the change in available water may be due to its overuse by humans in the area, or over population by another species.

The researcher will then predict what the expected outcome is. This prediction can be based on previous experimental work, or established scientific fact that supports a specific potential outcome (the principle underpinning the hypothesis). Both the prediction and hypothesis can be proven incorrect.

When planning the methods that are going to be used, the researcher needs to consider the following:

- How will the primary data be collected?
- Is secondary data available to further develop the results?
- What equipment is needed?
- What are the costs involved?
- Is it possible to collect relevant data with the resources that are available?

Once the researcher has collected the data, they need to analyse the results to determine what they reveal. They then draw conclusions based on both the data collected and on existing information. If their hypothesis is proved incorrect then the investigation needs to start again to determine the cause of the identified problem.

Figure 2.39: A researcher carrying out soil sampling in a pineapple field.

ACTIVITY 2.10

Planning an environmental investigation

In pairs or groups of three, plan an environmental management investigation of your choosing. The investigation must include both independent and dependent variables, and an identified control. The results produced by the experiment must be quantitative. In other words, it must produce data that is numerical.

By the end of this task, you must produce a written or typed report with the following headings:

Question: a question based on an identified area of concern, such as pollution in a river.

Hypothesis: a precise, testable statement predicting the outcome of a study that is designed to answer a specific question.

Introduction: an outline of the topic of discussion so that it is clear what is being investigated. State the aim of the research here too.

Principle underpinning the hypothesis: The reasoning for the hypothesis that has been proposed.

Methods: Methods need be well laid out and planned. Someone else may wish to replicate your experiment in the future.

Steps:

1. Identify the environmental problem or area of interest.
2. Refine the research problem as a question and set the hypothesis.
3. Carry out research to determine the findings of previous similar studies. Write an introduction, with the principles underpinning the hypothesis (what knowledge supports the hypothesis you have proposed).

> ### CONTINUED
>
> 4 Plan the experimental study.
>
>   a Decide on the independent variable and range of values you are using.
>
>   b List the variables that need to be controlled or considered.
>
>   c Decide how to monitor the control variables, or how to keep them the same (keeping them the same in field work is not possible).
>
>   d Determine the frequency and duration of the sampling.
>
>   e Plan which equipment, materials and measurements you are going to need.
>
>   f Consider and list the potential limitations of the experiment.
>
> **Discussion:** It is not necessary to complete the entire experiment. Discuss what you think the next steps would be if you carried out the investigation.
>
> Then present your plan to the class. Have an open discussion about what you could have done differently. How could feedback from your classmates help to improve the data collected and so determine if the hypothesis is true or false?

2.5 The use of technology in data collection and analysis

New technologies have resulted in equipment which meets specific fieldwork sampling needs. These include, but are not limited to, geospatial systems, satellite sensors, radio tracking, computer modelling and crowd sourcing.

The use of technology in data collection and analysis requires both finance and skill. The cost of technology is often high, and the operation and interpretation of data captured using technology requires people who are highly trained.

Methods of environmental data collection using technology

Geospatial systems

A **geospatial or geographic information system (GIS)** is a system designed to capture, store, analyse and manage geographic data. This tool allows users to create interactive searches, analyse geographic information, edit data and present results. It uses maps and 3D scenes to organise layers of information and assists in revealing patterns and relationships between data sets.

Figure 2.40: An image showing how different layers are used to form a GIS map.

> **KEY TERM**
>
> **geospatial/geographic information systems (GIS):** electronic mapping systems designed to capture, store, analyse and manage geographic information

2 Environmental Research and Data Collection

Digital maps contain geographic information in layers of data (for example, road maps, rivers, topography, towns and forested areas). These can be overlaid to show how one layer fits in relation to another.

Figure 2.41 shows how these layers can be used. The hiking trail along a mountain includes information points along the trail. These points are accessed through the mapping system and give hikers important information, such as the location of camp sites, risk areas or communication points.

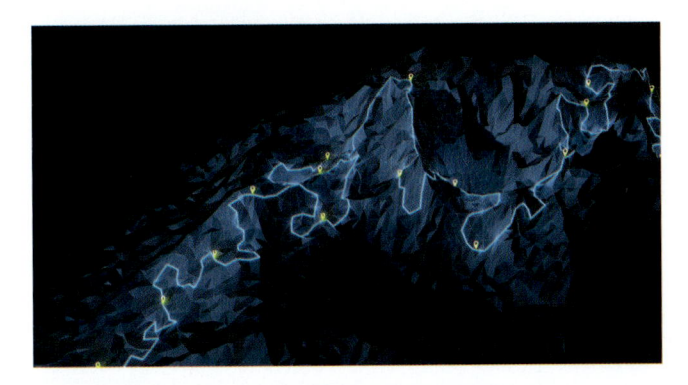

Figure 2.41: An image showing how a GIS map of a mountain can be used together with a map of a hiking trail to show the path the trail follows over the mountain range.

Figure 2.42 is an example of GIS mapping and satellite imagery to map the movement of an oil spill in the Gulf of Mexico. Oil is shown as the black marks on the map. The other information on the map includes the temperature of the oceans and the direction of ocean current flow. This information can be used to predict the movement of the oil.

Satellite sensors

Satellite sensing (also known as remote sensing) uses satellite information to track changes on Earth's surface or in the atmosphere. This technology can be applied to large-scale environmental monitoring. The use of medium- and high-resolution imagery, thermal infrared data, radar and laser technology makes it possible to record a wide variety of data. Many of the applications look at mapping and monitoring land cover changes over time. Some examples are:

- changes to vegetation/forest cover
- sea-level rises
- carbon dioxide and temperature changes in the atmosphere, oceans and on land
- land and sea ice monitoring.

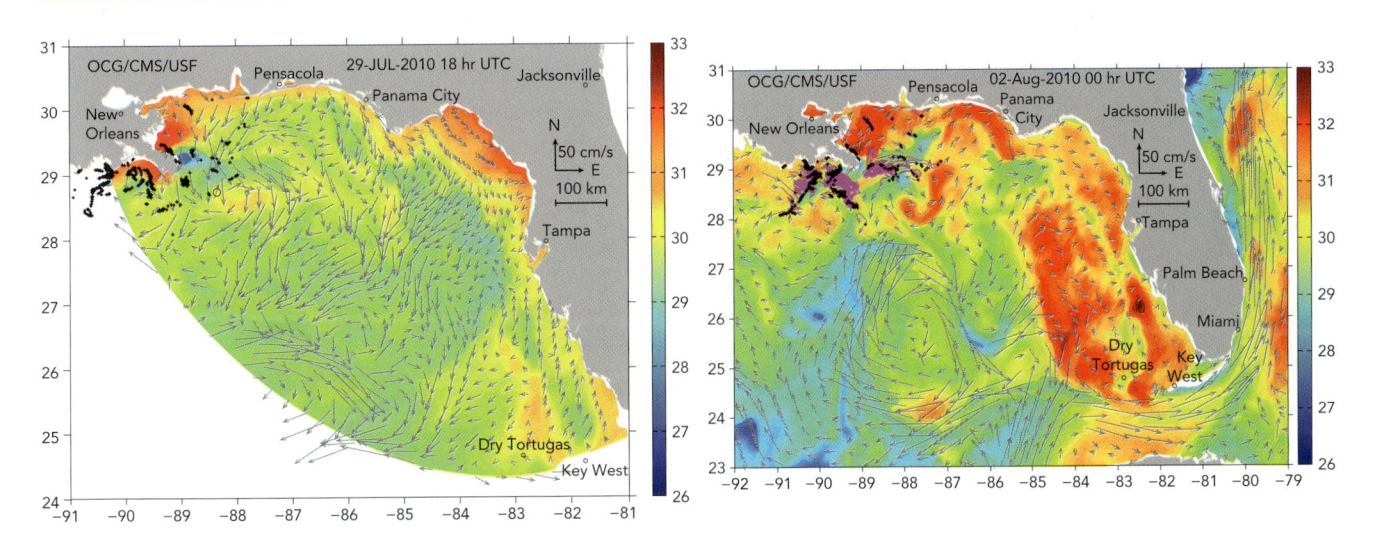

Figure 2.42: Images showing the movement of the Deep Water Horizon oil spill between 29 July 2010 and 2 August 2010.

97

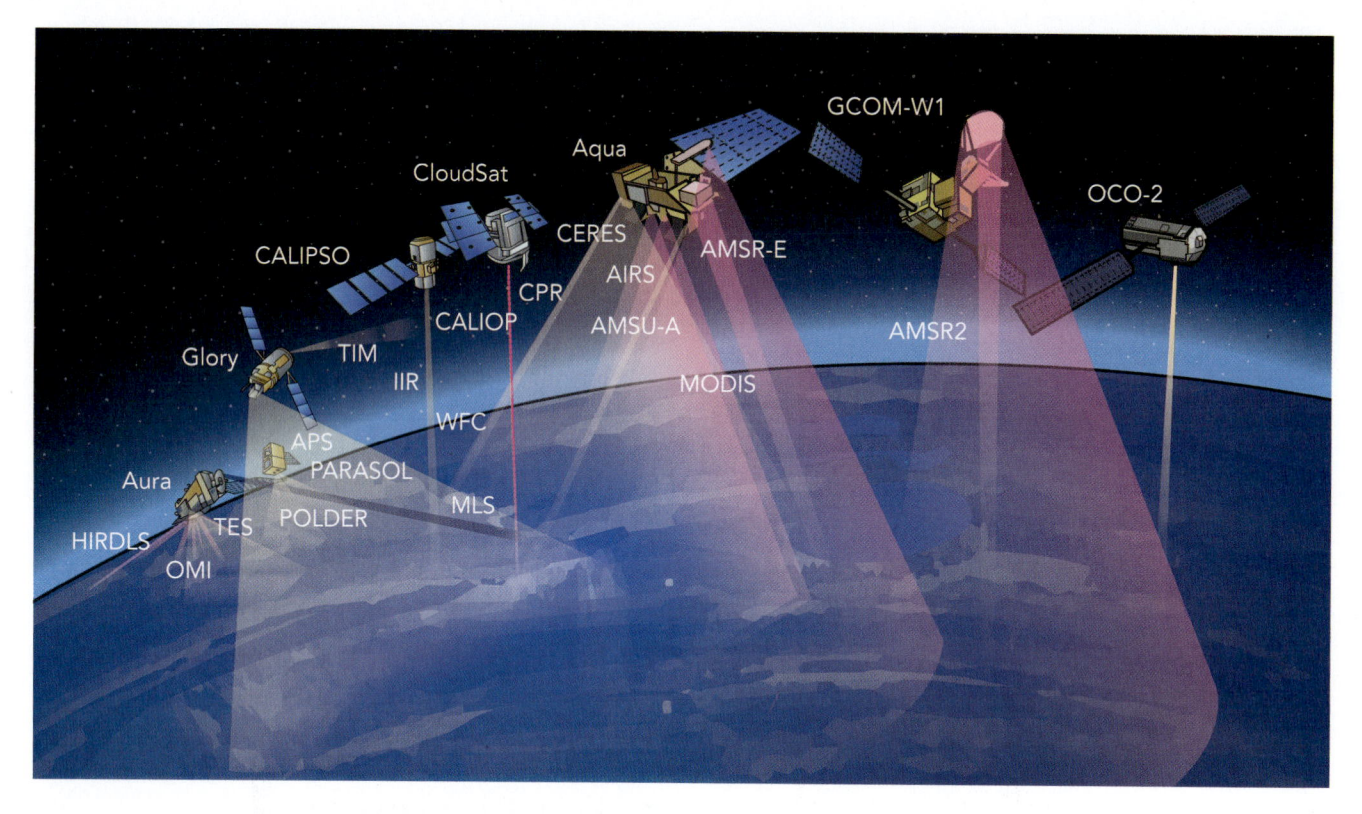

Figure 2.43: Various satellites, and the area that they are able to scan on the surface of Earth from space.

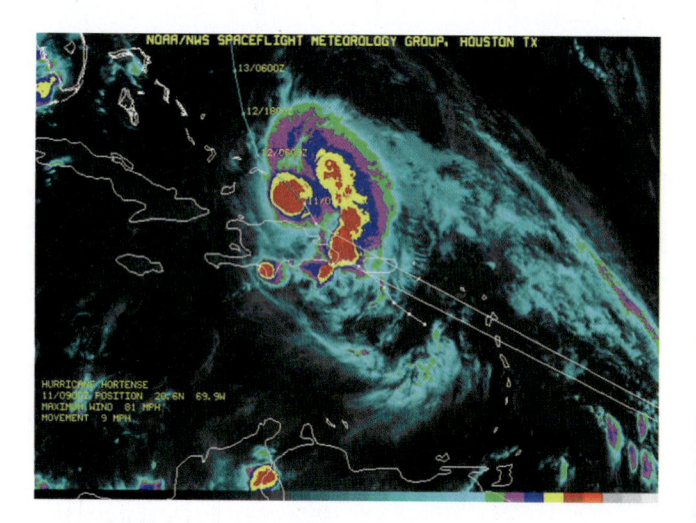

Figure 2.44: An infrared satellite image of a hurricane, showing estimated wind speed and rainfall intensity.

Radio tracking

Radio tracking (Figure 2.45) is a technology used to collect information about an animal through radio signals transmitted from a device attached to the animal. Radio tracking systems use radio frequencies to ensure that individual animals can be identified and tracked. Radio tracking gathers data about location, movement patterns and seasonal habits.

Figure 2.45: An African elephant wearing an anti-snare/poaching collar, which is used to track it.

Computer modelling

Computer modelling uses computer programs to analyse data and predict outcomes through mathematical analysis. Modelling allows for integrated environmental assessment, which is the assessment of an environmental question using a wide range of studies across disciplines which have a common environmental link. For example, an environmental issue may arise from numerous interacting factors – economic sectors, types of emissions, time, space, location, the ecosystem affected, impact on human health, biodiversity and financial costs, to name a few. Computer models determine the best solutions to complex problems.

Crowd sourcing

Crowd sourcing is the practice of engaging a group of people, or 'crowd', to collect data. In the environmental field, this may mean a farmer who has observed changes in how his crops are growing. Crowdsourcing collects information from scientists and the population in a region about their observations, changes they are experiencing and how they have managed to cope with the changes. This data can identify changes occurring in the environment and can be used to solve the problem.

Figure 2.46: Crowd sourcing collects data from a group of people in order to help answer a scientific question.

ACTIVITY 2.11

Satellite remote sensing research

Research the use of satellite and remote sensing in the study of climate change. Write 500–600 words explaining:

1. What satellite and remote sensing is.
2. What information it collects.
3. How the data reveals changes to the global climate.

Big data

Big data is extremely large sets of data that are collected using technology. This data is analysed using computers to reveal patterns, trends and associations. It is not just the amount of data that is important, but the information that can be obtained from analysing the it. Analysing big data allows for better planning and strategic decision making.

Big data is determined using five metrics:

1. **Volume**: the amount of data being gathered every second, measured in **petabytes** or **exabytes**, which are much larger than **terabytes**.
2. **Value**: the type of data being collected must have a use. If the information collected has no use, then it is of low value.

KEY TERMS

big data: extremely large sets of numerical information collected using technology and analysed using computers

*****volume:** an amount or quantity of something

*****petabyte:** a unit of information equal to one thousand million million (10^{15})

*****exabyte:** a unit of information equal to one quintillion (10^{18})

*****terabyte:** a unit of information equal to one million million (10^{12})

*****value:** to have a use, or a worth

3 **Variety**: this gives more **data streams**, in the form of continuous flows of data, to analyse for patterns.

4 **Velocity**: to be relevant, big data must be processed rapidly so that any trends the data reveals are seen quickly (for example in early warning systems ahead of a natural disaster).

5 **Veracity**: the data must be verifiable (possessing accuracy and integrity) to be trustworthy, and the data must add value. There is no point in collecting data that does not find answers to the questions posed.

Where any of these five metrics are absent, the use of big data becomes questionable and therefore less useful in analysing large complex problems.

Big data can be used to identify the root causes of problems in real time. It can help to understand complex concepts like climate change. It enables businesses and governments to optimise their resources for example, cutting back or managing the use of water or energy.

Big data has limitations. These include finding meaningful relationships between the data, securing and processing it. The volume of data is often so large that it can be difficult to determine what it indicates, or even what meaningful question to ask of it. Another limitation is the need for large, complex computer systems, with storage banks and speeds that can cope with the volume of information. These systems are expensive to design, install, manage and secure.

Data is collected using different variables or criteria. This can result in differences in collected data, which makes it difficult to compare. The data may be collected with different software, in different formats and in different languages. This makes the transferral of data from one source to another, and making it compatible, both challenging and time consuming. The skills required to carry out the management of big data tools and run the computer models that process the data are limited to highly trained professionals. Increasing the number of people that can process these types of data sets takes both time and money. Big data has the potential to reveal information that is currently not fully understood. However, the ways of getting that information are being developed alongside the technology required to process the data sets.

KEY TERMS

*variety: diverse data

*data stream: the process of transmitting a continuous flow of data, typically via data processing software

*velocity: speed with magnitude and direction

*veracity: the ability to verify or confirm data

conservation: the protection and scientific management of natural areas to protect biodiversity in a sustainable manner

2.5 Questions

1 Describe the main limitations of using technology in data collection and analysis.

2 Briefly describe the five metrics used in determining big data.

3 Summarise the main limitations of big data.

4 As a researcher, you are producing a GIS map for the **conservation** of rhinoceros in a game reserve.

   a **Suggest** the layers that you would include on your map to manage the reserve and potential threats.

   b What other technology could you employ to try and protect the rhinoceros in the park?

5 How can radio tracking help with nature conservation?

COMMAND WORD

suggest: apply knowledge and understanding to situations where there are a range of valid responses in order to make proposals / put forward considerations

2 Environmental Research and Data Collection

EXTENDED CASE STUDY

The climate change controversy: when misleading data blurs the facts

The climate change debate has faced several challenges. These challenges include bias, unreliable data, false reporting and even disagreements within the scientific community. All of these can lead to misleading and confusing messages being released. As a result, there has been a rise in scepticism along with confusion as to how we should manage our environmental impact. Conspiracy theories surrounding climate change have also gained momentum.

Unreliable use of data

Industry bias

The electric vehicle industry is a technological revolution that benefits economically from the climate change discussion. Public anxiety resulting from climate change has helped electric vehicles to become marketable as a climate-friendly alternative to gasoline vehicles (Figure 2.49). The industry often uses phrases such as 'zero emissions' and makes climate-friendly claims backed up by data. However, this is an example of unreliable data by omission, as there are emissions involved in producing and running electric vehicles.

Figure 2.47: Cars and electric vehicles are the subject of much debate.

The science

Electric vehicles have no exhaust pipe. When the engine is running, the vehicle does not release greenhouse gases into the atmosphere. Using electric vehicles in congested urban areas has the additional benefit of lowering the levels of air pollution around the town. The industry benefits if they highlight the lack of atmospheric pollutants released from the running of an electric vehicle. From Figure 2.48, we can see that petrol vehicles have the highest rate of emissions when considering both the manufacture and the combustion of petrol.

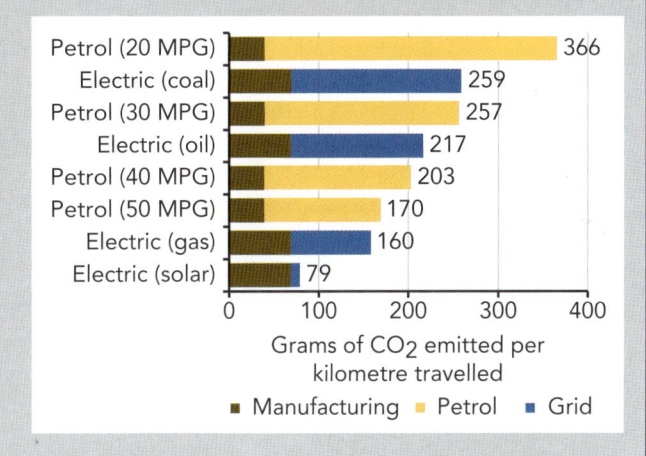

Vehicle type	Grams of CO_2 per km
Petrol (20 MPG)	366
Electric (coal)	259
Petrol (30 MPG)	257
Electric (oil)	217
Petrol (40 MPG)	203
Petrol (50 MPG)	170
Electric (gas)	160
Electric (solar)	79

Grams of CO_2 emitted per kilometre travelled
■ Manufacturing ■ Petrol ■ Grid

Figure 2.48: Carbon emissions for different types of vehicles, including both petrol and electric vehicles.

However, consideration needs to be given to the source of the electricity that is charging the vehicles. Running an electric vehicle off an energy source that is polluting will not result in net-zero emissions.

Another factor that needs to be considered is the fact that electric vehicles use rechargeable batteries. Battery manufacturing contributes significantly to atmospheric pollution. Figure 2.49 shows that carbon emissions during the manufacture of all electric vehicles are higher than emissions caused by petrol vehicles. Electric vehicles that are charged via coal or

CONTINUED

oil power stations actually have higher carbon emissions than some petrol-powered vehicles. Often, adverts for electric vehicles only highlight the benefits in relation to climate change. They do not reveal the ways in which electric vehicles contribute to climate change challenges, by requiring batteries, for example. This misuse of data or **greenwashing** of a product can lead to distrust in the science behind the electric vehicle.

Additionally, a climate change sceptic may exploit this misleading information to undermine transparent and robust scientific data showing a change in the quantity of greenhouse gases and their associated heating of the atmosphere. Sceptics may argue that the vehicle companies are only claiming zero emissions to increase their sales.

Figure 2.49: Electric vehicles are charged directly from the power grid.

KEY TERM

*greenwashing: the process of presenting misleading information (often to consumers) about how a product is more environmentally friendly than it actually is

Scientific disagreements

When unreliable data leads to false reports or climate researchers publicly disagree, there is a risk of confusing the public and strengthening climate change sceptics' arguments. When scientists disagree, the public may assume that that data is not consistent and that climate change is not real.

Figure 2.50: A group of scientists having a research discussion in a laboratory.

In 2016, a climate scientist accused his former colleagues of manipulating the data in a 2015 climate report, now known as the K15 Report. He claimed that the report had been rushed, and that data was manipulated to show higher global temperature increases. He added that the data had been handled unreliably and that key steps in handling the data had been omitted. He also said that some of the data had been adjusted. The scientist argued that the data had been changed in order to influence decisions at the 2015 Paris Accord climate talks. All of this suggests that the scientist who wrote the report had a climate change bias. It appears that the data was analysed to show results supporting the fact that

CONTINUED

climate change was occurring at an unexpectedly faster rate. This public disagreement between two scientists led to some climate deniers and politicians supporting climate scepticism. As a result, some politicans tried to limit the reduction in the burning of fossil fuels.

However, it is important to note that the scientist that made the accusations does believe in human-caused climate change. He was not saying that climate change does not exist. He was just questioning the way in which the data was analysed. He argued that there was misuse of data supporting the climate change discussion, and that this made the report unreliable. The public interpreted this to mean that the writers of an important report had manipulated data in favour of climate change. Other scientists have since replicated the report's findings to show that the K15 report was correct, and that the data is reliable.

The arguments put forward by both scientists are complex. Both involve the use of climate models and data archiving. However, the public argument gives climate deniers an opportunity to undermine any other climate change reports that are released. The argument is a technical discussion between two scientific viewpoints. The perceived unreliability of scientific data means that the public may assume that all data produced by scientists is unreliable. This casts doubt on whether climate change exists.

Extended case study questions

1. What is meant by the term 'climate change sceptic'?
2. Can climate change alarmism be as harmful as climate change scepticism to the climate change discussion? If so, why?
3. Explain how the 'greenwashing' of electric vehicles during marketing benefits vehicle companies.

Extended case study project

In small groups or as a class complete the following project:

Find three electric vehicle advertisements. These can be either film or newspaper advertisements.

a   Review the advertisements and compare the information given in each one.

b   What is the main message supported by each of the advertisements?

c   Create a summary table or poster of the information given in each advertisement.

d   Carry out research to find out more about the impact of the production of an electric vehicle on the environment.

- Do the advertisements 'greenwash' the product? How selective is the message being promoted?
- Add this research to your table or poster.

Figure 2.51: A group of people protesting with signs against climate change.

SUMMARY

The scientific method is a feedback procedure that seeks to prove or disprove a hypothesis.
Qualitative and quantitative data can be collected.
Testing a hypothesis must avoid scientific bias.
Scientific theory is formed out of the repeated scientific testing and proving of a hypothesis.
A variety of climate change data needs to be collected in a transparent and unbiased manner to ensure data reliability and to clearly support the climate change theory.
A number of theories argue against the existence of climate change.
Field work data collection is complex due to numerous variables. It can result in large data sets.
Sampling methods include both primary and secondary sampling, They also include the use of point sampling, line sampling and area sampling.
Various equipment is used to carry out field work. The correct techniques must be selected for each investigation.
Statistical analysis of data is required to prove or disprove a theory.
Constantly updated technology is being used to collect data, reduce bias and increase data reliability.
Big data is large sets of data that are collected using technology.

REFLECTION

What kind of activities in this chapter helped you further your understanding of new concepts?

Were there any concepts that you found particularly difficult to understand?
If so, what can you do to gain a better understanding of them?

PRACTICE QUESTIONS

1 Using the following sets of data and Simpson's index:

$$D = 1 - \left(\Sigma\left(\frac{n}{N}\right)^2\right)$$

   1 Calculate Simpson's diversity index for each location.
   2 Compare the differences in diversity between the two locations listed below.
      a An area of forest in Norway contains 122 pitch pines, 31 firs, and 42 red pines.
      b An African National Park contains 8 lions, 17 rhinos, 1219 impalas, 42 elephants, and 12 hyenas.

CONTINUED

2 An ecologist wants to determine if soil downhill from an illegal gold mine is contaminated with mercury.

The scientist collects 500 g soil samples from eight sites, at 25 metre intervals, along a transect that runs downhill from the mine.

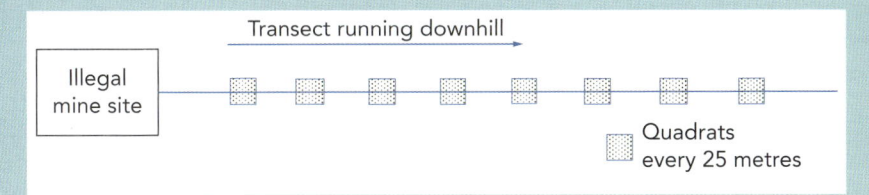

Figure 1

The scientist considers two methods:

Method 1
- Sieve each 500 g soil sample separately to remove any stones, plant and animal matter.
- Place 100 g from each sample into the same bag, and mix well.
- Test the mixed sample for mercury levels.

Method 2
- Sieve each 500 g soil sample separately to remove any stones, plant and animal matter.
- Test each sample individually for mercury content.
- Record the results for each sample.

   a Give **one** advantage for using method 1 over method 2. [1]

   b Using method 2 the scientist ran two transects of eight quadrats and got the following results:

Soil mercury content in mg/kg of soil		
Quadrat	Transect 1	Transect 2
1	73.8	81.4
2	45.8	59.0
3	36.1	47.1
4	20.4	22.3
5	9.8	11.2
6	4.5	6.1
7	2.2	3.4
8	0.7	2.1

Table 1: Mercury content.

> CAMBRIDGE INTERNATIONAL AS LEVEL ENVIRONMENTAL MANAGEMENT: COURSEBOOK

CONTINUED

    i   Plot the information in the table as a line graph.   [4]

    ii Describe the trend shown by the graph and suggest an explanation for the results.   [4]

    iii Calculate the range for transect 1:   [1]

c The scientist had proposed the following hypothesis: *Mercury concentrations decrease with distance from the mine site.* With reference to Table 1, state if this hypothesis is true or false.   [3]

> **Mercury fact sheet**
> - Mercury is a heavy metal that is toxic to living organisms.
> - Mercury is harmful if inhaled and may be absorbed through healthy skin.
> - Exposure can cause organ failure, cancers, brain damage, and even death.
> - Mercury can cause harm to eyes and lungs.
> - Action should be taken when concentrations of mercury reach 0.3 mg/kg in soil.

d With reference to the information in Figure 1, Table 1 and the mercury fact sheet, write a suitable conclusion about the health risks for both illegal miners and the researcher working on the site.   [6]

SELF-EVALUATION CHECKLIST

After studying this chapter, think about how confident you are with the different topics. This will help you to see any gaps in your knowledge and help you to learn more effectively.

I can	Needs more work	Getting there	Confident to move on	See Section
Describe what the scientific method involves.				2.1
Design an investigation.				2.1
Interpret data.				2.1
Determine whether data supports or refutes a hypothesis.				2.1
Explain the limitations of methods.				2.1
Understand and formulate hypotheses.				2.1
Define the terms reliable and bias.				2.1
Understand the development of historical data.				2.2

CONTINUED				
Explain the misuse of scientific data due to bias.				2.2
Outline the limits of unreliable data.				2.2
Describe strategies for collecting environmental data.				2.3
Understand the factors that influence the strategies used for collecting environmental data.				2.3
Use statistics to calculate estimated population size.				2.4
Use statistics to calculate biodiversity.				2.4
Use statistics to calculate frequency and abundance.				2.4

Chapter 3
Managing Human Population

LEARNING INTENTIONS

In this chapter you will:

- explore the dynamics and structure of human population
- calculate population density and dependency ratios and explain the factors that influence them
- identify ways in which birth, death and migration rate affect populations
- analyse the reasons for differences in population structure between high-income countries (HICs) and low-income countries (LICs)
- identify the impacts of an aging population on a country
- evaluate strategies for managing a changing population.

3 Managing Human Population

GETTING STARTED

Discuss the following questions as a class or in small groups.

1 Why do humans live in specific areas on the planet?

2 How do population structures differ between countries with high and low incomes? Why do you think this is?

3 Is the population in your country growing, shrinking or stable? What information can you find to support your answer?

4 In pairs, talk about the different ways that population growth can be managed. Share your ideas with the class.

5 As a class, discuss how managing population growth may be difficult.

COMMAND WORD

discuss: write about issue(s) or topic(s) in depth in a structured way

ENVIRONMENTAL MANAGEMENT IN CONTEXT

The North Sentinel islanders: people in balance with their environment

Populations that are in balance with their environment are often those with little or no modern medical or technological intervention. As a result, people may suffer from higher death rates, due to lack of access to safe water, modern medical care and a reliable food supply. They are likely to be subsistence farmers (people who produce enough food for themselves with nothing extra or left over) who live in small tribes. Medical care is usually in the form of natural remedies, and treatment is often administered by people whose skills have been passed down through the generations. An example of such a population can be found on North Sentinel Island in the Bay of Bengal, off the coast of India.

The Sentinelese are a small group of indigenous people (population size is estimated to be 15–400, but is not actually known). They are one of six native peoples of the Andaman and Nicobar Islands (Figure 3.1). The Sentinelese have a subsistence based survival that relies on the hunting and trapping of wild animals, and gathering wild plant foods, fish and shellfish. They are not believed to farm any plants or animals that they use as a source of food, fuels or tools. As food sources decrease, the population decreases, and as food sources increase, the population increases. The population lives in harmony with nature and in balance with the environment.

To protect the islanders' way of life, North Sentinel Island was declared a tribal reserve by the

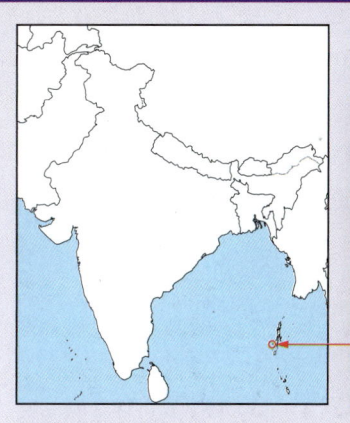

Figure 3.1: Aerial photograph of North Sentinel island with a map of India showing the location of the island.

Government of India in 1956. It is illegal to travel within 5.6 km of the island, and the Indian government maintains a constant armed patrol to prevent intrusion by outsiders.

Discussion questions

1 Why do you think the Indian Government protects the North Sentinelese population from outsiders with armed patrols?

2 Research a recent incident (from 2006 to the present day) involving North Sentinel Island. Share your findings.

3 Discuss how the Sentinelese population is considered to be in balance with the environment in which they exist.

3.1 Human population dynamics and structure

Various factors influence the growth of human populations and their impact on the environment. Factors include birth rates, death rates, increased population densities and access to resources such as water, food, shelter and minerals for mining. Understanding population growth trends is critical to environmental science because these trends help to determine the environmental impact of human activities. Rising populations put increased pressure on natural resources such as land, water, energy and biodiversity (Figure 3.2). As humans use resources, the risk of water and air pollution increases.

Population dynamics is the study of factors that influence population size (birth rates, death rates and **migration**) and population structures around the globe.

Figure 3.2: Dense crowds of urban population in Times Square, Manhattan, New York, USA.

> **KEY TERMS**
>
> **population dynamics:** the study of how and why populations change in size and how they can be managed
>
> **migration:** the movement of peoples from one place in the world to another

Factors influencing population distribution and density

Population distribution is the pattern of where people live, whereas population density is the number of people living in an area. Global population is uneven with some areas being densely populated (such as London or Singapore) and others sparsely populated (the Sahara Desert or Antarctica). For example, the population density of people living in New York City is ten times more than that of Alaska. One of the factors which causes this difference is climate.

> **KEY TERM**
>
> **population distribution:** the way in which the population is spread out across a given area

Environmental factors

Environmental factors that influence population density include geographical factors (features related to the shape of the land), climatic factors (weather) and natural resources.

Relief, landforms and accessibility

Certain types of land makes food production easier. Land which presents opportunity for food production includes lowland plains, flat river valleys and fertile river deltas or volcanic areas. These areas tend to have higher population densities. Mountainous areas with poor soils and steep slopes tend to have low population densities.

Landforms such as mountains, oceans and rivers can act as a barrier to accessing an area. However, mountain passes and narrow or shallow parts of a river may allow the establishment of transport routes and the growth of a population. On the other hand, large, flat lowland plains are easier to access and so have higher population densities.

Weather and climate

Areas which have few extreme weather events are more attractive than areas that experience extreme weather. Very hot or cold, or dry or wet climates, are more challenging for humans to thrive in (Figure 3.3). They tend to be more sparsely populated than areas with moderate climate and rainfall that is evenly distributed throughout the year.

3 Managing Human Population

Vegetation and natural threats

Vegetation and natural threats affect population density. Historically, humans are more likely to settle in grasslands and open areas than areas of dense forest. Areas with pests, dangerous animals, natural hazards or a high risk of disease are less appealing for human settlement. They are therefore likely to have lower population density.

Raw materials/natural resources

Areas that have a wealth of natural resources such as fish, wood, coal, oil, and minerals, may have higher population densities than areas without. However, some resources may be found in remote locations with harsh climates which therefore have low human population densities.

> **KEY TERM**
>
> **leaching:** when water soaks into soils, removing the minerals and nutrients and reducing their ability to support plant life

Human factors

Historical

Vegetation and natural threats affect population density. Historically, humans are more likely to settle in grasslands and open areas than areas of dense forest. Areas with pests, dangerous animals, natural hazards or a high risk of disease are less appealing for human settlement. They are therefore likely to have lower population density.

Social

Areas where extensive environmental degradation has occurred (toxic spills, deforestation, soil erosion, water pollution), are either unable to support human life or are harmful to human life. These areas may have previously supported human life, but now have low population density. An example of environmental degradation is the nuclear accident at the Chernobyl Nuclear Power Plant, Ukraine, in 1986 (Figure 3.4). Two or more explosions released large amounts of radioactivity into the environment. Radiation is not visible, unlike other forms of pollution, such as plastic in water. It is not possible to see the pollutant that is causing the damage. The area around the power plant was evacuated and an exclusion zone of 2600 km^2 created. It is estimated that it will take 20 000 years for the Chernobyl area to become habitable again.

Figure 3.3: Extreme climates can pose a challenge for water supply. Both too much water or too little water are problematic and affect population densities.

Soil type and quality

Areas with rich, fertile soils, such as river deltas or volcanic areas, which support agriculture often have higher population densities than areas with poor soils. Areas on steep slopes tend to have thin poor soils, while areas with high annual rainfall (for example, the equatorial rainforest) can have poor, leached soils. Very cold areas with permanently frozen soil (permafrost) can limit the amount of food that can be grown. Humans can also negatively impact fertile soils through deforestation, overgrazing and overcropping. These activities result in soil degradation and reduced agricultural output.

Water supply

A secure water supply is essential for human survival. Areas which have sufficient water (but not excessive amounts) can be more densely populated than areas that are too dry or too wet.

111

Figure 3.4: An abandoned school room in Chernobyl. Harmful radiation caused the area to be evacuated.

Economic

Agriculture is a **primary industry**. Raw materials such as wheat, corn and rice are grown and harvested. These materials may be processed, or made into something else such as when wheat is ground into flour to make bread. Processing raw materials, or manufacturing, is known as **secondary industry**. In places where farms are well developed and manufacturing is established, there are job opportunities and possibly a better standard of living. For this reason, areas with well-developed farms and manufacturing industries tend to be more densely populated. Even when agriculture and industry decline, areas can remain densely populated as the population has become established in the area.

Accessibility is affected by both physical and human factors. Human factors include well-developed transport routes, links and infrastructure (roads, railways, shipping, canals or airports). Areas with good transport links are more likely to be densely populated than poorly connected areas. Transport routes allow for the development of trade between populations, and the movement of goods.

Political

Government policies can affect population density, too. For example, policies might encourage or discourage investment in an area, encourage migration into an area or build transport links to remote places. Wars or conflict can result in significant **out-migration** and a decrease in population density in one area while simultaneously increasing population density in others.

In areas of human conflict (see the South Sudan case study later in this chapter), people leave their homes because they are unsafe. This results in the out-migration of population from one area, and **in-migration** into another (the displacement of the population). Conflict also results in decreased food production as farmers abandon their farms. This puts further pressure on the population and results in even greater out-migration, as people move in hope of finding a food source.

The patterns of human distribution we see today (Figure 3.5) have developed due to the physical and human factors discussed above. However, with technology such as refrigeration, transport, communication, commercial farming, heating and cooling, humans can now live in areas previously considered unsuitable. Examples of such extreme locations include the scientific research stations in Antarctica, Kuwait City (one of the hottest cities on the planet) or La Rinconada, which is located 5090 metres above sea level in the mountains of Peru.

Population distribution is not only uneven between countries or regions, it is also uneven within countries. For example, the east coast of South Africa is more densely populated than the west coast, which is dryer and hotter.

> **KEY TERMS**
>
> *primary industry:* industry such as mining, agriculture, fishing or forestry that involves harvesting raw materials
>
> *secondary industry:* industry that converts raw materials such as farming or mining products into products for sale. The manufacturing industry
>
> *out-migration:* to leave one community or area in order to settle in another area
>
> *in-migration:* to move into an area or region in order to settle down and live

3 Managing Human Population

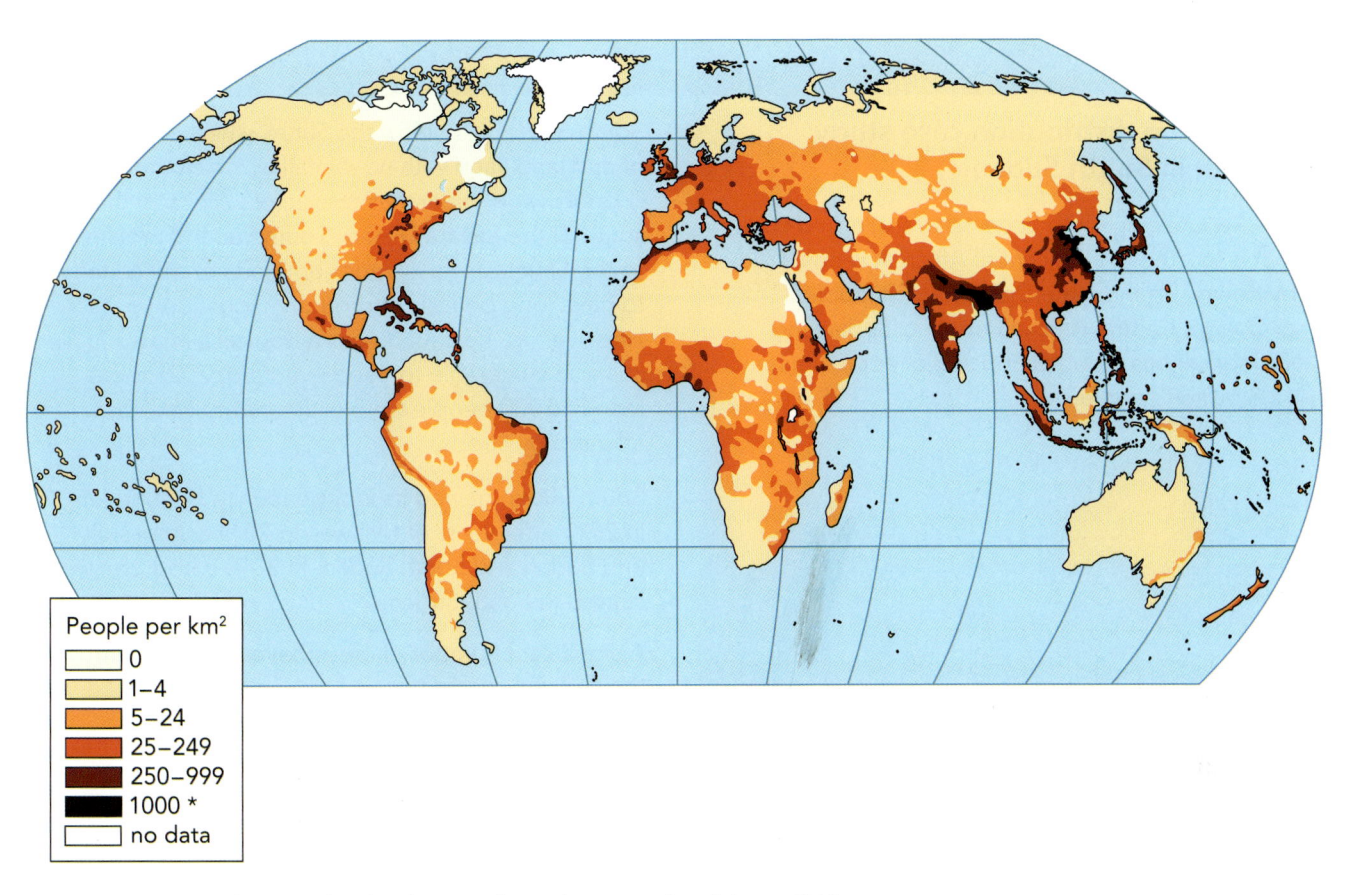

Figure 3.5: A map showing the distribution of people across the globe in 2015.

Key — People per km²:
- 0
- 1–4
- 5–24
- 25–249
- 250–999
- 1000 *
- no data

ACTIVITY 3.1

Population distribution poster

1 In pairs, create a map showing the population distribution of a country of your choice. Use Figure 3.5 as an example of how to show areas of greater density. You could use different colours or shading to indicate the density in different areas. If you use this option, make sure you include a key indicating the density represented by each colour.

2 Select a country from a continent/island other than your own.

a Suggest reasons for the distribution you observe. You could do this through annotating a map, or you could have a class discussion on the topic.

b How does the distribution differ from that found in your home country?

c **Compare** and **contrast** the reasons for the distribution of the population in the country you selected with that of your own country.

COMMAND WORDS

compare: identify/comment on similarities and/or differences

contrast: identify/comment on differences

CAMBRIDGE INTERNATIONAL AS LEVEL ENVIRONMENTAL MANAGEMENT: COURSEBOOK

CASE STUDY

Population distribution case study: conflict in South Sudan

The Republic of South Sudan is a landlocked country that is located in east/central Africa. It is bordered by Kenya, Uganda, the Democratic Republic of Congo, Central African Republic, Sudan and Ethiopia. It has a population size of approximately 12.8 million people.

Figure 3.6: In the city of Juba, in an arid region of South Sudan, the rates of precipitation are low.

South Sudan has a tropical climate, which means it has a rainy season with high humidity and large amounts of rainfall, followed by a dry season. Temperatures range between 20 °C and 30 °C through the year. The environment includes tropical forest, swamps and grasslands.

South Sudan lies below the Sahel, which is a band of arid land (land with no rainfall which cannot support vegetation) that runs across Africa to the south of the Sahara Desert. South Sudan's proximity to the Sahel can cause northern parts of the country to have long periods of drought (Figure 3.6). However, as the country has a tropical climate, it is also vulnerable to flood events. In 2021 more than 835 000 people were estimated to have been affected by floods.

South Sudan gained its independence from Sudan in 2011 after years of civil war. In 2011 Sudan split into Sudan and South Sudan. In 2013 a new civil war started in South Sudan.

A significant number of crises have resulted from the civil war, including:

- the collapse of infrastructure
- political instability
- food insecurity
- water insecurity
- lack of access to sanitation
- lack of access to education or medical care
- human rights violations
- loss of natural areas, habitat and wildlife
- displacement of population.

CONTINUED

The refugee crisis includes the international displacement of 2.1 million refugees out of South Sudan and into neighbouring countries such as Uganda, Kenya and Ethiopia (see Table 3.1 and Figure 3.7). To date as many as 1.87 million people have been displaced internally.

Countries people have migrated into from South Sudan	The number of migrants recorded
Kenya	110 377
Uganda	1 034 106
Democratic Republic of Congo	85 426
Central African Republic	2 057
Sudan	447 287
Ethiopia	416 886

Table 3.1: The number of people recorded as migrants out of South Sudan to surrounding countries since 2013.

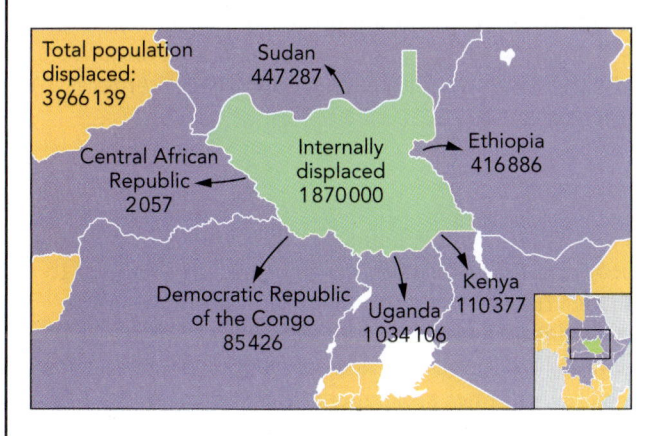

Figure 3.7: Migration out of South Sudan into the surrounding countries.

Total population displaced: 3 966 139
- Sudan 447 287
- Central African Republic 2 057
- Internally displaced 1 870 000
- Ethiopia 416 886
- Democratic Republic of the Congo 85 426
- Uganda 1 034 106
- Kenya 110 377

The spread of disease, famine and ongoing conflict have resulted in ongoing pressure on South Sudanese people, the natural environment and surrounding countries that take in refugees from South Sudan. These factors are **push factors**, causing people to leave an area.

Case study questions

1 **Calculate** percentages for the displaced South Sudanese population. Give your answers to two decimal places. Consider:
   a people displaced internationally
   b people displaced internally
   c the total displaced population

2 Make a mind map to show the causes for Sudan's displaced population (push factors). Decide what the main cause is and say why you think this.

3 As a class, discuss the potential impacts of this displacement on receiving countries and the environment.

4 Investigate another location where either a natural or a human-made disaster has resulted in the displacement of the population. Write a short newspaper article or blog explaining the causes of the displacement and its impact on the population.

KEY TERM

*push factors: these are the factors that cause people to leave an area. They include war, drought, floods, lack of housing, food, education, lack of jobs or a poor standard of living

COMMAND WORD

calculate: work out from given facts, figures or information

Calculating population density

Population density is an average measurement of the number of people in an area.

The formula for calculating population density is:

$$Dp = N/A$$

Dp = Population density

N = Total population

A = Land area covered by that population

> ### WORKED EXAMPLE 3.1
>
> #### CALCULATING POPULATION DENSITY
>
> **Question**
>
> 10 000 people are living in an area of 32 km². What is the population density?
>
> Step 1   Dp = N/A
>
> Step 2   DP = 10 000/32
>
> Answer   DP = 312.5 people per km²
>
> Remember to always include the units in your answer. In this example it is people per km².
>
> Now calculate the population density for the following countries:
>
> | Country | Area (km²) | Population size |
> |---|---|---|
> | Macau | 30 | 631 636 |
> | Bahrain | 785 | 1 501 635 |
> | Taiwan | 36 193 | 23 604 265 |
> | Maldives | 300 | 515 696 |
> | Mali | 1 240 192 | 19 077 749 |

Factors affecting changes to a population

Population change is defined as how the number of people in a population changes between the start and the end of a given time period, usually a year. Population change is the complex interaction of a number of factors. These factors include:

- **birth rates**
- **death rates**
- **child mortality rates**
- **natural increase**
- **net migration**.

When any of these factors change, population numbers are affected.

> ### KEY TERMS
>
> **birth rate:** the number of live births per thousand people in the population, per year. Also known as the crude birth rate as it does not take age or gender into account
>
> **death rate:** the number of deaths per thousand people in the population, per year
>
> ***child mortality rate:** the number of children, per 1000 live births, that die under the age of five in a population in a year
>
> ***natural increase:** the difference between the birth rates and death rates in a population; natural increase differs from overall increase
>
> ***net migration:** the difference between the number of people entering a country (immigration) and the number of people leaving a country (emigration). Net migration is negative when more people leave a country than enter it

Birth rate

The birth rate is the number of live births per thousand people in the population per year. It is also referred to as the crude birth rate and is the most basic measure of fertility. The crude birth rate applies to the total population. It does not take gender or age into account. Where birth rates are higher (40–50 births per 1000), populations tend to increase in size. In contrast, where birth rates are lower (10–20 births per 1000), the population may decrease in size.

A high death rate, especially **infant mortality rate**, in low-income countries (LICs) often results in high birth rates as families try to have children that survive to adulthood. In such countries, larger families are important, as they work together to harvest crops, raise children and care for older people. Figure 3.9 shows birth rates around the world.

3   Managing Human Population

Figure 3.8: In Niger, family wealth is reflected in the number of children that a couple has.

KEY TERM

*infant mortality rate: the number of infant deaths for every 1000 live births, of children under the age of one

Even when death rates fall, birth rates may remain high as it takes a long time for cultural values to change.

Government policies can play a significant role in birth rates. Some countries have policies to slow or reduce population growth. For example, China used a population policy to stabilise its rapid population growth. Concerned that there would be insufficient resources to support the population, China limited families to having one child each in 1979. This became known as the One Child Policy. This policy was effective at slowing the growth rate of the Chinese population, helping the Chinese government to balance both population and available resources more effectively (Figure 3.10).

Countries with declining populations may use policies to encourage population growth.

These strategies are discussed in more detail in the section on managing human population change later in this chapter.

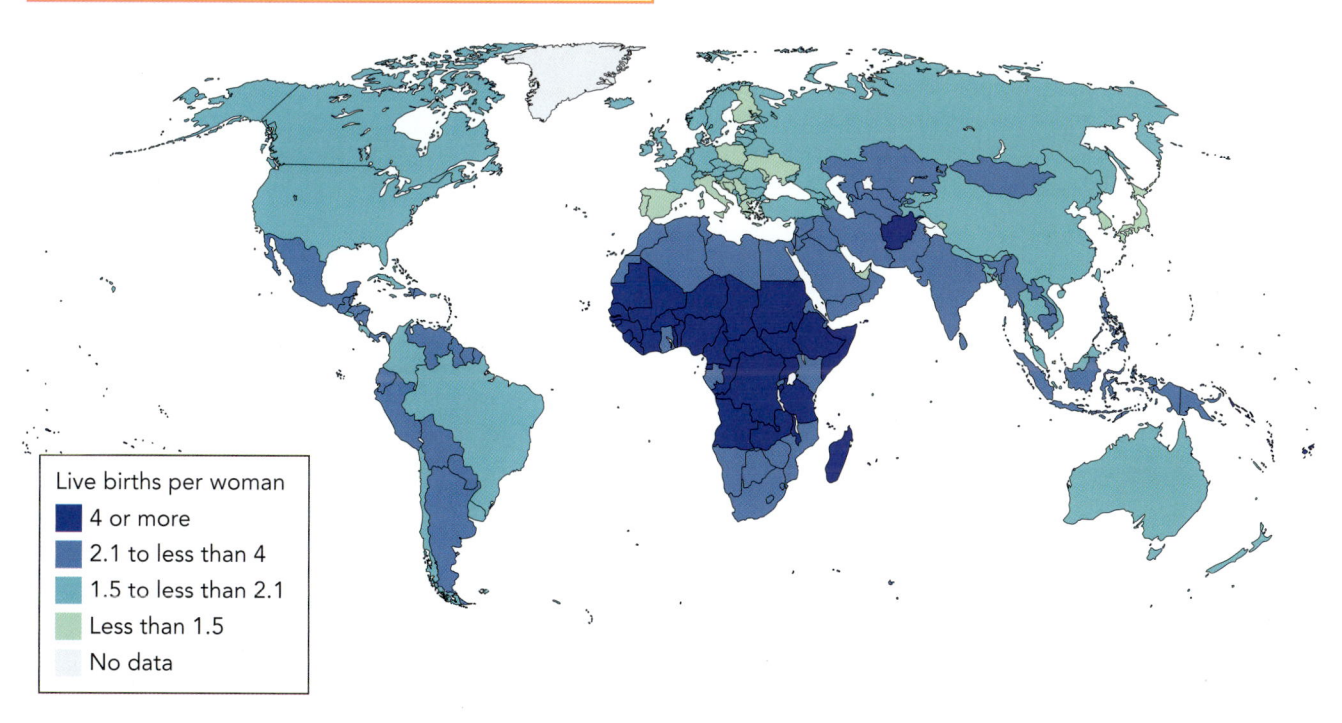

Live births per woman
- 4 or more
- 2.1 to less than 4
- 1.5 to less than 2.1
- Less than 1.5
- No data

Figure 3.9: Live births per woman by region.

117

CAMBRIDGE INTERNATIONAL AS LEVEL ENVIRONMENTAL MANAGEMENT: COURSEBOOK

Figure 3.10: The One Child Policy in China was effective in reducing population growth.

Death rates

The death rate is the number of deaths per 1000 people in the population in any given year. Death rate varies greatly between countries and is influenced by factors including climate, medical facilities, living standards, access to clean drinking water, hygiene levels, social conflict and crime rates.

Child mortality rates (Figure 3.11) are usually higher in LICs. Key causes for this tend to be lack of medical care for mother and child, lack of access to clean water and nutritious food, and spread of disease due to lack of sanitation. All of these factors increase the chance of a child dying before the age of five.

> ### ACTIVITY 3.2
>
> **Comparing graphic data**
>
> With reference to Figures 3.9 and 3.11, which show global live births and child mortality, write 150–200 words to **describe** and **explain** the patterns you observe.

> ### COMMAND WORDS
>
> **describe:** state the points of a topic / give characteristics and main features
>
> **explain:** set out purposes or reasons / make the relationships between things evident / provide why and/or how and support with relevant evidence

> ### SELF ASSESSMENT
>
> - Were you able to see how the maps in Figures 3.9 and 3.11 showed a similar trend, reflecting the close link between birth rate and child mortality?
> - Did you explain, as well as describe, what you observed?
> - How comfortable do you feel about discussing key concepts related to population change?

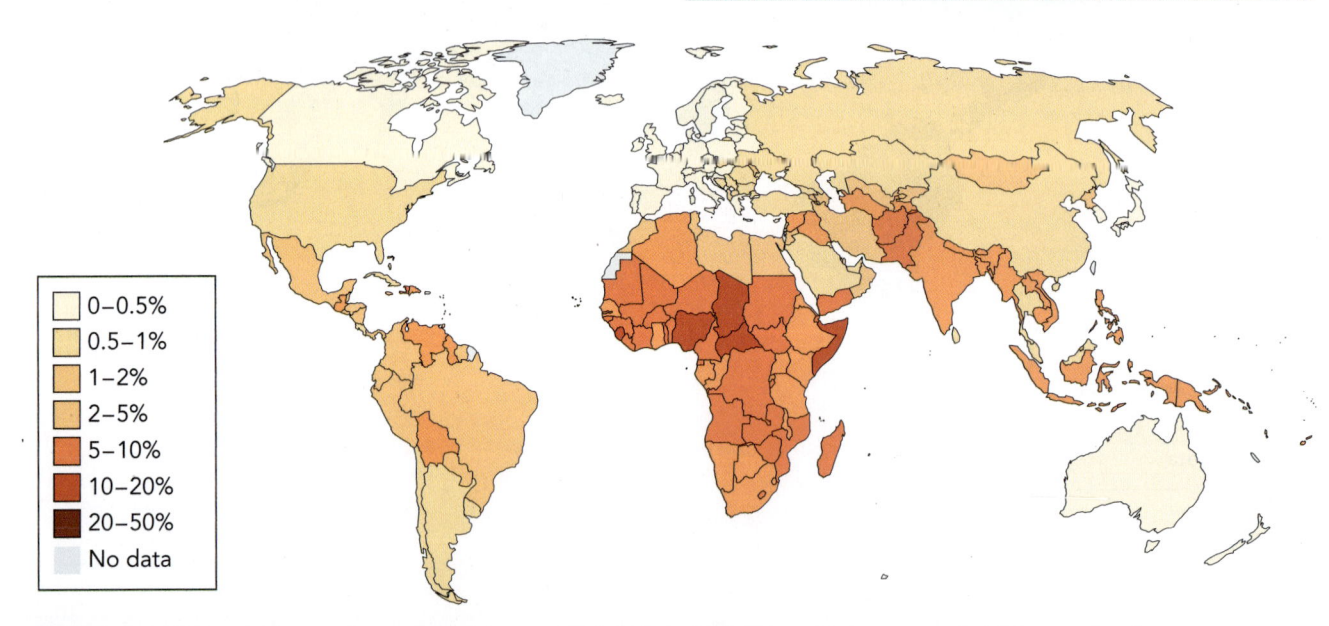

Legend:
- 0–0.5%
- 0.5–1%
- 1–2%
- 2–5%
- 5–10%
- 10–20%
- 20–50%
- No data

Figure 3.11: Global child mortality rates by country, 2019. LICs in Africa show higher rates of child mortality than high-income countries (HICs) such as Australia and countries in the European Union.

Life expectancy

Life expectancy is the age that a new-born child would, on average, be expected to live to. Life expectancy is rising around the world, with more people expected to live beyond 60 years, and, in some countries, beyond 80 years. However, a big gap still remains between HICs and LICs. Life expectancy is as low as mid-50s in some African countries compared to 80s in Japan.

Life expectancy acts as a measure of the general health of the population. When medical care, standard of living, access to clean water, access to nutrition and education improve, life expectancy increases.

Migration rates

Natural population change (increase or decline) is the change in the size of a population caused by the difference between the birth rate and death rate. Where the birth rate exceeds the death rate, the population grows. Where the death rate exceeds the birth rate, the population declines.

However, with globalisation, migration plays a role in population change. Net migration is the balance between **immigration** and **emigration**. Where immigration is higher than emigration, the population grows. If emigration is higher than immigration, the population declines.

Overall population change is the annual population change of an area. It takes both natural increase and net migration into account.

Overall population change = (birth rate − death rate) + net migration

Note that net migration can be negative when more people leave a country than enter it, for example, during times of war.

> **KEY TERMS**
>
> *life expectancy:* the average age that a new-born child is expected to live to
>
> *natural population change:* the change in the size of a population due to birth and death rates
>
> *immigration:* people migrating into a country
>
> *emigration:* people migrating out of a country
>
> *overall population change:* the change in the size of a population due to birth rates, death rates and net migration rates

Understanding population differences in high and low income countries

Population structure is the number of males and females in different age groups within a population. It is represented as a population pyramid. Population pyramids are a useful tool for conducting a detailed analysis of the country's population. The data allows governments to plan for future needs, and it also allows comparisons of population structures over time and between countries. It is possible to assess information on population structure in relation to current events such as economic growth, war or natural disasters.

On an age/gender pyramid, the vertical axis divides ages into five-year intervals apart from the uppermost bar which indicates population of a certain age and older (65+ or 80+). The horizontal axis indicates either total numbers or percentages of the population.

Population pyramids change significantly in shape as a country develops. Five main population pyramids have been developed to show population structure at different levels of economic development. These population pyramids reflect how many children are being born, how many survive to adulthood, how long people live and the balance between males and females within a population. They can be interpreted in connection with factors such as child mortality rates, birth rates, death rates, access to medical care and contraception.

The first population pyramid is rarely found. It reflects a population with a high birth and death rate and has zero population growth. Birth and death rates are high, and life expectancy is low. Such populations tend to be limited to small tribes in remote locations- the population of North Sentinel Island, discussed in the case study in this chapter, for example.

> **KEY TERM**
>
> population structure: the number of males and females within different age groups in a given population

The second population pyramid structure (Figure 3.12) reflects a stage of population growth, which is characterised by a rapid decrease in the death rate. The introduction of medicines, better nutrition, clean water and septic waste removal result in lower mortality rates. However, and this is often due to cultural influences, birth rates still remain high. As a result, natural population growth is high, with a large proportion of the population under the age of 15. Examples of countries at this stage of population development include Niger in Africa and Guatemala in Central America.

Average life expectancy increases and urbanisation starts to slow. Mortality rates stabilise as people live longer. During stage 3, countries start to become more industrialised.

Bangladesh, 2019
Population: **163 046 173**

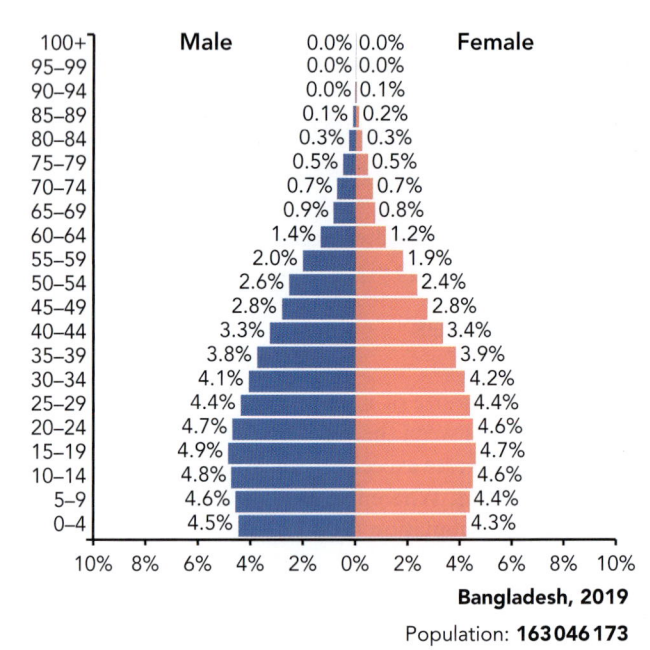

Figure 3.13: The Stage 3 population pyramid for Bangladesh. It shows a narrowing of the base and widening of the middle section of the pyramid due to higher life expectancy.

Niger, 2019
Population: **23 310 718**

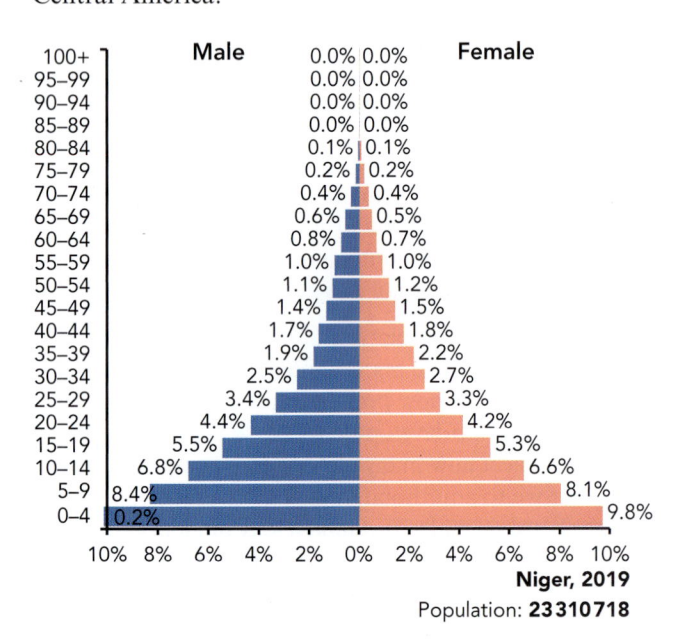

Figure 3.12: The Stage 2 population pyramid for Niger, showing the typical triangular shape with a wide base and narrow top.

The population pyramid for Niger, (Figure 3.12) an LIC, has a wide base, reflecting a high birth rate (45.8/1000 people). A marked decrease in numbers between each age group indicates relatively high death rates between each age group. There is also low life expectancy (62 years).

In the third population pyramid (Figure 3.13), the birth rate is lower than in the second pyramid, while the death rate continues to decline or remains low. During this stage, the population continues to grow, but at a slower rate than Stage 2. This is because birth rates are decreasing along with death rates. A change in culture may be reflected in lower mortality rates and declining child mortality. As a result of this cultural change, birth rates also start to decline. Contraception is available, making it possible for birth rates to be controlled.

The Stage 3 pyramid in Figure 3.13 shows the population structure in Bangladesh, an LIC. This pyramid has a narrower base than countries at Stage 2, reflecting a steady decrease in birth rates. The relatively wide bars in the teenage and young adult age groups show falling death rates, increased life expectancy and larger numbers of people living to the age of 72.

In the fourth pyramid (Figure 3.14), both birth and death rates are low and the population size fluctuates due to economic conditions. Population growth is slow, and death rates increase slightly as the aging population reaches full life expectancy.

The pyramid (Figure 3.14), the USA (an HIC), shows a narrowing base, reflecting an ongoing decrease in birth rates. Falling death rates and increased life expectancy (78.8 years in 2019, 76.6 years in 2000) are reflected in the reduced narrowing of each successive bar, and a taller pyramid overall.

3 Managing Human Population

the population replacement level and older people are reaching full life expectancy. Natural population decline is occurring but this can be counterbalanced with inward migration.

The pyramid's narrowing base reflects a rapid decrease in birth rates. Each age bar is wider than the one below it up until the age of four. This shows a low death rate in people under the age of 70. More people are living longer, resulting in a life expectancy of 84.7 years.

Analysing a population pyramid

Population pyramids can be analysed to discuss changes that are taking place within a population. Figure 3.16 is an example of how a pyramid can be annotated to suggest what those changes might be. In order to do this, it is necessary to understand the five population pyramids. You can then review a population pyramid, using your knowledge to suggest possible causes for the you see.

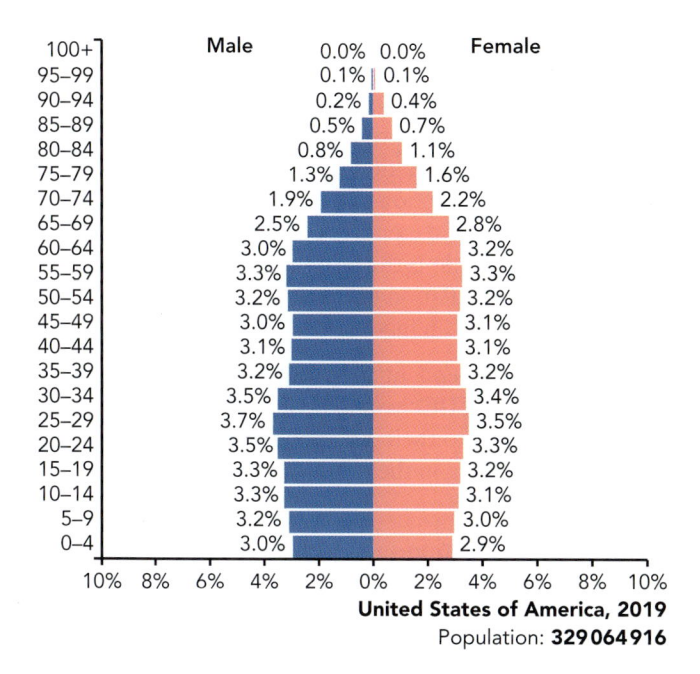

Figure 3.14: The Stage 4 population pyramid for the USA. This pyramid has developed straight sides with a narrow base. It is wider towards the top.

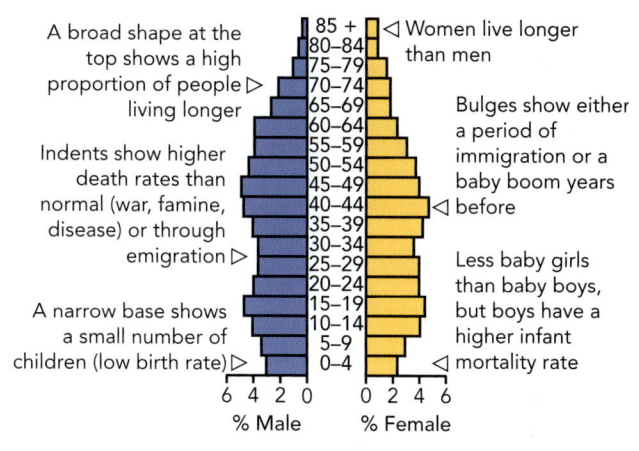

Figure 3.16: Annotating an HIC population pyramid.

Figure 3.16 gives guidance for understanding or analysing population pyramids.

These steps can be used to carry out the analysis of a population pyramid.

1. Divide the pyramid into three sections (0–14 young dependents, 15–65 independents and 65+ old dependents). This way, you determine if there is a youthful or aging population and how the dependency in the population is distributed.

2. Note how long people are living by looking at the highest point in the diagram. This gives you the population's life expectancy and an idea of its economic development.

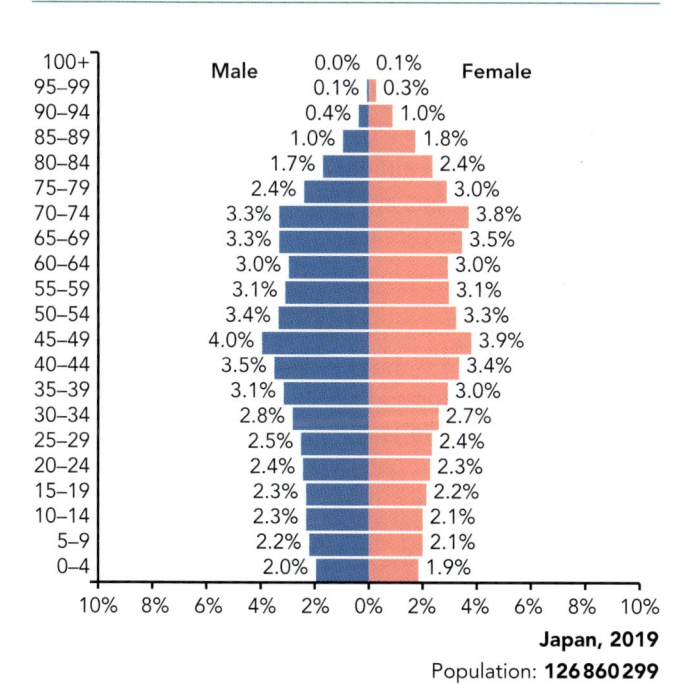

Figure 3.15: The Stage 5 population pyramid for Japan, with a very narrow base and wider mid to top section.

Japan, which is an HIC, is the fifth population pyramid structure (Figure 3.15). In Japan, birth rate has fallen below the death rate. Birth rates are below

3. Look for unusual bulges in the pyramid. Bulges can indicate events like emigration or immigration, war, or natural disasters. Suggest possible causes for a bulge or a narrowing of the pyramid.

4. Note the shape of the pyramid. Does it have a wide or a narrow base? Suggest what this indicates.

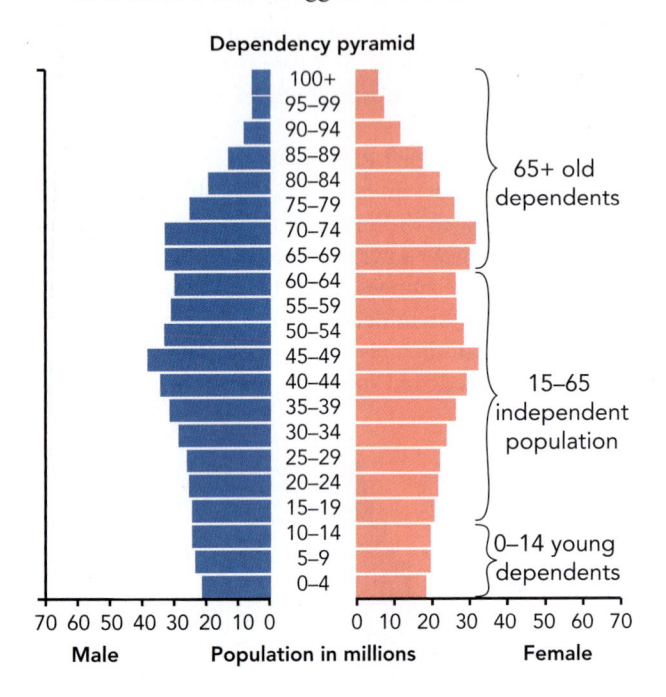

Figure 3.17: A population pyramid showing the age groups within a population that are considered dependent and independent.

Dependency ratio

The **dependency ratio** shows the relationship between a country's working population (economically independent) and non-working population (economically dependent). The dependent population can be split into young (0–14 years) and old dependents (65+ years). The dependency ratio is therefore defined as the portion of the population that are not economically active or producing an income.

Young and old dependent age groups are a crude generalisation. In HICs young people may be dependent on their parents into their 20s, while the older population may still be working in their 70s.

The **total dependency ratio** is a relative measure of the working age population supporting the non-working age population. The total dependency ratio is calculated by adding young and old dependent populations together and calculating a single figure, rather than splitting them into the two age categories. The ratio reflects how much of the overall population is dependent on the working portion of the population.

Total dependency ratio =

$$\frac{[\text{Number of people (0 to 14)} + \text{number of people (65+)}]}{\text{Number of people aged 15 to 64}} \times 100$$

Note: The smaller the difference between the numerator (number above the line) and the denominator (number below the line) the higher the dependency ratio.

ACTIVITY 3.3

Population pyramid investigation

1. In pairs, investigate different population pyramids from countries around the world. Look for one that has an unusual shape and suggest reasons for this.

2. Make a small poster by drawing and annotating the population pyramid you have selected. Share your poster with your class.

3. What can you learn from the population pyramids that other pairs investigated? Discuss your findings as a class.

KEY TERMS

dependency ratio: the dependency ratio is the measure of the dependent (non-working) portion of the population (age groups 0–14 and 65+) compared to the total independent (working) portion of the population (15–64 years). The ratio is expressed as the number of dependents per hundred people in the workforce

total dependency ratio: the total dependency ratio is a measure of both young (age 0–14 years) and older dependents (age 65 and older) added together to show their total versus the independent population (15–64 year olds). The ratio is expressed as the total number of dependents (young and old) per hundred people in the workforce

3 Managing Human Population

> **WORKED EXAMPLE 3.2**
>
> **CALCULATING TOTAL DEPENDENCY RATIO**
>
> **Question**
>
> Assume that a country has a total population of 10 000 people. There are:
>
> - 2500 children under the age of 15
> - 5500 people between the ages of 15 and 64
> - 2000 people age 65 and older.
>
> What is the total dependency ratio?
>
> Step 1: = ((2500 + 2000)/5500) × 100
>
> Step 2: = (4500/5500) × 100
>
> Step 3: = 0.82 × 100
>
> Answer: the total dependency ratio = 82%
>
> Calculate the total dependency ratio for the following states in the USA.

State	Population size by age group		
	0–14 yrs	15–65 yrs	65 yrs +
Utah	869 907	1 663 748	250 917
Colorado	1 228 089	3 262 718	533 941
Florida	4 057 419	11 281 585	3 198 965
Alaska	183 512	463 018	51 943

What does the dependency ratio tell you?

A total dependency ratio is considered to be high when it exceeds 62%. In other words, when there are more than 62 dependents for every 100 workers. A high total dependency ratio indicates that the working-age population and the economy of a country face the burden of financially caring for a large number of people.

Figure 3.18: Young people in Nairobi, Kenya, East Africa.

The **youth dependency ratio** measures the ratio of young dependents (people younger than 15) to the working-age population (those aged 15–64). The **old age dependency ratio** measures the ratio of older retired people (over the age of 64) in relation to the working-age population (those aged 15–64). Youth and old age dependency ratios are calculated using the following formulae.

Youth dependency ratio =

$$\frac{\text{number of people 0 to 14}}{\text{number of people 15 to 64}} \times 100$$

Old-age dependency ratio =

$$\frac{\text{number of people 65+}}{\text{number of people 15 to 64}} \times 100$$

> **KEY TERMS**
>
> ***youth dependency ratio:** the youth dependency ratio is a measure of the young dependents (age 0–14) in a population, in relation to the working-age population (15–64 years old)
>
> ***old age dependency ratio:** the old-age dependency ratio is the number of older dependents (age 65+) in a population, in relation to the working-age population (15–64 years old). The ratio is expressed as the total number of older dependents per hundred people in the workforce

CAMBRIDGE INTERNATIONAL AS LEVEL ENVIRONMENTAL MANAGEMENT: COURSEBOOK

> **WORKED EXAMPLE 3.3**
>
> **CALCULATING YOUTH AND OLD AGE DEPENDENCY RATIOS**
>
> **Question**
>
> Assume that a country has a total population of 10 000 people. There are:
>
> - 2500 children under the age of 15
> - 5500 people between the age of 15 and 64
> - 2000 people aged 65 and older.
>
> What are the youth and old-age dependency ratios?
>
> | Youth dependency ratio | Old-age dependency ratio |
> |---|---|
> | Step 1: = (2500/5500) × 100 | Step 1: = (2000/5500) × 100 |
> | Step 2: = 0.45 × 100 | Step 2: = 0.36 × 100 |
> | Answer Youth dependency ratio = 45% | Answer Old-age dependency ratio = 36% |
>
> For the following states in the USA, calculate:
>
> **a** the old-age dependency ratio
>
> **b** the youth dependency ratio
>
> | State | Population size by age group |||
> |---|---|---|---|
> | | 0–14 yrs | 15–65 yrs | 65 yrs + |
> | Utah | 869 907 | 1 663 748 | 250 917 |
> | Colorado | 1 220 009 | 3 262 718 | 533 941 |
> | Florida | 4 057 419 | 11 281 585 | 3 198 965 |
> | Alaska | 183 512 | 463 018 | 51 943 |

In 2020, Niger was an example of a highly dependent young population, with a youth dependency ratio of 104%. Meanwhile, Qatar had a low youth dependency ratio of 16.1%. The global average was 38%. In contrast, a highly dependent older population was seen in Japan, with an old-age dependency ratio of 48%, whereas this figure was just 1.5% in the United Arab Emirates. Compare these figures with the average global old-age dependency ratio, which was was 10% in 2020. The same year, the youth dependency ratio in Afghanistan was 75.3%, while the old-age dependency ratio was 4.8%.

A total dependency ratio of 80.1% indicates that for every 100 people working there are 80.1 people that need financial support. A high number of young dependents indicates a young growing population. In 2020, the Japanese youth dependency ratio was 21%; Japan has an aging population, as seen by the old-age dependency figure given above. The country with the highest dependency ratio in 2020 (according to the CIA World Factbook) was Niger, with an overall dependency ratio of 109.5%. In Niger, for every 100 people of working age, there are 109.5 people who depend on them.

Figure 3.19: A group of older men in a park. Aging populations such as the USA and Japan have a high ratio of older people.

3.1 Questions

1. Explain the difference between natural population change and overall population change.

2. Copy the population pyramid for Bahrain (2012) in Figure 3.20, and annotate it. Use the guidance shown in Figure 3.16.

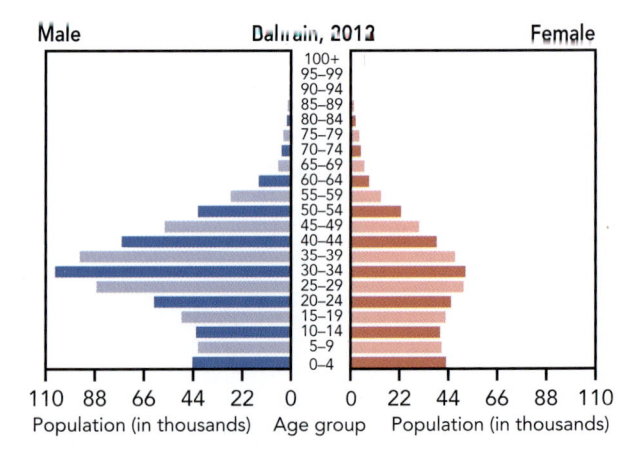

Figure 3.20: A population pyramid showing the structure of the population in Bahrain in the Middle East in 2012

3   Population densities:
   a   Calculate the population density for the following populations.

Country	Area (km²)	Population size
Australia	7 692 024	24 898 152
Venezuela	912 050	28 887 118
Nepal	147 181	28 095 714
Greenland	2 166 086	56 564

   b   Explain three possible factors for the differences in the population densities calculated.
   c   Discuss three factors influencing birth rates.

3.2 Impacts of human population change

Changes in human population growth rates can result in populations that have very high percentages of either young or old people. A population with a high percentage of older people is known as an **ageing population**.

> **KEY TERM**
>
> **ageing population:** a population with a high percentage of old people (aged 65 years or older)

Ageing populations

In an ageing population, the median age is higher than in a population with fewer old people/more young people. Aging populations occur when birth rates decline while life expectancy remains constant or increases. This population structure is reflected in the Stage 5 population pyramid and in some HICs (for example, Japan, the UK and Italy).

Older people now have access to better nutrition, sanitation, standards of living and care. This means that the average age of many populations is increasing. According to the United Nations, fewer than 25% of men lived to the age of 60 in Western Europe during the early 1800s. This figure has now risen to 90%. The global life expectancy was 46 years in 1950, and it was 72.6 years in 2019. The number of people over the age of 60 has tripled since 1950. It is expected to rise from 700 million in 2006 to 2.1 billion in 2050.

The problem of ageing populations has, for some time, been a concern for many HICs. It is now starting to be a concern for many LICs. In LICs, the population is aging at a far faster rate than it did in HICs. This rapid rate of change gives populations less time to adapt to or manage the problem.

Figure 3.21: A medical doctor caring for an older couple.

An ageing population presents a number of challenges and economic difficulties, such as:

- A shrinking workforce (when more people are retiring from than joining the workforce) puts pressure on the economy's productivity.
- Older people need to be taken care of, which requires the provision of housing, nutrition, health care, sanitation and carers.
- If the independent portion of the population (the working population) is smaller than the older, dependent portion (non-working), financial pressures are greater. In such countries the taxes collected from the workforce may not supply sufficient income for a country to care for its older people.
- Pensions may be insufficient to care for the older population as they live for longer.
- Health care systems come under greater pressure, as older people tend to require more medical care than the young.
- Poverty amongst older people is a considerable global problem.
- The longer people live, the more likely it is that governments will need to increase the retirement age from 60 to 65. This means that people work for longer, and that governments can collect tax from

them for longer (Figure 3.22). This extra tax income from the workforce contributes towards the care for the older population.
- Care homes need to be built to care for the older population. With a shrinking workforce and shrinking government income, this can be difficult.

Figure 3.22: A skilled worker who has continued to work past retirement age.

3.2 Questions

1. **Describe** what is meant by the terms 'ageing population' and 'youthful population'.
2. Which population pyramid matches that of an ageing population?
3. **Explain** why an LIC is less likely to have an ageing population.
4. **Discuss** the issues that arise out of an ageing population.

COMMAND WORDS

describe: state the points of a topic / give characteristics and main features

explain: set out purposes or reasons / make the relationships between things evident / provide why and/or how and support with relevant evidence

discuss: write about issue(s) or topic(s) in depth in a structured way

3.3 Managing human population change

Governments and global organisations may consider managing population change to cope with rapid population growth or decline. A rapidly growing population could employ **antinatalist** population policy to slow down this growth. A country which is experiencing population decline could implement a **pronatalist** policy to encourage population growth. In addition to decisions made by a country's government, groups of nations may take a global stance. The Club of Rome and UN Agenda 21 are examples of organisations that were formed to investigate and suggest possible solutions to global population change.

KEY TERMS

*antinatalist: a policy that discourages human reproduction

*pronatalist: a policy that promotes human reproduction

The Club of Rome

The Club of Rome was created in 1972 in order to address the multiple crises facing humanity and Earth. Its underlying goal was to alert the world to the consequences of rapidly growing populations and their impact on the planet's health. One of the club's key concerns was the extent to which pollution was harming the environment. It saw population growth and the resulting environmental deterioration, poverty, ill health and crime as 'problematic'. One hundred member countries called on scientists, economists, business leaders and former politicians to help address the 'problem of humankind'.

The Club of Rome's underlying belief was that economic and human growth could not continue indefinitely due to the planet's limited resources. The question was how growth could be made sustainable. The club's agenda was to limit population growth and promote sustainable economic development in order to prevent humankind exceeding the carrying capacity of the planet.

3 Managing Human Population

Figure 3.23: A steel plant in the Netherlands.

The Club of Rome produced its first report, 'The Limits to Growth', in 1972. This attracted considerable public attention. The club stated that its goal was 'to act as a global catalyst for change through the identification and analysis of crucial problems facing humanity, and the communication of such problems to the most important public and private decision makers, as well as the general public'.

The Club of Rome assisted with a model for global development that investigated five major trends. These were accelerating industrialisation, rapid population growth, widespread malnutrition, depletion of non-renewable resources and a deteriorating environment. Although this model did help to contribute towards change, economic complexities made it difficult for the club to achieve its ultimate goal. Political, social and economic change all take time and global policies cannot be applied equally to countries at different stages of economic or population development.

The UN Agenda 21

The UN Agenda 21 was a global plan of action which was needed to curb the impact of humans on the environment. Agenda 21 included the Rio Declaration on the environment and development as well as the Statement of Principles for Sustainable Management of Forests. It was adopted by more than 178 governments at the UN Conference on the Environment and Development in Rio de Janeiro, Brazil, in 1992 (Figure 3.24).

Figure 3.24: A poster for the UN Agenda 21, reflecting the broad scope of the agenda and the drive towards sustainability around the globe.

127

The agenda is a document of 40 chapters that outlines an action plan for sustainable development. It covers a wide range of specific natural resources and the role of different groups within issues of social and economic development.

The main goals of Agenda 21 were:

- international cooperation to accelerate sustainable development in LICs
- combating poverty
- changing consumption patterns
- demographic dynamics and sustainability
- protecting and promoting human health conditions
- promoting sustainable human settlement development
- integrating environment and development in decision-making
- protection of the atmosphere.

Both Agenda 21 and the Club of Rome are concerned with underlying issues of global human population growth and resource management.

Pronatalist policies

Some HICs with high rates of economic development face the problem of a shrinking population. This occurs when birth rates fall below death rates. A country facing with this challenge may employ a **pronatalist policy**. With such a policy, the government encourages people to have children in a bid to encourage population growth.

Pronatalist countries like France have employed policies that encourage couples to have children (Figure 3.25).

This can be done through monetary incentives, where rewards are given to women having more than one child.

The country could offer childcare assistance, free schooling, medical support, free public transport, monthly grants, and even housing for families that have large numbers of children.

> **KEY TERM**
>
> **pronatalist policy:** a population strategy designed to encourage people from having children and to increase birth rates

A country may implement an immigration policy that encourages skilled workers or fertile families into the country in order to increase the population size.

However, despite being pronatalist and increasing the birth rates in 2007, France still has a declining population.

Figure 3.25: Couples in France were encouraged to have more than one child.

Antinatalist policies

Rapidly increasing populations with high birth rates and declining mortality rates may face high rates of poverty, unemployment, disease and malnutrition along with low standards of living (Figure 3.26). Countries facing such population challenges may employ **antinatalist policies**, which are policies designed to reduce birth rates. A country can employ various strategies to try to control its growth rate.

> **KEY TERM**
>
> **antinatalist policy:** a population strategy designed to discourage people from having children and to decrease birth rates

Figure 3.26: Densely populated streets with traffic congestion in Lagos, Nigeria, which is an LIC in Africa.

Antinatalist policies may include either education of the population or laws to help control population growth.

Thailand used voluntary family planning education to decrease the birth rate between 1970 and 1990. This was achieved through the rapid spread of family planning services and information. Both small families and contraception were promoted in a way which broke taboos and clarified misconceptions.

In China, the One Child Policy was enforced as a law in 1979 to control the rapid population growth that China was experiencing. Couples who had only one child got better benefits, housing, education, healthcare and longer maternity leave. Couples who had a second child were fined and denied access to benefits that one child families were given.

Successful antinatalist policies have various factors in common. These include:

- Encouraging marriage later in life (voluntarily or by law)
- Making contraception and family planning easily available and educating the population about how contraceptives work (Figure 3.27 shows the parts of the world where contraception is widely used.)
- Improving education and employment opportunities for women.
- Improving health care, ensuring that both mother and child survive childbirth and that children live to adulthood.
- Free sterilisation for couples who have had a child/children and do not wish to have more.

In both Thailand and China, population policies successfully reduced the rate of population growth. It is estimated that China's policy resulted in 400 million children not being born. As a result, resources under pressure from overpopulation were protected. This policy ended in 2016 and was replaced by a two-child policy.

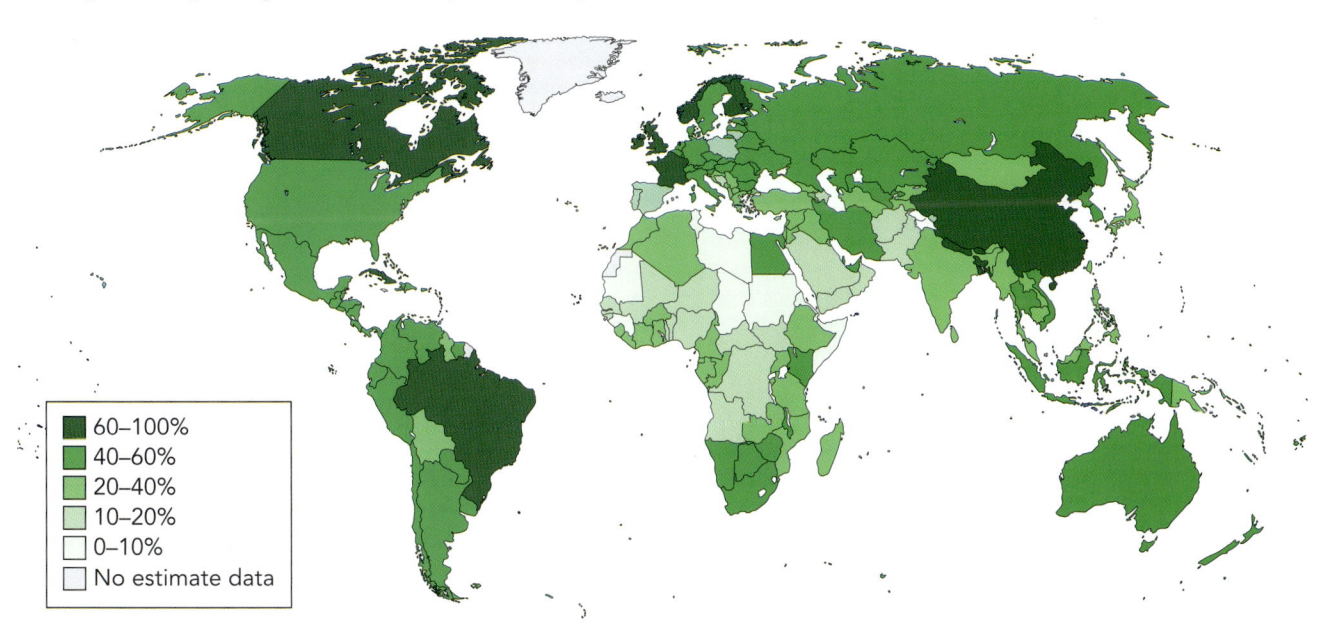

Figure 3.27: Prevalence of modern contraception use among women aged 15–49, by country or area, 2017.

CAMBRIDGE INTERNATIONAL AS LEVEL ENVIRONMENTAL MANAGEMENT: COURSEBOOK

ACTIVITY 3.4

Antinatalist and pronatalist policy change

Between 1972 and 1987, Singapore had an anti-natalist policy in place, using the slogan 'stop at two' (Figure 3.28). This policy included access to:

- low cost contraception
- easy access to family planning
- free education
- access to low cost healthcare for small families.

The government also promoted sterilisation.

This policy was too successful and resulted in a shrinking population, as men and women had chosen to pursue careers instead of having families. As a result, Singapore had to change its policy to a pro-natalist one.

The new, pro-natalist policy was targeted at well-educated young women, as they were more likely to be able to afford children. Mothers with four or more children were offered 12 weeks' maternity leave. To promote the 'have three' policy, posters and slogans were put up that read 'three or more'. Larger child benefits and tax breaks were offered for each child that the family had.

1. Analyse how successful the 'have three' policy was. Conduct your own research to help with this.

2. What do you think the Singaporean public felt about the change in policy? Support your answer with a brief argument.

Figure 3.28: A poster from Singapore promoting small families.

The limitations of anti-natalist population policies are only seen one to two generations later (20–30 years) when the population has stablised.

- As the number of children declines and the standard of living improves due to economic growth and medical care, the death rate also declines. This results in an ageing population.
- The workforce declines as there are too few young people to fill the jobs as older people retire.
- There are not enough young people to care for older people.

- A population in decline is difficult to stabilise, especially when people have embraced the cultural shift to smaller families.
- Children from single child families in rural areas are more likely to live in poverty as they may have four grandparents and two parents to care for. Urban families may also struggle. However, they have greater earning potential in a city than their rural counterparts.

3.3 Questions

1. **Define** the meaning of the terms antinatalist and pronatalist.
2. **Describe** the goal of the Club of Rome.
3. What are the problems faced by a government trying to manage an ageing population?
4. Why does it take so long to slow population growth?
5. What is the limitation to population control in an LIC with a strong cultural or religious influence?

> **COMMAND WORDS**
>
> **define:** give precise meaning
>
> **describe:** state the points of a topic / give characteristics and main features

EXTENDED CASE STUDY

The young and old: considering the population age gap

Countries of the world are classified according to their level of social and economic development. LICs (low income countries) and HICs (high income countries) have distinctive population structures which are driven by a number of factors, such as:

- cultural values or traditions (for example, large families representing wealth or success).
- the level of industrialisation.
- geographical location.
- natural resources.
- levels of education.

Two countries that are at the opposite ends of the population spectrum (youthful compared to ageing) are Uganda and Japan.

Uganda's youthful population

Uganda is an LIC in central east Africa that has a Stage 2 population pyramid. According to UK Aid (a British-based charity), Uganda's population is projected to grow from 40 million people in 2015 to 141 million by 2065. This increase is driven by a high birth rate (Figure 3.29). The age group of 15–34 years will more than double between 2015 and 2065. Uganda is one of ten countries that is projected to collectively account for more than 50% of the projected world population increase by 2050.

Figure 3.29: Young mothers in Uganda at a community clinic.

Decades of high birth rates have resulted in a high dependency ratio of approximately 103 young and old dependents (0–14 and 65+ years) for every 100 people in the independent age group (15–64 years). This puts significant pressure on household and country resources. In the last 30 years, the birth rate has reduced only slightly, while the infant mortality rate has decreased by 64%.

The high birth rate is caused by a combination of a number of factors. Childbearing and marriage start early. Further causes include low levels of female education. Women have limited access to contraception (especially young women and women who live in rural and poorer areas). Also, having more children is seen as a sign of wealth. Death rates are decreasing while birth rates have not slowed significantly. In addition to this, people are

CAMBRIDGE INTERNATIONAL AS LEVEL ENVIRONMENTAL MANAGEMENT: COURSEBOOK

CONTINUED

living longer and there has been a rapid decline in infant mortality. Both of these factors have resulted in the population increasing from 9.5 million in 1969 to 40 million in 2015.

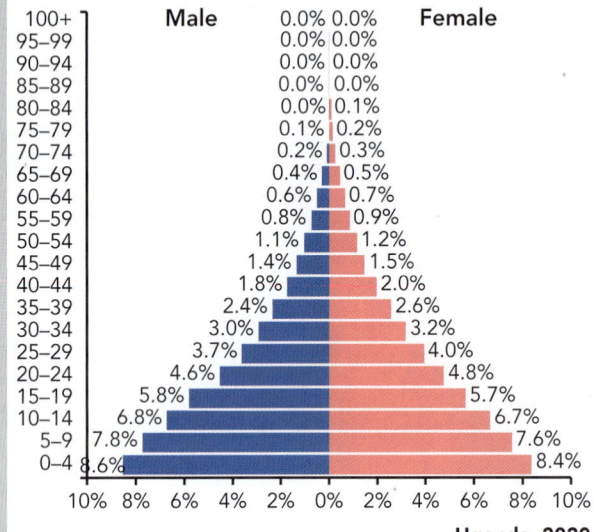

Uganda, 2020
Population: **45 741 000**

Age	Male	Female
100+	0.0%	0.0%
95–99	0.0%	0.0%
90–94	0.0%	0.0%
85–89	0.0%	0.0%
80–84	0.0%	0.1%
75–79	0.1%	0.2%
70–74	0.2%	0.3%
65–69	0.4%	0.5%
60–64	0.6%	0.7%
55–59	0.8%	0.9%
50–54	1.1%	1.2%
45–49	1.4%	1.5%
40–44	1.8%	2.0%
35–39	2.4%	2.6%
30–34	3.0%	3.2%
25–29	3.7%	4.0%
20–24	4.6%	4.8%
15–19	5.8%	5.7%
10–14	6.8%	6.7%
5–9	7.8%	7.6%
0–4	8.6%	8.4%

Uganda, 2065
Population: **109 932 490**

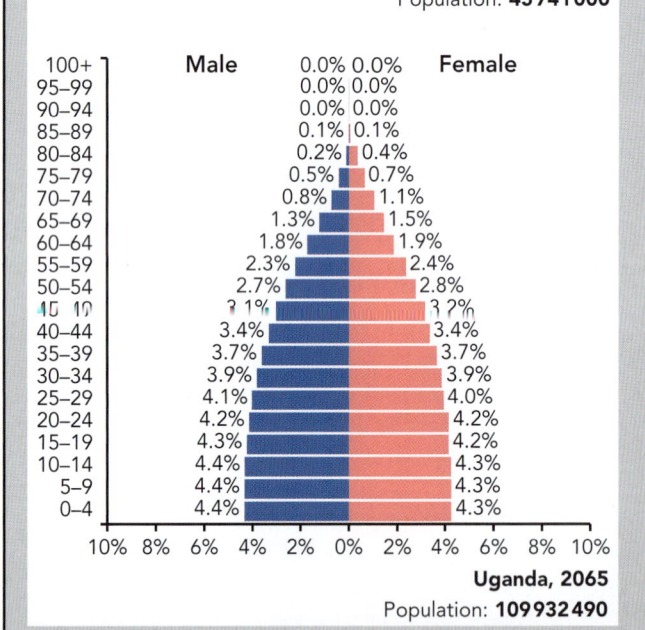

Age	Male	Female
100+	0.0%	0.0%
95–99	0.0%	0.0%
90–94	0.0%	0.0%
85–89	0.1%	0.1%
80–84	0.2%	0.4%
75–79	0.5%	0.7%
70–74	0.8%	1.1%
65–69	1.3%	1.5%
60–64	1.8%	1.9%
55–59	2.3%	2.4%
50–54	2.7%	2.8%
45–49	3.1%	3.2%
40–44	3.4%	3.4%
35–39	3.7%	3.7%
30–34	3.9%	3.9%
25–29	4.1%	4.0%
20–24	4.2%	4.2%
15–19	4.3%	4.2%
10–14	4.4%	4.3%
5–9	4.4%	4.3%
0–4	4.4%	4.3%

Figure 3.30: Uganda population pyramids for 2020 and the projected population in 2065. The population is projected to more than double in size in that time period.

The population pyramids in Figure 3.30 show how the population structure will change by 2065. It is important to note the change in population size as well as the shape of the pyramid. The change in population structure will result in an increased demand for housing, schooling, sanitation and healthcare. For example, between 2015 and 2065, primary schools places need to increase from 8 million to 20 million. This increase puts pressure on all other resources. As young people reach working age, employment will need to be available for them. Otherwise, the level of poverty will increase and put further pressure on the country's resources.

Figure 3.31: Children playing outside under a tree at school in Uganda.

The Ugandan government face a challenge in meeting the demands of a young population. It will need to finance the building of more schools (Figure 3.31) and clinics, along with the provision of sanitation and water supply systems. In order to meet these demands, the country has put a National Youth Policy in place. The major challenges are: poverty, limited access to education or skills, lack of work opportunities, disease and civil unrest. The country's goal is to accelerate its demographic transition from LIC to HIC in order to achieve economic and social stability.

3 Managing Human Population

CONTINUED

This can be done through:

- Addressing the shortages in health care, making sure maternity hospitals are prioritised.
- Making contraception more readily available and promoting it to reduce the birth rate.
- Educating the female portion of the population.
- Delaying the age at which marriages are allowed to occur (the 'later, longer, fewer' concept, with women not having babies so young, having bigger gaps between babies, and therefore having fewer children in their life).
- Developing industry and commercial-scale farming to increase job availability and food for the population. This can be done through encouraging TNCs (transnational corporations) to invest in Uganda and set up factories. However, corruption and political instability can prevent TNCs from choosing to invest in a country.
- Encouraging foreign aid to develop education, sanitation and infrastructure.

Japan's ageing population

Japan is an HIC in Asia that is at Stage 5 in the population pyramid. Japan's population is expected to decline 20% by the year 2055. The country has an increasing proportion of older dependents due to declining birth rates and increasing life expectancy.

Fewer women are having babies due to changes in lifestyle. Women tend to start families later, and have fewer babies in their life time.

Several factors are responsible for the ageing of the Japanese population. Figures 3.32 and 3.33 show how the population has changed and may change over time. Firstly, there has been a decline in birth rate. Since the 1950s, Japan's birth rate has been below the population's replacement level.

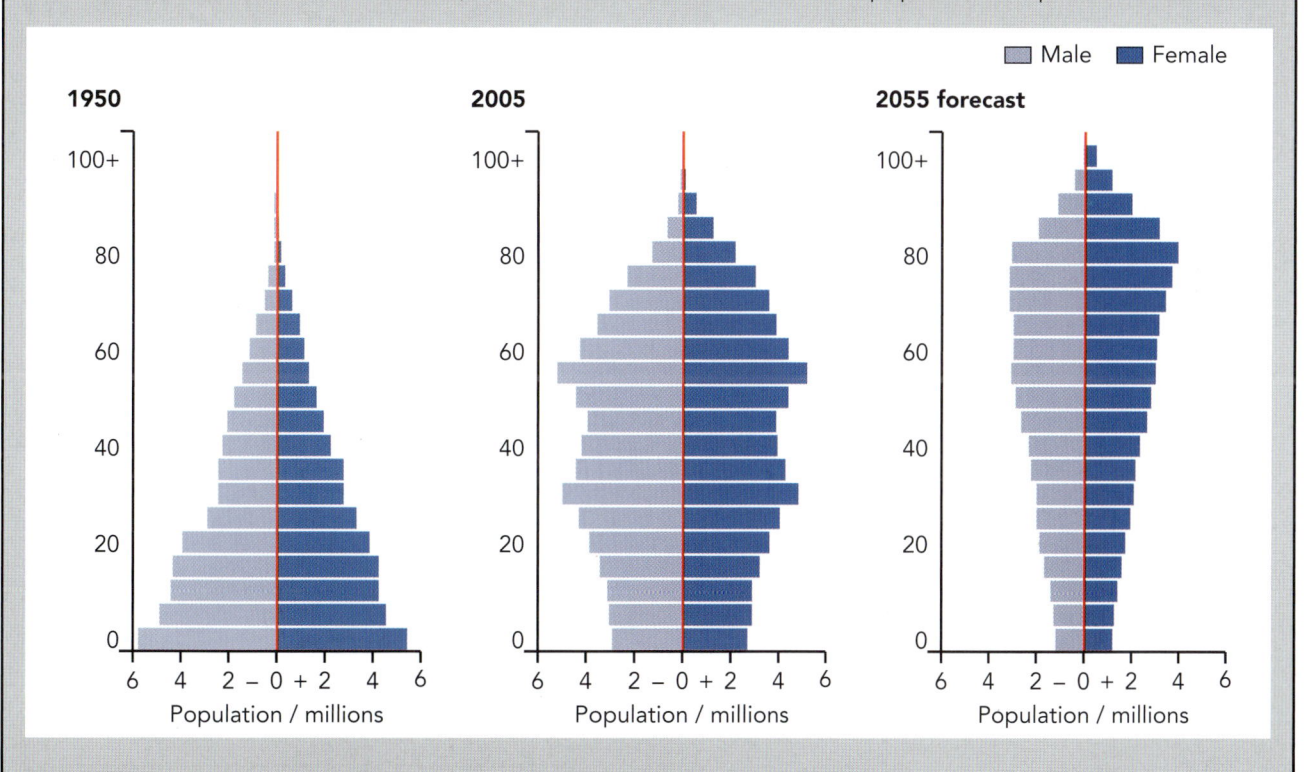

Figure 3.32: Japan's population pyramids for 1950, 2005 and 2055, showing the marked change in the shape of the population structure. The 2055 population pyramid is projected based on current birth and death rate trends.

133

CONTINUED

Most women in Japan marry later in life. Some women expect to marry after the age of 50 and do not intend to have any children. Secondly, working married couples choose to have fewer children due to work pressure, the cost of raising a child and the cost of living.

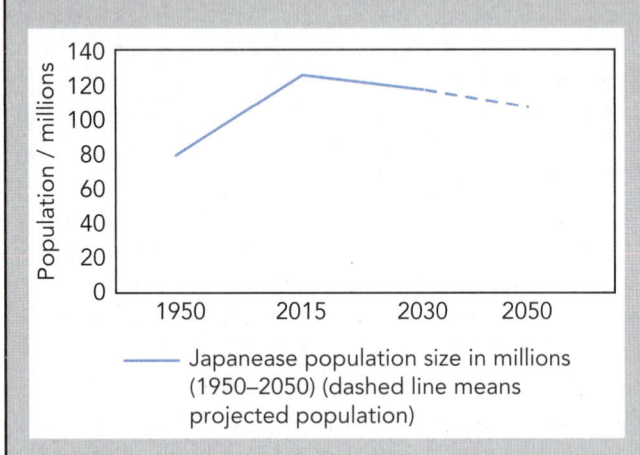

Figure 3.33: The past and predicted change to Japanese population size over time.

The old-age dependency ratio in Japan was estimated by the UN Statistics division to be 35.7% in 2022. The impacts of this high ratio include the following:

- Lower tax revenues – a smaller workforce means less money is collected as tax. Retired people also pay lower income tax. Therefore, there is pressure on the amount of money being collected by the government.

- Higher government spending – the government is committed to paying pensions and benefits to guarantee a minimum income and standard of living for older people. There is greater demand for indirect spending on older people, with a higher need for access to government health care and old age homes.

- Higher tax rates – with fewer workers, the government needs to increase taxes in order to cover costs.

- Lower pension funds – because people are living longer, pensions funds have to stretch further. Many pension funds were not planned to manage this expansion in the old-age portion of a population, therefore the average income for retired people will decline.

- Pressure to raise the retirement age – because of the pressure on pension funds, older people need to work longer. Some companies have raised the retirement age from 60 to 67 years. Older Japanese people tend to remain active members of the workforce.

- Higher national debt – Japan's debt has increased to more than double the country's economic output due to the health and social security costs of the ageing population.

- Greater pressure on health and social care services.

- Increased pressure on public transport systems as older people are less likely to drive.

Solutions to the challenge of an ageing population might include encouraging more older people to remain in the workforce. This would increase the amount of tax being collected. Japan could also try to increase its birth rate by encouraging couples to have children. Another option could be encouraging migration to increase the size of the workforce. However, Japan is resistant to migration, as it is likely to result in the dilution of the country's culture and beliefs.

Case study questions

1. Calculate the percentage change in population for Japan between 1950 and 2015, and then between 2015 and 2050 using the following information:

Change in Japanese population size over 100 years	
Year	Population (millions)
1950	82
2015	127
2030	120
2050	107

CONTINUED

2 How can the age-sex structure of a population help a government determine the needs of a population?

3 What is the main problem that the Japanese government is facing due to the decline in population?

4 Explain how the 'later, longer, fewer' concept can help to reduce the rate of population growth within a country.

5 The challenges for managing an ageing population and youthful population differ. Create two spider diagrams which compare strategies to manage youthful and aging populations.

6 Case study project: Work with a partner to create a presentation that evaluates pro- and antinatal population policies.

- Consider the following statement: *pronatalist policies are more difficult to implement than antinatalist policies.*
- Use this case study and additional research into other population policies to support your argument. Look for both strengths and weaknesses when investigating the policies and making comparisons between them.

PEER ASSESSMENT

Once other groups have presented their project findings, give feedback. Look at other groups' arguments and think about:

1 How much evidence did they give?
2 How clearly did they communicate that evidence?
3 Did they make good use of their additional research?
4 Provide feedback to the teams, describing two things that you liked, one thing that could have been improved, and one thing that you did not know before.

SUMMARY

Human population growth, distribution and density are influenced by physical factors and human factors.
Population density is calculated as the number of people per square km.
Population change is influenced by birth rate, cultural or traditional factors, levels of education, religious influence, government policy, and death rates.
The dependency ratio is the number of working vs non-working people within a population. The smaller the working population, the greater the pressure on the economy.
The impact of population change includes ageing populations. These cause problems such as shrinking workforces and economic stress.
Population change can be managed through either pronatalist or antinatalist policies.
The Club of Rome and the UN Agenda 21 were implemented to address human population growth and its impact on the environment.
Pro-natalist polices encourage families to have babies, whereas anti-natalist policies discourage this.
The high dependency of youthful and aging populations put economic and social strain on a country.

> CAMBRIDGE INTERNATIONAL AS LEVEL ENVIRONMENTAL MANAGEMENT: COURSEBOOK

> REFLECTION
>
> Did this chapter contain any new terminology that you found difficult to understand or remember? Did you try writing the words out along with their definitions to reinforce your learning?
>
> How did you find the mathematical questions? Did you use the activities and exam style questions to practise using the formulae? What could you do to remember the formulae more easily?

> PRACTICE QUESTIONS
>
> 1 Rwanda is a low income country (LIC) in Africa and The Netherlands is a high income country (HIC) in Europe.
>
> Table 1 shows population data for Rwanda and The Netherlands in 1980 and 2022.
>
> | Country | Area km² | Population 2022 | Population density km² 2022 | Population density km² 1980 | Percentage increase in population density |
> |---|---|---|---|---|---|
> | Rwanda | 26 338 | 13 600 464 | | 194 | 165.98% |
> | Netherlands | 41 850 | 17 211 447 | 411 | 342 | |
>
> Table 1
>
> a Calculate the population density for Rwanda in 2022, giving your answer in people/km² to one decimal place. [2]
>
> b Calculate the percentage increase in population density in The Netherlands. Give your answer to two significant figures. [2]
>
> c Suggest strategies that the government could employ in Rwanda to address the problems faced by their population change. [6]
>
> d Figure 1 shows two population pyramids representing the percentage of the population in different age groups in country X and country Y.

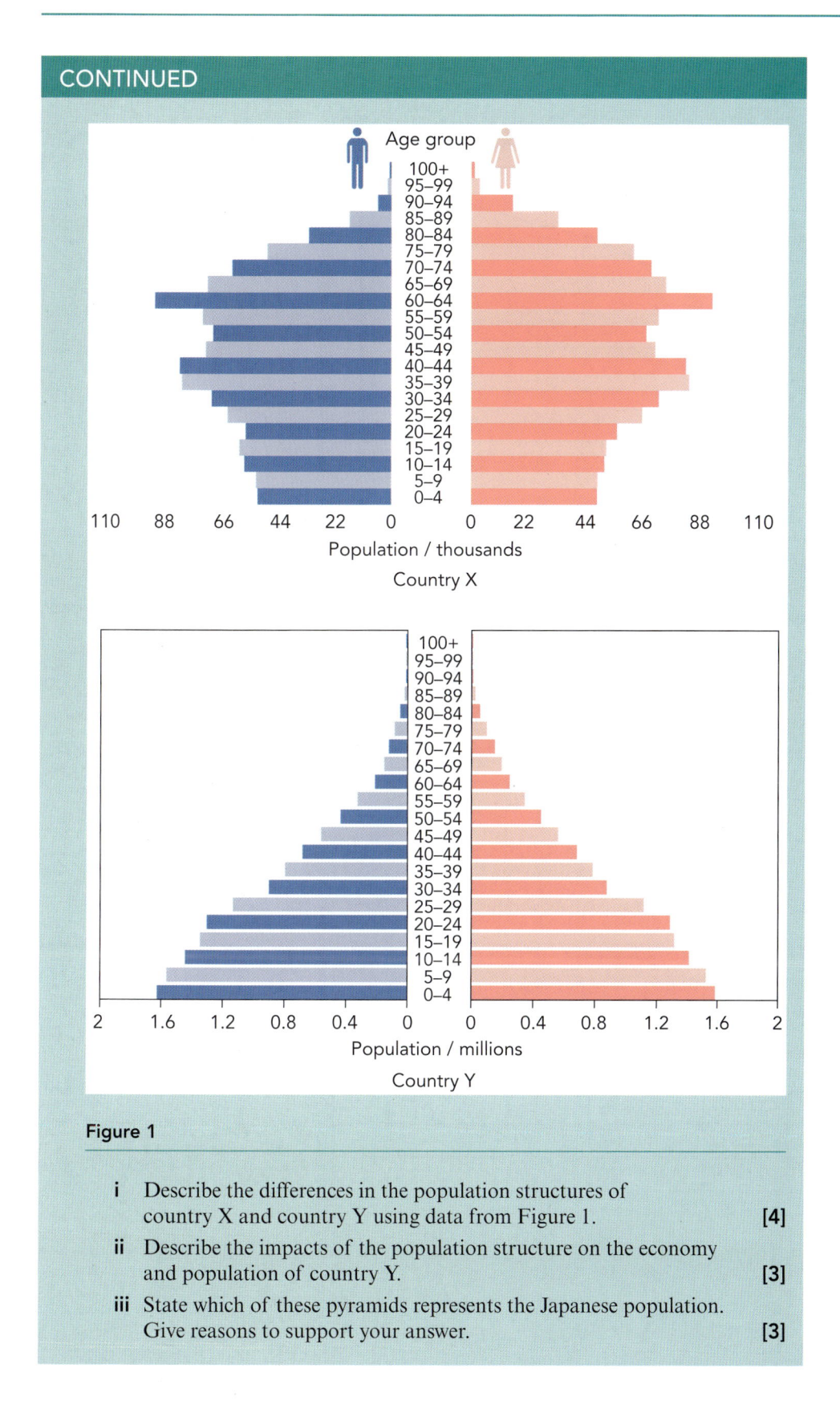

CONTINUED

Figure 1

Country X — Population / thousands (axis: 110, 88, 66, 44, 22, 0, 0, 22, 44, 66, 88, 110); Age groups 0–4 to 100+

Country Y — Population / millions (axis: 2, 1.6, 1.2, 0.8, 0.4, 0, 0, 0.4, 0.8, 1.2, 1.6, 2); Age groups 0–4 to 100+

i   Describe the differences in the population structures of country X and country Y using data from Figure 1. [4]

ii Describe the impacts of the population structure on the economy and population of country Y. [3]

iii State which of these pyramids represents the Japanese population. Give reasons to support your answer. [3]

CONTINUED

2 Figure 2 shows dependency ratios for Canada (1971–2051).

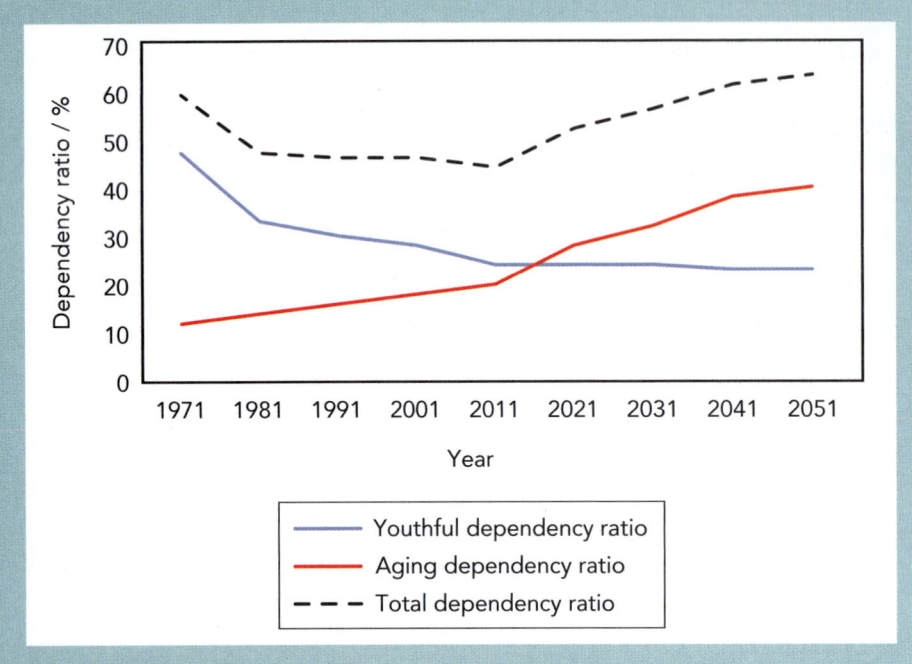

Figure 2

a Using figure 2 state:
  i The aging dependency ratio in 2011. [1]
  ii The year in which the aging dependency ratio is predicted to be 40. [1]
b Describe the trend in the youth dependency shown in Figure 3. [3]
c Explain how an aging population can be a challenge for an HIC. [5]

3 Table 2 shows population data for Belarus, Chile, Japan and Sri Lanka in 2020.

Country	Total population	Young population 0-14 years	Old population 65+ years	Young dependency ratio	Old dependency ratio
Belarus	9 379 952	1 617 104	1 500 792	17	16
Chile	19 116 209	3 677 959	2 293 945		12
Japan	125 836 021	15 654 001	35 234 086	12	
Sri Lanka	21 919 000	5 190 419	2 411 090	24	11

Table 2: Population data 2020.

a Calculate the young dependency ratio for Chile. Give your answer to one significant figure. [2]
b Calculate the old dependency ratio for Japan. Give your answer to one significant figure. [2]

CONTINUED

    c Which country has the largest percentage of old dependent population? [1]

    d Calculate the total dependency for each country and state which one has the highest total dependency ratio. [3]
- i Belarus
- ii Chile
- iii Japan
- iv Sri Lanka

    e Describe and suggest reasons for the differences observed in the old dependency ratios. [3]

4 The challenges for managing an aging population are greater than those posed by a youthful population. To what extent do you agree with this statement? Give reasons and include information from relevant examples to support your answer. [20]

SELF-EVALUATION CHECKLIST

After studying this chapter, think about how confident you are with the different topics. This will help you to see any gaps in your knowledge and help you to learn more effectively.

I can	Needs more work	Getting there	Confident to move on	See Section
Calculate population density				3.1
Describe and explain factors influencing population density and distribution				3.1
Describe population size and composition				3.1
Explain the impacts of changes in birth rates, death rates and migration on population structure				3.1
Define and calculate a dependency ratio				3.2
Suggest reasons for differences between LIC and HIC population structures.				3.2
Describe the impacts of an ageing population				3.2
Define and evaluate strategies for managing a changing population				3.3
Compare and contrast the population dynamics of a LIC and a HIC				3.3

Chapter 4
Managing Ecosystems and Biodiversity

LEARNING INTENTIONS

In this chapter you will:

- explore the major terrestrial biomes
- outline the characteristics of primary and secondary succession
- find out about ecosystem productivity
- identify the impact of invasive species on biodiversity
- evaluate the benefits of conserving biodiversity
- consider legislation and protocols as methods for biodiversity conservation

4 Managing Ecosystems and Biodiversity

CONTINUED

- identify the role of various conservation methods
- explore the impact of human activity on tropical rainforests and Antarctica, and evaluate strategies for managing them

GETTING STARTED

As a class or in small groups complete the following tasks.

1. What are soils composed of? How do they support plant growth?
2. **Describe** the main causes for the differences between hot deserts and tropical rainforests.
3. **Discuss** ways of protecting biodiversity. Consider examples of conservation you are familiar with.

COMMAND WORDS

describe: state the points of a topic / give characteristics and main features

discuss: write about issue(s) or topic(s) in depth in a structured way

ENVIRONMENTAL MANAGEMENT IN CONTEXT

Conservation of rare or unusual species

Rare and unusual species tend to be different both physically and genetically to the other species which share their ecosystems. This means they play a unique role in ecosystem function. They also offer a kind of insurance for the future of a planet adjusting to climate change.

In addition to their survival potential, rare and unusual species offer other benefits. Scientists have found that some species contribute to their ecosystems and help to support human life in a variety of ways, for example:

- The Cantor's giant softshell turtle lives in Indian rivers, where it balances the aquatic ecosystem. If the turtle becomes extinct, the river ecosystem will collapse . People in India depend upon these rivers for food and water, so the loss of these turtles would have a direct impact on people's lives.

Figure 4.1: The Hispaniolan solenodon (*Solenodon paradoxus*), also known as the agouta, is an example of an unusual species.

- Many species offer potential medical opportunities. For example, the Madagascan rosy periwinkle (*Catharanthus roseus*) contains chemicals that are used to treat cancer.

> CAMBRIDGE INTERNATIONAL AS LEVEL ENVIRONMENTAL MANAGEMENT: COURSEBOOK

> **CONTINUED**
>
> As well as benefitting people, many species play an important role in the function of their ecosystem. They could be a key element in their food chain or they might help to stabilise or regenerate ecosystem habitats.
>
> Some species may be rare or under threat due to over exploitation (capturing and selling) by humans or destruction of their habitat. Others may have a limited range, relying on very specific habitat conditions, like the agouta in Figure 4.1, which are considered a priority species. By protecting these species, many other species that share their habitats are protected.
>
> **Discussion questions**
>
> 1. How does protecting a priority species help to protect other species?
> 2. Does it matter if a single species becomes extinct? Explain your answer.
> 3. 'Current protection measures in place for rare and unusual species are effective.' How far do you agree with this statement? Discuss as a group.

4.1 Ecosystems

Ecosystems are made up of all the organisms (biotic factors) and the physical environment (abiotic factors) in which they interact. Biotic and abiotic factors are linked through the nutrient cycle and energy flows.

Ecosystems exist on different scales, ranging from small (e.g. a small wetland) to large (e.g. tropical rainforests). The biosphere is the largest scale, as it includes the entire envelope around Earth that supports life. Biomes are smaller than the biosphere. They consist of large zones on Earth that have similar climatic and ecosystem characteristics.

Biomes

A biome is a large zone characterised by its soil, climate, vegetation and wildlife. The type, amount and frequency of rainfall impacts the soils, plants and animals that exist in such an area. Biomes around the world are defined by the factors they share. For example, tropical rainforests in South America have similar characteristics to tropical rainforests in Africa. Hot deserts in Africa have similar characteristics to hot deserts in Australia (Figure 4.2).

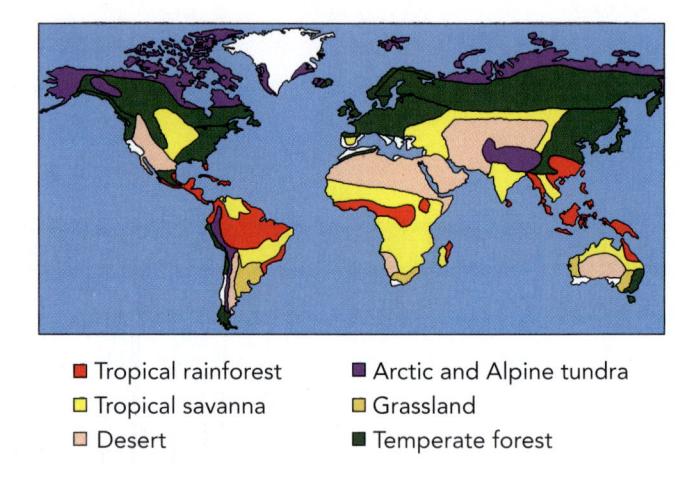

- ■ Tropical rainforest
- ■ Tropical savanna
- □ Desert
- ■ Arctic and Alpine tundra
- ■ Grassland
- ■ Temperate forest

Figure 4.2: Global biome distribution showing tropical rainforests in regions with the highest rainfall, deserts in regions with low rainfall and tundra in regions of very low temperatures.

There are four major terrestrial biomes: **desert** (hot and cold), **grasslands**, forest and **tundra**. The fifth major biome, that is not discussed here, is the aquatic biome. Biomes can be broken down into more specific categories. Within each biome, there is a wide variety of ecosystems which are not mapped at this scale.

> **KEY TERMS**
>
> **desert:** a hostile, barren landscape where less than 250 mm of precipitation occurs annually, and biodiversity is low
>
> **tundra:** a biome found far north in Asia and Alaska, characterised by long cold, dark winters, and short cool summers. Permanently frozen ground limits vegetation growth to short shrubs and grasses
>
> **grasslands:** a biome with grassy plains and few trees, in the tropics and subtropics, typically referred to as savanna

Figure 4.4: Savanna grasslands in Kenya, showing trees with their characteristic canopy shape.

Soil

Soil develops in relation to the climate of a region. Rainfall impacts the types of plants that can grow in an area and the rates of leaching. This, in turn, influences the processes that bring about the formation of soil.

Different biomes have different soil types. Table 4.1 outlines the general characteristics of the soil found in each biome.

Terrestrial biomes

Figure 4.3: Vertical stratification in a tropical rainforest.

Figure 4.5: Vertical layers in the rainforest.

> **KEY TERMS**
>
> ***diurnal temperature range:** a variation between high and low air temperatures that occurs during the same day (e.g. changes between night and daytime temperatures)
>
> ***convectional rainfall:** rainfall that occurs when the energy of the sun heats up Earth's surface and causes water to evaporate and become water vapour. This then condenses to form clouds at higher altitudes

Terrestrial biome	Location/ distribution	Examples of plant adaptation	Climate	Soil type	Human impact
Tropical rainforest	Equatorial zones of Central America, Central Africa and Asia.	Trees are tall and thin (Figure 4.3) to allow for them to reach the light. An example would be the Ficus. The bark is usually smooth to allow stem flow to guide water to the roots. The bark is thin as there is no need to conserve water. Leaves of the plants have a waxy surface with drip tips to allow excess water to runoff and prevent algae from growing on the leaves. Leaves are also often angled and different angles to capture as much of the light filtering through as possible and optimize photosynthesis. The trees are shallow rooted and supported by buttresses or stilted roots (Figure 4.5). The widespread roots absorb nutrients from the shallow soils. Woody vines grow up around the trunks of trees to reach the forest canopy, where they spread out to catch the light.	Temperature range 22–31 °C, depending on location. Local **diurnal temperature** is low. Precipitation can reach 2000 mm per year, and humidity levels are high. High rates of insolation cause daily **convectional rainfall**.	Usually red or yellow in colour. High levels of rainfall all year around result in nutrients and clays being leached out of the soil and replaced with aluminium oxides.	Deforestation. Mining. Urbanisation. Agricultural development. Desertification.

Terrestrial biome	Location/ distribution	Examples of plant adaptation	Climate	Soil type	Human impact
Grassland	Middle of continents, away from coasts and between 5° and 15° north and south of the equator.	Savanna vegetation is characterized by a variety of grasses and umbrella shaped trees, like the acacia. The trees are usually spread far apart as water is limited. The root systems can be either widespread to absorb water from a large area or deep tap roots that reach the ground water during dry season. Some plants store water in their stems or bulbs underground, while other plants can go dormant during the dry period. The leaves are long and narrow to reduce water loss through transpiration. Many have developed thorns and grown tall to protect themselves from primary consumers. The deep root systems also allow the trees to regrow after a fire. In addition, some trees have very thick bark to protect them from fire damage (Figure 4.4).	Temperature range 20–35 °C. Seasonal precipitation with distinct dry and wet seasons. Up to 750 mm precipitation per year with wet summer. Rainfall very unpredictable.	Tend to be red in colour due to high iron content. Deep, highly weathered, porous soils with rapid drainage. Thin, organic layer on surface decays rapidly due to high temperatures. Soils not very fertile and support limited vegetation growth due a hard sub-layer.	Extensive use of grasslands for grazing and urbanisation. Reduced biodiversity due to loss of habitat and hunting.

Terrestrial biome	Location/ distribution	Examples of plant adaptation	Climate	Soil type	Human impact
Hot desert	Between latitudes 18° and 28° north and south of the equator. Middle of continents and along western coasts. Deserts cover approximately 14% of Earth's surface.	Cacti have thorns and succulent, deep-rooted or wide root balls. Cacti have short reproductive cycles and drought tolerant seeds. Cacti leaves are small or non-existent. Thick waxy cuticles prevent water escaping. Sunken stomata help to reduce water loss (Figure 4.7).	High daytime and low night-time temperatures. Less than 250 mm precipitation per year. Winds generally light. Evaporation rates high.	Usually light in colour. Dry soils, generally sandy. Can vary in depth, texture, porosity, mineral and salt content. Top layer may be absent or very thin due to lack of vegetation for provision of organic material. Lower layers poorly developed (Figure 4.6). Deserts with no significant soil horizons are called Entisols.	Mining. Removal of rare species. Use of ground water and drying out of surface water resources.
Tundra	In the far north of Asia and South America. South coast of Greenland.	Tundra plants are typically small (mosses, grasses and shrubs). They grow close together and low to the ground. Many plants have waxy hairs coating them to shield them from the cold and wind, helping them to retain heat and moisture. They typically have shallow root systems due to permafrost. They are perennials that do not die off in winter, and have long life cycle with a short growing season (Figure 4.8).	High annual temperature range: as low as −34 °C with summer highs reaching 12 °C. Less than 250 mm precipitation per year. Stong polar winds. Short summers, with 50–60 days of 24-hour daylight.	Dark brown in colour. Sub layer is permanently frozen (permafrost) within 100 cm of the surface. Tundra soils have an accumulation of organic material at surface due to cold temperatures and low level of decomposition.	Melting of permafrost, release of large volumes of methane. Solifluction (slope failure). Drilling for resources.

Table 4.1: Key characteristics of four different terrestrial biomes: tropical rainforest, grassland, hot desert and tundra. Refer to Figure 4.2 for the map of the biome distribution.

4 Managing Ecosystems and Biodiversity

> ### ACTIVITY 4.1
>
> **Biome animal adaptations**
>
> Your teacher will assign you a biome to research.
>
> 1. Research two to three animals that have adapted to live in that biome. Choose unusual animals that not many people know about. Focus on how the animals have adapted to the climate and habitat of the biome.
> 2. In pairs, spend three minutes talking about your animals. Do not mention the name of your biome.
> 3. Guess the name of the biome that the animal your partner is talking about comes from. Explain why you think the animal would live in that biome.
> 4. Change partners and repeat the process. Find out about more animals that have adapted to different climates.

Figure 4.8: Tundra vegetation showing low lying grasses and bushes and a lack of trees.

Figure 4.9: Buttress roots of two different Dipterocarp trees supporting the tall tree trunks in the rainforest in Borneo's Gunung Mulu National Park.

Figure 4.6: A desert in Utah, USA, showing low vegetation cover and exposed rocks and soils.

> ### SELF ASSESSMENT
>
> - Were other students able to correctly guess the biome from your description of the animals?
> - Could you have improved on the information you provided about your chosen animals?
> - What were the other students' animal descriptions like? Were you able to correctly guess the biome from the clues that they gave for their animals?

> ### KEY TERMS
>
> ***stomata:** pores in the leaf or stem of the plant. These form a slit which allows the movement of gases in and out of the spaces between the cells; found mainly on the underside of leaves
>
> **permafrost:** areas of permanently frozen ground

Figure 4.7: Desert plant adaptations include retaining water in stems, replacing leaves with thorns to reduce water loss and offer protection, and having root systems that maximise access to water resources.

147

Primary and secondary succession

Ecological succession is the process through which an ecosystem changes and develops over time (Figure 4.10). No ecosystem is static: there is constant change occurring. Change is sometimes on a large scale, due to a significant event (e.g. volcanic eruption, tsunami, fire, human activity) or on a small scale (e.g. if a small mudslide or rockslide disturbs a mountain slope).

> **KEY TERMS**
>
> *****ecological succession:** the process by which the structure of a biological community changes over time
>
> **primary succession:** the gradual process by which an ecosystem develops and changes in a region that has not previously been colonised, for example new lava flows
>
> **secondary succession:** the gradual process by which an ecosystem develops and changes in a region that has previously been colonised, however, it has been disturbed, damaged or removed

There are two main types of succession: primary and secondary succession.

Primary succession occurs in areas that are essentially uncolonised, where the soil is not capable of sustaining life. It occurs on an entirely new landscape which has never been colonised before. For example, on a new lava flow, newly formed sand dunes, or a newly quarried rock face.

Secondary succession occurs in an area that was colonised but which has since been disturbed or damaged. This is usually a smaller scale disturbance that has not eliminated all life or removed all the nutrients from the environment. For example, it could occur in an area where a mudslide has occurred (Figure 4.11), or some trees have been felled.

As the physical and chemical environment in an area change over time, so do the species living there. Through the process of living, growing and reproducing, organisms change the environment they exist in, resulting in succession. In both primary and secondary succession, there is a continually changing mix of species as disturbances of different frequency, intensity and size alter the landscape.

However, the sequence of succession is not random. At each stage, certain species have evolved to exploit very specific conditions. Each species has adapted to thrive and compete against other species in a very

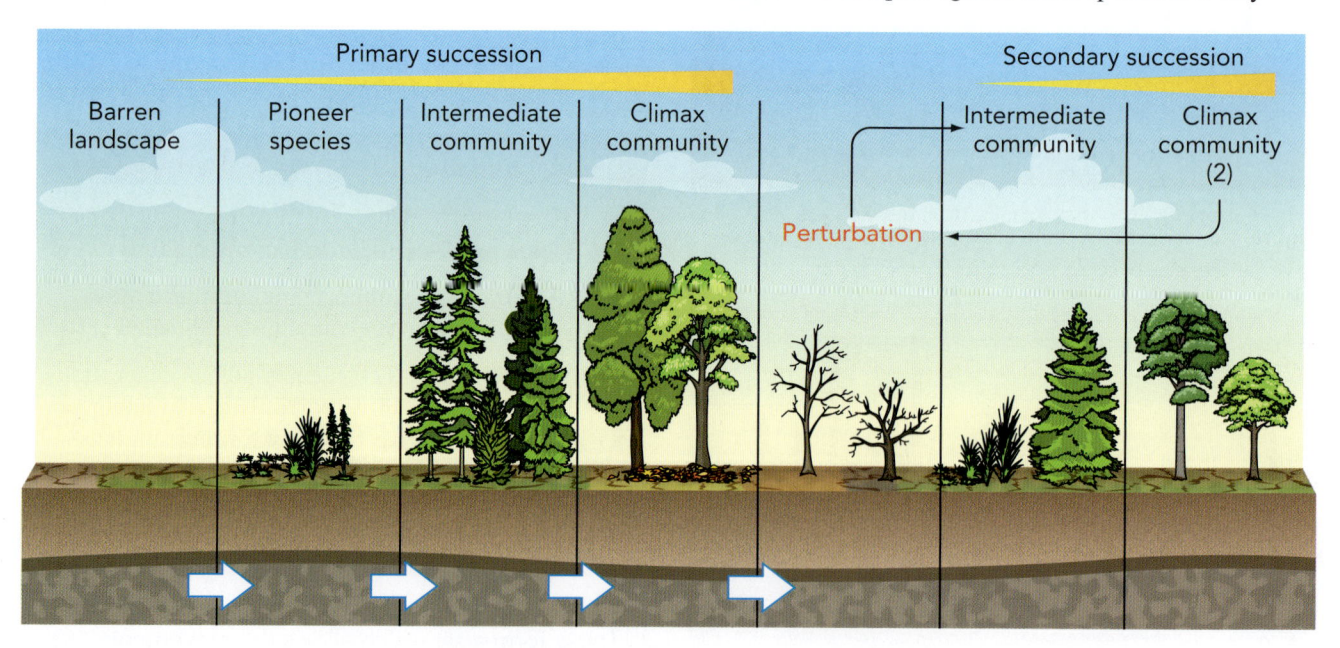

Figure 4.10: Succession of an ecosystem from bare rock through to a climax community.

4 Managing Ecosystems and Biodiversity

specific set of environmental conditions. Therefore, the process of succession within an area is partially predictable. Initially, only a small number of more robust species from surrounding habitats are able to thrive in the disturbed area (Figure 4.12). Where there is bare rock, the **pioneer species** that first establish themselves are lichens and mosses. They start the formation of soils by breaking down the rock. As soils develop, larger grasses and shrubs can start growing. Over time, larger and more complex ecosystems develop (see Figure 4.10).

The greater the disturbance, the hardier the species that carry out the first stages of succession. Pioneer species may cause a change in the conditions, such as creating soils or providing shade. When that happens, more species can grow in the new area, resulting in a change to the existing species. A different set of species better adapted to survive in the new conditions out-compete the initial species that populated the area. This phase (Figure 4.10), where change from the initial community has occurred, but the final stage has not yet been reached, is referred to as the 'intermediate community'.

The final stage in succession is known as the a 'climax community'. A climax community takes on different forms depending on the prevailing climate. For example, a tropical rainforest compared to savanna grassland, which are both climax communities, but very different to each other. Reaching a climax community does not mean there is no further change; landslides, fires or trees falling in a climax community allow for secondary succession to take place.

This process of succession involves the whole community, not just the plants. However, changes in plant species are one of the driving forces behind changes in animal species, because each plant species has an associated animal or group of animals that feed on it.

The actual species involved in a succession in a particular area are controlled by factors such as the geology, climate and other abiotic factors of the area. Succession occurs on many different time scales (a few days to hundreds of years), for example, a climax woodland may take hundreds of years while the succession of invertebrates and fungi in cow dung may be over within three months.

> **KEY TERM**
>
> **pioneer species:** a hardy species which is capable of being the first to colonise disturbed or newly formed environments

Figure 4.11: A landslide in Sierra Leone. Here the existing ecosystem has been disturbed. The area will now undergo secondary succession, as the ecosystem recovers from the disturbance.

Figure 4.12: Lichen and small plants colonising a rocky area with little soil present. These pioneer species are some of the first to inhabit an area during the process of succession.

> **ACTIVITY 4.2**
>
> **Succession case study**
>
> Research an example of secondary succession.
>
> 1 Write a short (600–800 word) case study on your chosen example.
>
>   Include:
>
>   - the event/s that caused the initial change
>   - examples of affected species and the impact of the change
>   - the path of secondary succession that the ecosystem followed.
>
> 2 Choose a format for your case study (word document, informative poster or presentation).
>
> 3 Share your example with the class.

149

Primary productivity

Primary productivity is the rate at which energy is converted into organic material through photosynthesis by plants. The total amount of biological productivity within an ecosystem or biome is known as the **gross primary productivity**. However, each photosynthesising organism within an ecosystem uses some of that energy to respire. The amount of energy captured, minus the energy used to respire, is the **net primary productivity**.

The amount or mass of living organic material a particular ecosystem can support is determined by:

- The amount of energy captured and stored as chemical energy by the primary producers in that ecosystem. This is high in tropical zones where rates of insolation and precipitation are high, while it is low in areas where rates of insolation or precipitation are low.
- How rapidly the plants can store the energy.

Figure 4.13 shows how primary productivity varies around the world. This is influenced by many factors, however, they include the key factors of sunlight and water availability. In terrestrial ecosystems, primary productivity ranges from approximately 2000 g/m²/year in the tropical forests to less than 100 g/m²/year in some of the deserts. The grams in these units refer to the amount of dry organic matter produced.

Gross primary productivity (GPP) is shown as g/m²/year, and is the rate at which ecosystem's producers convert solar energy into biomass, measured as:

$$\text{Gross primary productivity (GPP)} = \frac{\text{Energy production per unit area}}{\text{Units of time}}$$

Net primary productivity (NPP) is shown as g/m²/year. It is the rate that the producers produce and store energy minus the loss of energy through respiration (R).

$$NPP = GPP - R$$

NPP differs between different climates and ecosystems. It decreases as distance from the equator, and proximity to the poles, increases (as available sunlight decreases, there is reduced insolation).

The rate at which biomass is produced within an ecosystem is the **ecosystem productivity**. This rate varies between different ecosystems depending on the numbers of producers, decomposers and consumers within the ecosystem.

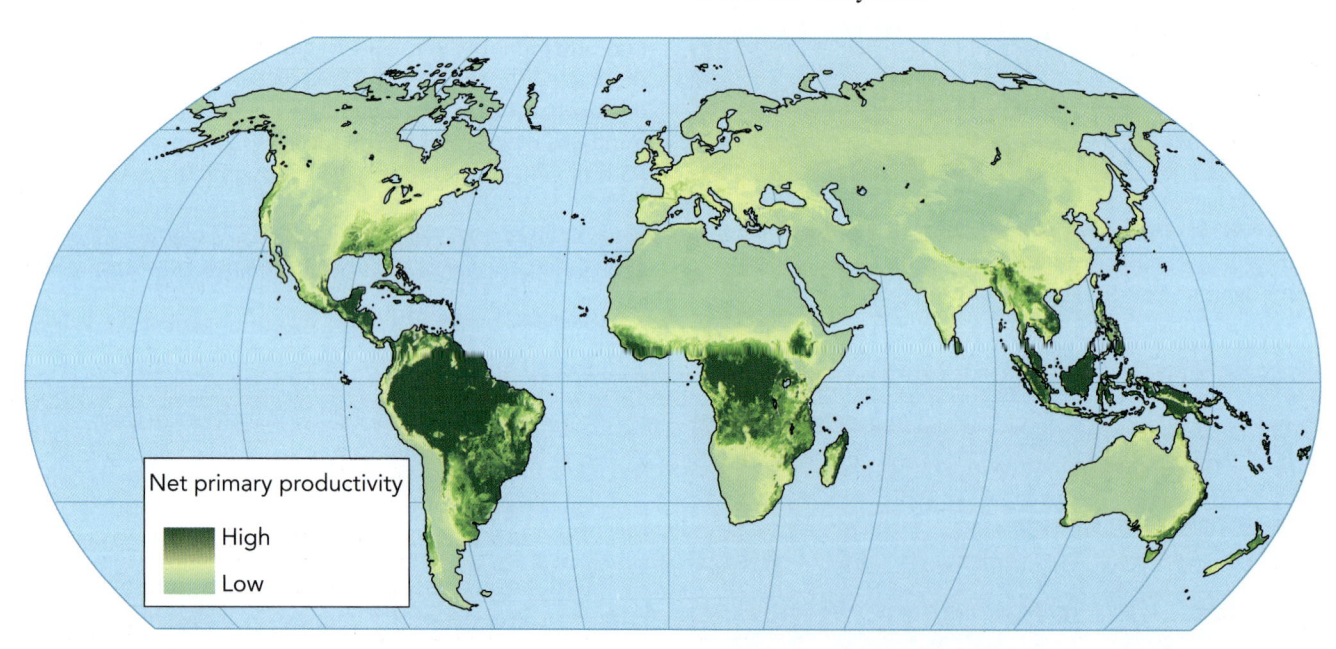

Figure 4.13: Spatial distribution of global net primary productivity of carbon per unit area per unit time. The highest rates of primary productivity are found in regions of high annual insolation rates.

4 Managing Ecosystems and Biodiversity

> **KEY TERMS**
>
> *****primary productivity:** the rate at which energy is converted into organic material through photosynthesis by plants (producers)
>
> **gross primary productivity:** the rate at which producers convert solar energy into biomass
>
> **net primary productivity:** the rate as which producers convert solar energy into biomass minus the loss of energy through respiration
>
> **ecosystem productivity:** the rate of production of biomass for an ecosystem

Ecological pyramids

Ecological pyramids, also known as trophic pyramids, are used to compare communities within or between ecosystems by analysing their trophic levels. The pyramids show the energy structure of an ecosystem as a diagram representing the energy, number of individuals or biomass in each trophic level. The following are some examples of ecological pyramids and energy flow in ecosystems.

Examples of ecological pyramids and energy flow within an ecosystem

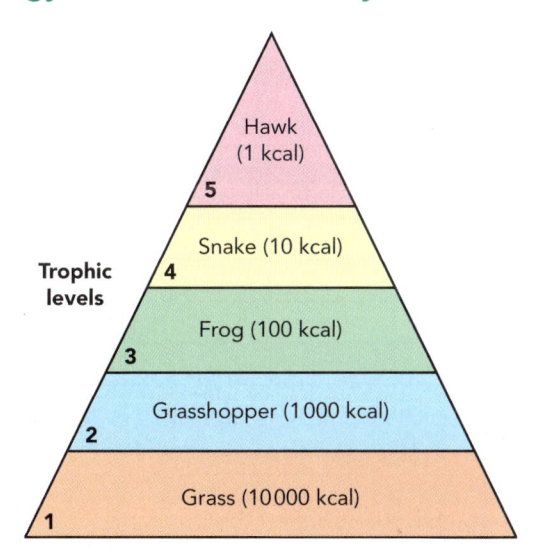

Figure 4.14: Pyramid of energy.

A pyramid of energy

A pyramid of energy is the best way to show the feeding relationships of an ecosystem. The amount of energy available decreases further up the pyramid. The greatest amount of energy is available at the bottom of the pyramid. This is due to the use and loss of energy at each level (Figure 4.15).

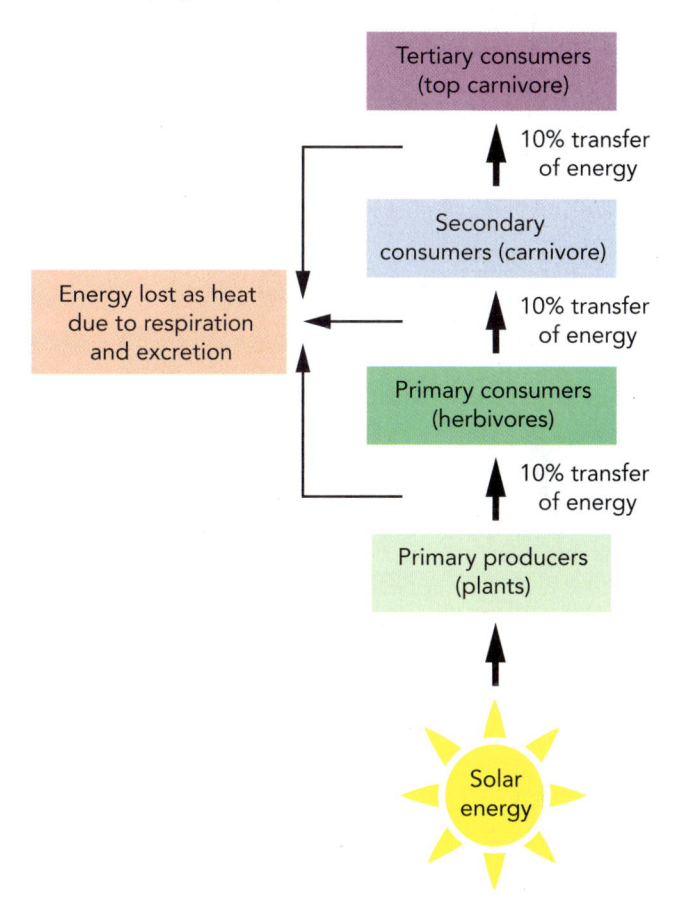

Figure 4.15: Energy transfer up the trophic levels.

Energy transfer up the trophic levels

Energy flows into the ecosystem from the sun. It then continues to flow through the different trophic levels along food chains. Energy is used at each trophic level by organisms, and it is lost from the ecosystem at each trophic level, through respiration and excretion. Approximately 10% of the energy available to one trophic level is passed onto the next level when it is eaten.

151

A normal pyramid of numbers

In a normal pyramid of numbers (Figure 4.16), the largest number of organisms is at the bottom, These are the producers, such as plants on a forest floor. The smallest number is at the top, which are the tertiary consumers, such as a few large carnivores. There are limitations in this model as it does not consider the size of organisms. For example, one rainforest tree can support thousands of other organisms.

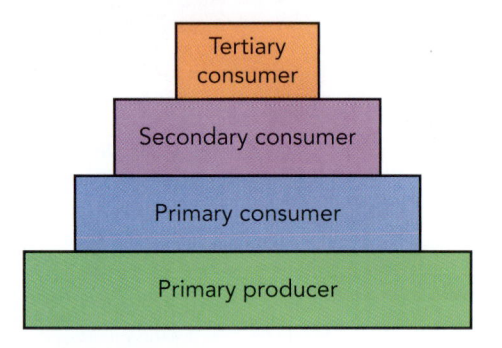

Figure 4.16: A normal pyramid of numbers.

When organism size is considered, unusual pyramid shapes are likely to occur. For example, in Figure 4.17, one rose bush can support a large number of greenfly. This results in the lowest layer of the ecological pyramid being smaller than the layer above it.

Like the pyramid of numbers, a pyramid of biomass can also be an unusual shape, with a lower level smaller than the level above. For example, the biomass of phytoplankton may be lower than the mass of organisms that feed on them (fish or insects). Phytoplankton reproduce quickly and have a shorter life span. This means their biomass is able to support a larger group higher up in the trophic pyramid.

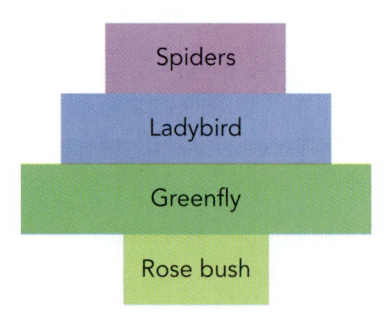

Figure 4.17: A distorted pyramid of numbers, where the producers are fewer in number than the organisms above them.

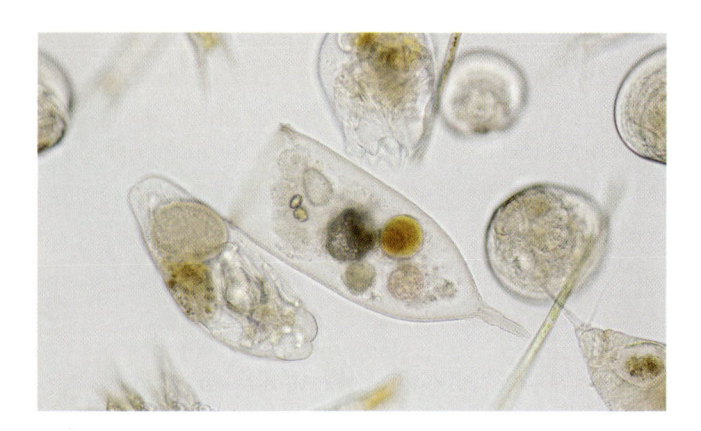

Figure 4.18: Phytoplankton seen through a microscope.

ACTIVITY 4.3

Interpreting ecological pyramids

Using the data from Tables 4.2 and 4.3, draw the energy, biomass and number diagrams for the ecosystem represented.

Organism	Pyramid of biomass (g/m^2)	Pyramid of numbers
Grass	4000	1000
Antelope	3000	3
Ticks	2000	800
Birds	1000	100

Table 4.2: Data for drawing of a pyramid of biomass and pyramid of numbers.

Organism	Pyramid of energy (cal/m^2/year)
Water plants	135 000
Tadpoles	21 000
Small fish	4000
Larger fish	1

Table 4.3: Data for drawing an aquatic pyramid of energy.

Once you have completed the diagrams:

1. Suggest the reason for these ecological pyramids' shapes.
2. Explain why the amount of energy decreases so rapidly up the aquatic pyramid of energy.

4 Managing Ecosystems and Biodiversity

4.1 Questions

1. **Explain** the difference between a biome and an ecosystem.
2. **Compare** primary and secondary succession.
3. Explain how desert organisms have adapted to live in harsh conditions.
4. **Describe** how soils found in the tundra differ from other soils around the world.
5. Explain why the tundra has low biodiversity.

COMMAND WORDS

explain: set out purposes or reasons / make the relationships between things evident / provide why and/or how and support with relevant evidence

compare: identify/comment on similarities and/or differences

describe: state the points of a topic / give characteristics and main features

Figure 4.19: The cane toad is an invasive species in Australia.

Figure 4.20: The Australian green tree frog (*litoria caerulea*): a species **indigenous** to Australia.

Cane toads were introduced into Australia to control pests in sugar cane fields. They will eat almost anything of the appropriate size and have no natural predators in Australia. They lay aproximately 30 000 eggs, while indigenous green tree frogs produce between 1000 and 2000. Not only is the cane frog able to eat the green tree frog the large number of eggs produced by the cane frog means they easily outnumber and outcompete their native rivals.

4.2 Managing the conservation of biodiversity

Native and invasive species

Native species are species that originated and developed in a specific ecosystem or region and adapted to living in that area. An **invasive species** is able to outcompete other species in the area it has invaded. The invasive species causes changes to the balance in an ecosystem that it invades. An invasive species can be either native or non-native (a species that originated somewhere other than its current ecosystem).

KEY TERMS

native species: aspecies that originated and developed in a specific ecosystem or region and has adapted to living in that area

invasive species: a species that is able to outcompete other species, causing changes to an ecosystem's balance

*indigenous: originating or occurring naturally in a specific area; a species that is native to an ecosystem

Invasive species cause harm to indigenous wildlife in a number of ways. When an invasive species is introduced into a new ecosystem, it may not have any natural predators or controls. It is therefore able to spread rapidly. Native wildlife may not have mechanisms to defend itself against the invader and so is unable to compete against it. Invasive species pose direct threats to native species numbers through carrying disease, predation and outcompeting. Indirectly, invasive species can change the food web within an ecosystem because they replace native food sources (Figure 4.22).

Figure 4.21: A sign asking hikers to protect the region from invasive species by brushing off their shoes before leaving a trailhead.

The loss of indigenous flora results in reduced biodiversity in an area. This is because indigenous vegetation has the seeds, flowers, fruit and habitat that native fauna has adapted to for survival. Invasive species of flora do not offer the same plant characteristics as the native species, and organisms dependent on them may disappear from the area.

Many human economic, commercial, recreational and agricultural activities depend on healthy native ecosystems. Invasive species degrade native ecosystems and can therefore be harmful to human health and local economies. The impact of invasive species is estimated to cost billions of dollars globally each year.

Invasive species are primarily spread through human activities, both intentional and unintentional. An example of intentional human activity is the fast-growing softwood trees from Australia and the Americas (such as blue gum and pine trees) which were planted in Africa to supply wood. The trees became invasive species, outcompeting local vegetation. An example of unintentional human activity is when ships carry rodents on board, unintentionally spreading them to new locations. This happened in the Galapagos islands in the 1700s and 1800s, where the black rat was accidentally introduced to many of the islands by whaling ships. The rats ate tortoise eggs, impacting tortoise numbers. In 2012, rat eradication was used to undo the damage done by the invasive species.

> **KEY TERMS**
>
> *flora: the plants of a particular area, region or environment
>
> *fauna: the animal life characteristic of an area, region or environment

Invasive species				
The potential impacts				
Grazing	Predation		Parasitism	Poisoning
Competition	Flammability		Disease transmission	Suppression of growth
Examples of potential impacts				
Environmental			Social	
Soil erosion			Loss of agricultural land to invasive species	
Excessive water consumption			Reduced access to water	
Decreased biodiversity			Damage to infrastructure	
Habitat degradation			Reduction in tourism opportunities	

Figure 4.22: Some of the potential environmental and socio-economic impacts of invasive species.

Native species can become invasive when they outcompete other organisms in the ecosystem. An example of this is when **eutrophication** occurs in a water body (Figure 4.23). Blue-green algae native to the area blooms, becoming harmful to other native species in the ecosystem.

> **KEY TERM**
>
> ***eutrophication:** an increase in nutrients in a body of water results in a rapid growth of algae. Algal blooms cover the surface of the water, forming a green layer. When the algae decay and die, a decline in oxygen level occurs, causing significant ecological degradation

Figure 4.23: A stream with a thick algal bloom, evidence of eutrophication occurring in the water.

Benefits of conserving biodiversity

Conservation biodiversity involves protecting and preserving the variety of species, ecosystems, habitats and genetic diversity, both in a specific area and globally. The benefits of conserving biodiversity include, but are not limited to, the following:

- Food security: this is dependent upon natural resources as the basis of food production. Biodiversity conservation protects plants, animals, microbes and genetic resources that support soil fertility, nutrient recycling, pest and disease regulation, prevention of soil erosion and pollination of crops.

- Economic growth and poverty reduction: many poor populations live in rural areas and depend upon natural areas for their livelihoods. Forests provide wood and non-timber products (fruit, medicines, animals), while fresh water and marine areas are an opportunity for fishing. Scarce or mismanaged natural resources are also often a source of conflict. This can further increase poverty and political instability.

- Combatting climate change: by conserving forested or vegetated areas, the amount of CO_2 released into the atmosphere can be reduced. By conserving coastal mangroves and coastal ecosystems, the effects of increased storm surges and coastal inundation can be lessened.

- Medical resources: many plants have medicinal qualities (Figure 4.24) in their natural form and are harvested by indigenous populations to treat illness. The development of pharmaceutical medicines also relies on the qualities that plants offer. Protecting plant biodiversity ensures that the medicine people need can continue to be produced.

> **ACTIVITY 4.4**
>
> **Case study research**
>
> Research and write an information leaflet on an invasive species of your choice.
>
> Include:
>
> - Information on the location
> - How did the invasive species spread to the area?
> - The impact the invasive species is having on native flora and fauna
> - Actions being taken to manage the problem
>
> Include images so your reader has a clear idea about the species you are discussing.

Figure 4.24: Common medicinal herbs dried for use. Some are used in pharmaceutical medicines, while others are believed to have inherent medicinal qualities.

- Genetic diversity: this refers to the range of different characteristics within a species that are a result of specific **genes** in their **DNA**. The greater the number and types of genes, the greater chance the species has of being able to adapt to changes in the environment, and the less likely it is to face extinction. Examples of environmental changes include increasing or decreasing precipitation and increases or decreases in temperature. Protection of genetic diversity is therefore central to the conservation and restoration of biodiversity as it ensures that the organisms are able to survive.
- Resources: items such as food, wood, fibres, oils, clean water and fuels can all be obtained from nature. Protecting biodiversity ensures that these resources are used sustainably and will be available for future generations.
- Ecological security: biodiversity ensures ecosystem stability and maintains an ecological balance. Within food webs, plants and animals are linked to each other through food chains. The loss of one or more species from an ecosystem results in ecosystem decline and ultimately ecosystem fragility and failure.
- Cultural and recreational value (Figure 4.25): biodiversity reflects many social values and local beliefs, as specific plants or animals often carry spiritual or cultural importance. Many cultural practices and ceremonies depend on biodiversity and specific species or ecosystems. For example,

a culture may use rivers for baptism and hillsides or forests for prayers. From a recreational perspective, biodiversity makes beautiful landscapes, which also encourages tourism and the creation of eco-tourism venues. Well-managed beaches and waterways create opportunities for outdoor activities and support local economies through recreation and tourism.

Figure 4.25: Protection of biodiversity in sites such as this this one in Switzerland also protects an area's recreational and cultural value.

> **KEY TERMS**
>
> **genes:** the basic units of heredity (characteristics) passed down from parent to young. For example tall parents are more likely to have tall children as their genes carry that characteristic
>
> **DNA:** the material in cells that carries information about how a living organism will look and function. Genes make up portion of the DNA

Legislation and protocols that conserve biodiversity

The evolution of plants and animals over millions of years has resulted in the global ecology that supports life as we know it today. These plants and animals help to stabilise the climate, renew soils, protect water resources and maintain the chemical balance of Earth.

We uses the planet's natural resources to meet our ever-changing needs. If the planet is to continue supporting us, biodiversity is needed. This is why legislation and protocols have been put in place to conserve and protect biodiversity.

The planet functions as a single system. Each biome relies on other biomes and climates to remain stable. However, modern politics has fragmented the planet into individual self-governing regions. This means that each country has the rights to all the natural resources within their borders. Governments can control, conserve, exploit or destroy these resources as they wish. Global, local, regional and international treaties are an attempt to manage the ongoing problem of depleted resources and declining biodiversity. However, membership of these treaties or agreements is voluntary, so they are not necessarily effective worldwide.

Protection of species

Global concern about conversation has resulted in the awareness that we need to protect not only individual species, but biodiversity as a whole. In cases of species which are hunted or collected, direct protection may be required to ensure their conservation. National laws such as the Endangered Species Act (USA) make the collection or killing of species on the endangered species list illegal and prosecutable. Offenders risk fines or even jail time. An example of a species with total protection in the USA is the American bald eagle (Figure 4.26).

Figure 4.26: Students in Boston, USA learn about the American Bald Eagle, which is an example of a protected species.

International laws are in place to protect various species. These usually fall under the management of organisations like CITES (Convention of International Trade in Endangered Species of Wild Fauna and Flora) and the IWC (International Whaling Commission), both of which are discussed later in this chapter. However, it can be difficult to enforce these laws in the face of international poaching and trafficking.

The African black rhino's (Figure 4.27) numbers fell to about 2400 by 1995. This was for two reasons. Firstly, the black rhino suffered loss of habitat and reduced range due to agricultural expansion, urbanisation and sport hunting. It was also hunted for its horn. In some cultures, the horn is believed to have medicinal value and is used to treat illness. Hunting continues to put pressure on the species, which is now protected under the CITES agreement. The World Wildlife Fund took action in the three main countries that still have African black rhino (Namibia, Kenya and South Africa). The number of black rhino is estimated to have recovered to more than 5000 today. However, there are still extreme poaching pressures on the species.

Figure 4.27: The African black rhino is under severe threat from poaching. This species is protected under CITES and appears on the Red List.

Sustainable harvesting

Sustainable harvesting is the use of a resource in a way that ensures its constant supply without harming future yields or causing irreversible damage to the ecosystem. In order for sustainable harvesting from an ecosystem to be effective, both educational and legislative solutions need to be put in place.

The law that covers the South African fynbos ecosystem is an example of legislature governing the sustainable harvesting of resources from a biome. Plants found in this biome have uses from feather dusters to teas, medicines, dyes, sugars, and decorative flowers. The plants' usefulness has resulted in their over harvesting. For this reason, permits are required for harvesting products, and there are regulations about how much can be harvested and when (Figure 4.28). Table 4.4 shows how sustainable harvesting can be determined.

Figure 4.28: Harvesters need permits to harvest in the area, and the number of flowers they can pick is restricted. This limts disturbance to the fynbos while still allowing the harvesters to earn a living.

Description	Sustainable/ Unsustainable	Reason
Total harvest		
All available material is harvested	Unsustainable	All material is removed. So no reproduction or regrowth can occur.
Maximum harvest		
Regrowth or reproduction is considered when harvest occurs	Unsustainable	Viability of restocking is not fully considered. This means there is a potential for species or ecosystem failure.
Random harvest		
No limits are set	Unsustainable	The frequency and quantities taken determine the survival of the species or ecosystem.
Optimal harvest		
A set % of the resource is harvested, leaving the rest	Sustainable	Allows for continued ecosystem function and replacement of resources over time.
Systematic random sampling		
A set % of available material is harvested from a defined part of the ecosystem, not the whole	Sustainable	Ensures full sustainability and biodiversity of the ecosystem as the disturbed area is minimised.

Table 4.4: The sustainability or unsustainability of difference har.vesting strategies.

4 Managing Ecosystems and Biodiversity

ACTIVITY 4.5

Graphing biodiversity threats

The data in this table shows the major threats to biodiversity. Use the data to complete a 100% stacked bar chart. Then answer the questions.

1. What is the greatest overall cause of loss of species?
2. Which group is suffering from the highest level of overexploitation and why?
3. Explain why you think climate change is still ranked relatively low as a threat to biodiversity.

Biodiversity category	Habitat loss (%)	Exploitation (%)	Invasive species (%)	Pollution (%)	Climate change (%)
Birds	48	12	10	13	17
Reptiles & amphibians	46	20	15	11	8
Mammals	42	40	10	4	4
Fish	28	54	4	3	11

Organisations and treaties

Below are some of the key legislation and protocols used to conserve and protect biodiversity.

CITES: Convention of International Trade in Endangered Species of Wild Fauna and Flora

Purpose: To protect endangered plants and animals, and ensure sustainable trade in species.

Goal: An international agreement that involves approximately 180 countries. Designed to ensure that the international trade in wild animals and plants does not threaten the survival of the species.

Methods: Organisations adhere to the treaty rules voluntarily. All export, import and re-export of species on the CITES list has to be authorised through a licensing system. Each member country must have an authority in place to manage the licensing system. Both the country exporting a species and the one importing it must follow the CITES rules.

Limitations: Rhino, tiger and elephants are well-known examples of CITES species, but there are many other species that require protection. Despite CITES, trafficking of endangered species and their products has not stopped and there is still a large international market in the illegal trade of the protected species. However, being caught results in legal action. Some countries have not entered into the CITES agreement, which gives illegal traffickers areas that they can operate more freely in.

Figure 4.29: Globally, the trade of any tiger products is prohibited under CITES.

IWC: International Whaling Commission

Purpose: The management and conservation of whale species around the world.

Goal: Established in 1946 to manage the rapidly declining whale numbers internationally. To establish protected whale sanctuaries, as no-fishing and safe breeding zones.

Methods: The IWC puts regulations in place for the management and conservation of global whaling. It currently has 89 member governments, which abide by its whaling rules. The IWC also addresses a wide range of whale conservation issues, such as oceanic noise pollution, debris, whale entanglement, bycatch, whale/ ship collisions and sustainable whale watching. In 1986, the commission introduced zero catch limits for commercial whaling, but continues to set catch limits for communities whose survival depends on whaling.

Limitations: Catch limits are still considered too high. The IWC lacks an enforcement programme to punish members who do not comply. Illegal fishing is also very difficult to monitor due to the scale of the oceans.

Figure 4.30: A humpback whale. The IWC is responsible for managing and protecting global whale populations.

EU CFP: European Union Common Fisheries Policy

Purpose: To set the rules for fishing fleets in EU waters.

Goal: To ensure the sustainable management and conservation of fish stocks in EU fishing waters, which are shared by a large number of countries. To esnure fishermen can compete fairly for fish stocks.

Methods: The EU CFP sets policies that:

- manage maximum catch sizes (catch limits) to maintain fish stocks in the long term
- prevent illegal fishing
- apply rules and sanctions across the EU
- trace and check fishery products throughout the supply chain
- protect Red List marine species.

Limitations: EU vessels seeking more fish are targeting fragile fishing grounds in West Africa. It is estimated that more than half of West African waters are already over-fished. For the EU CFP to be truly effective, it needs to apply to the fishing practices of EU vessels wherever they are fishing.

Figure 4.31: A fishing catch on the deck of a factory trawler, showing the large number of fish harvested at sea.

ITTO: International Tropical Timber Organisation

Goal: To develop international policy guidelines to encourage sustainable forest management and sustainable tropical timber harvest and trade.

Methods: Assists member countries in adapting the guidelines to suit local industry. It also collects data on timber trade to track the harvesting and movement of timber globally. The ITTO has helped to implement projects associated with forest restoration, law enforcement and conservation.

Limitations: Illegal trade in tropical timber results in deforestation in areas that are supposed to be protected under the ITTO. The large areas that the forests cover make it difficult to catch the illegal harvesting of timber.

IUCN Red List: The International Union for Conservation of Nature

Goal: A partnership that provides scientific information and tools to guide international actions in conservation. The protection of plant, animal and fungi species and the habitats needed for their survival is central to the purpose of the list of species the organisation produces.

4   Managing Ecosystems and Biodiversity

Methods: Uses scientific research to determine the risks of extinction for species. Each species on the Red List is allocated to one of the following categories:

- Extinct or Extinct in the wild (only found in captivity).
- Critically Endangered, Endangered and Vulnerable (threatened with extinction).
- Near Threatened (species that are close to the threatened thresholds and require conservation).
- Least Concern (species with a lower risk of going extinct).
- Data Deficient (in sufficient data to categorise the species).
- These lists are used to guide the protection of species in different conservation efforts around the world.

Limitations: There may be species that should be on the list that are not. Documentation regarding sources of data not always properly maintained and so may not be valid.

Figure 4.32: A pangolin mother and baby in the wild. Pangolins are one of the most illegally trafficked animals on the planet and appear on the IUCN's Red List.

Evolutionarily Distinct and Globally Endangered species (EDGE) programme

The EDGE of Existence Programme was launched in 2007 by the Zoological Society of London. Its purpose was to raise funds and awareness of the need for conservation of endangered species such as the vaquita (a type of porpoise) or the bumblebee bat, which is argued to be smallest mammal in the world. The EDGE Programme is specifically interested in protecting species on the verge of extinction, and species that are distinct in the way they look, live and behave, as well as in their genetic structure. The programme represents unique and irreplaceable species that have been previously overlooked by conservation programmes.

EDGE is a global conservation initiative focusing specifically on species with unique evolutionary histories that are now threatened. Human activities, including deforestation and urbanisation, threaten these species which represent billions of years of evolution. In 2020, there were 106 species in 44 counties that were on the list for conservation action.

The EDGE score is a measure which combines endangered conservation status with the genetic distinctiveness of the species. Each species in scored according to its taxonomic group (like mammals or birds). This score is based on its evolutionary distinctivness (ED) and global endangerment (GE).

The ED score is determined by looking at the species' evolutionary tree. This tree represents the relationship between a group of species (Figure 4.33). Its branches represent the millions of years of evolution that have occurred since the species' original group. The fewer species there are on a branch, the higher the ED score. The species' GE score is based on IUCN Red List categories. The higher the Red List score, the higher the GE score. The ED and the GE scores combined form the EDGE score. Species on the EDGE list have a high 'EDGE score'. EDGE species are often the only surviving members of their genus. Their extinction would represent the loss of a unique evolutionary branch.

> **KEY TERMS**
>
> *****evolution:** the process by which living organisms have developed and adapted into different forms
>
> *****evolutionary tree:** a branching diagram or 'tree' showing the evolutionary relationships among various biological species based on similarities and differences in their physical and genetic characteristics

Some EDGE species, such as the giant panda or the elephant, are well known. However, many others on the list are not so well-known, such as the saw fish and the kakapo (Figures 4.34 and 4.35).

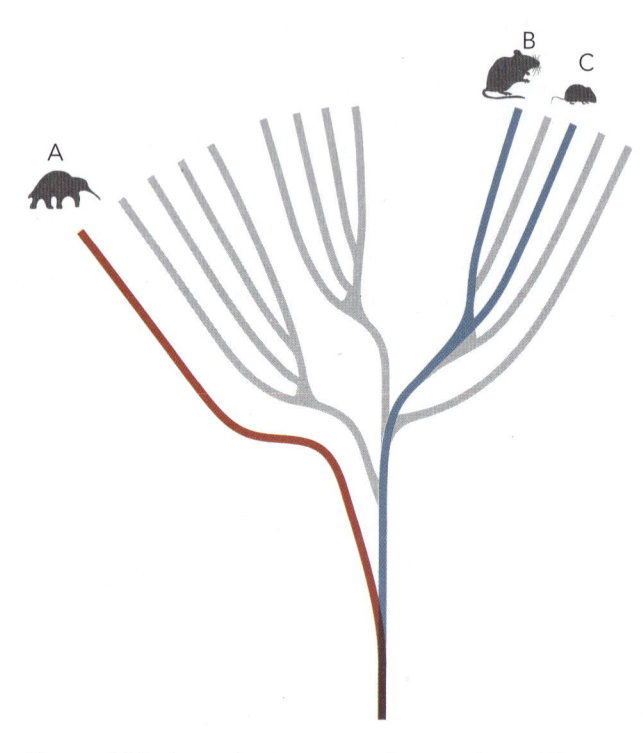

Figure 4.33: An evolutionary tree showing three different mammals that have evolved from a single group over millions of years. Mammal A represents an EDGE species.

Figure 4.34: A saw fish in an oceanarium. All five species of saw fish are on the IUCN Red list. They are hunted for food and their numbers are declining rapidly.

Figure 4.35: The kakapo is now extinct in its natural range in New Zealand due to loss of habitat and the introduction of rats and dogs by humans to the islands.

Captive breeding

Captive breeding is one option for the conservation of species, especially where wild populations are small and fragmented (spread far apart). Captive breeding is the breeding of endangered species in captivity, with the goal of releasing them back into protected wild areas in the future. This ensures the conservation of the species, and helps to prevent extinction.

The primary goal of captive breeding is to develop a self-sustaining captive population of a species, without the need to capture any other individuals from the wild. A second goal is to ensure maximum genetic diversity within the population.

Detailed captive breeding plans aim to maintain genetic diversity where possible, to increase the chances of long-term survival.

In species where numbers have fallen to below ten, saving the species becomes the goal. The northern white rhino is an example of such a species. In 2017, there were three females and one male. Now, there are just two females. In this instance, genetic diversity was no longer possible as there were too few individuals left. Captive breeding in this species was unsuccessful, and the last male died without successfully reproducing. For this species, reproductive material (semen and embryos) were harvested with the goal of producing viable young in the future using a surrogate species (the southern white rhino).

Captive breeding with a small population can lead to increased chances of inbreeding and result in a selection for weak genetic traits, such as cancers, hearing and sight problems. These traits lead to lower survival rates. For captive breeding to be viable, there needs to be a global effort to share genetic material between individuals found in different zoos and reserves around the world. Well-run captive breeding programs keep a close record the genetic lines of each individual (parents, grandparents and great grandparents) to prevent inbreeding and maintain genetic diversity.

If individuals are to be released back into the wild, it is important that tendencies allowing them to survive have not been lost during breeding in captivity. If these tendencies have been lost, individuals may be more at risk of predation, or unable to hunt when released back into the wild.

Individuals bred in captive breeding programmes need to be prepared for life in the wild. This can be a challenge with species that rely on complex social systems for survival-mating, communication, foraging, predator avoidance, offspring rearing and migration. Such behaviours are normally learned by the young in the wild from their parents. Captive individuals, on the other hand, have not been exposed to natural conditions. To get around this problem, survival skills are taught to captive-bred individuals before they are released into the wild.

After their release, individuals are monitored to determine if they are able to survive the challenges of living in the wild. Often individuals are given a 'soft release'. This might mean the provision of food and water at the release site until the individual is able to forage on their own. Successes and failures of each release are recorded and understood for use in future releases.

With species such as the giant panda, captive breeding has been very successful. With other, species, such as the Sumatran and northern white rhino, results have not been so encouraging. The captive breeding process requires an in-depth understanding of a species' reproductive biology. We do not yet fully understand the reproductive triggers in many species. Even when an individual successfully reproduces in captivity, there may be nowhere suitable to release them to if their habitat has not been preserved.

Figure 4.36: An orangutan relaxing in the sun. This species is bred in captivity at Chester Zoo in the UK in order to try and increase its numbers.

Figure 4.37: Captive breeding of the giant panda has been successful.

CASE STUDY

Captive breeding: Sumatran rhino (*Dicerorhinus sumatrensis*)

Sumatra is the second largest of the volcanic Indonesian islands. It is separated from the Malay Peninsula by the Strait of Malacca in the northeast, and from Java by the Sunda Strait in the south. The high Barisan Mountains run for approximately 1600 km in a northwest-southeast direction across the island, while in the east there are flat flood plains.

Figure 4.38: A map of the Indonesian islands, showing Sumatra in the west, south of Malaysia.

Sumatra's climate is tropical, with high rates of seasonal rainfall (3000–4000 mm per year). Its warm climate and good water supply has resulted in areas of dense forested vegetation which are high in biodiversity. The island's animal life includes orangutans, elephants, tigers and the two-horned Sumatran rhinoceros.

The climate and soils of Sumatra support the production of agricultural products such as rubber, tobacco, tea, coffee palm oil and peanuts for export.

Timber is harvested from the forests for wood, oils and fibres. In addition to this, the islands have reserves of minerals such as coal, gold, silver and tin. These activities have resulted in the deforestation of many of the island's areas of natural vegetation.

Figure 4.39: Hidden waterfalls in the forests of the Karo highland, northern Sumatra.

This island is also home to the Sumatran rhinoceros, the only two-horned rhinoceros in the Asian region. It lives in dense, tropical forest, and is a primary consumer that feeds from a wide variety of plants. The Sumatran rhinoceros lives for between 35 and 40 years, and its pregnancies last approximately 15–16 months. Females are believed to give birth to one calf every three years.

Decline of the Sumatran rhinoceros

The initial decline in the numbers of the Sumatran rhinoceros (rhino) was due to poaching, as its horn is used in traditional medicine. However, its continued decline was due to a number of factors. In addition to ongoing poaching, these included deforestation and habitat fragmentation. Between 1980 and 2022 the population of the Sumatran rhino went from 500 individuals to just 80.

4 Managing Ecosystems and Biodiversity

CONTINUED

The Sumatran rhino is considered the most endangered of all the rhino species. The species was declared extinct in the wild on mainland Malaysia in 2015, and then on the island of Borneo in 2019. The Sumatran rhino now only exists in the wild on the island of Sumatra, in areas that are protected and physically guarded by armed rhino protection units. This ongoing protection, combined with joining small isolated populations into larger populations, and continued captive breeding efforts are currently the best hope for saving the species from extinction.

The captive breeding programme

The Sumatran rhino is an example of a species where captive breeding has been less successful. This is due to difficulties associated with getting the females pregnant. The captive breeding program for this species started in 1984 when the American Association of Zoos and Aquariums (AZA) started to develop a Species Survival Plan for all rhino species. Due to a ban on exporting rhino from Malaysia, the country started its own program with 14 captive rhinos: four males and ten females. This programme ran from 1984 to 1994. However, it was not a success, and the last of these rhinos died from disease in 2003.

Figure 4.40: The endangered Sumatran rhino.

The first eight Sumatran rhinos captured in Indonesia for the captive breeding program were divided amongst four different breeding sites. Only in 1988 were another seven captured for export to four different breeding facilities in the United States of America. One site, Cincinnati Zoo, succeeded with three Sumatran rhino births (2001, 2004 and 2007). By the early 1990s, however, it was clear that the breeding program for the Sumatran rhino faced many challenges, it was at this point that discussions started to lead to the development of the Sumatran Rhino Sanctuary (SRS), located in the Way Kambas National Park in Sumatra, Indonesia. This facility was originally opened in 1998. In 2019 it was expanded to increase capacity for the captive breeding program.

The SRS homed three Sumatran rhinos that were transferred from other zoos in 1998, another two young females that were captured in Sumatra in 2005 and two of the young rhinos born in Cincinnati Zoo. One of the breeding pairs has produced young, one male and one female (2012 and 2016).

In total, 47 Sumatran rhinos have been captured for the captive breeding program since it started in 1984, with only six young being born between 1984 and 2022. This low number of births was due to the breeding program meeting a number of unforeseen challenges. These included behavioural and dietary differences between this species and other rhino species. In addition, many of the animals developed medical problems relating to reproduction. Even after 35 years of trying to make the captive breeding program a success, the females' reproductive problems are not fully understood by scientists.

CONTINUED

Figure 4.41: In 2012 a Sumatran rhino male calf was born to a mother known as Ratu. He was sent to the SRS in 2007 in the hope that he would eventually father calves of his own.

In this example, captive breeding has had very limited success. Time may be running out for the species as their numbers continue to decline.

Case study questions

1 With reference to the information in the case study, calculate the percentage change in the number of Sumatran rhinos between the year 1980 and the year 2022.

2 The physical characteristics of two rhino species are shown in the table below

Characteristics	Sumatran rhino	Southern white rhino
Weight:	600–900 kg	1 800–2 700 kg
Height:	1–1.5 metres tall at the shoulder	1.5–1.8 metres tall at the shoulder
Length:	2–3 metres long	3–5 metres long
Habitat:	Forest species	Grassland species
Predators of young or ill:	Forest Tiger	Lion, crocodile, hyena
Mouth shape:	Pointed or V-shaped upper lip	Wide square shape upper lip

   a Calculate the difference in the maximum weight of the two species.

   b With reference to the information in the table, describe and explain the differences observed in the two species.

3 With reference to the original geographic distribution of the Sumatran rhino species (Malaysia, Sumatra and Borneo) shown in Figure 4.37, explain how the islands act as a barrier to reproduction in this species.

4 The Sumatran rhino is a solitary species, which means that it prefers to live alone. What challenges do you think this presents for the captive breeding program?

Habitat conservation and creation

Habitat conservation and creation is an environmental management practice that seeks to conserve, create, and protect habitats in order to prevent the extinction of species. Habitat destruction is not the only area of concern. Sometimes a habitat has been reduced or fragmented (broken into patches that are separated to such a degree that individual members of species become isolated from each other).

A variety of techniques can be used to assist in the conservation of habitats. These include extractive reserves, the protection of habitats, nature reserves, protected areas, conservation zones and national parks.

Rewilding

Rewilding is the process of restoring an area of land to its natural uncultivated state (Figure 4.42). A combination of different techniques can be carried out in rewilding, depending on how significant the damage is to an area.

Figure 4.42: Wildflowers planted along fields in an attempt to encourage rewilding.

> **KEY TERM**
>
> **rewilding:** restoring an area of land to its natural undisturbed state, specifically through the reintroduction of species of wild animals that have been driven out or hunted to extinction in the area

By creating the correct conditions (such as removing dykes and dams to free up blocked rivers or reducing the active management of wildlife populations), nature can revert to its wild state.

Rewilding gives reintroduced species and communities the space to thrive, while the population is enhanced by reintroduction of key native species. As a result, biodiversity increases and the ecosystem's health improves.

Extractive reserves

An extractive reserve is an area of land (usually state owned) where access and resource use rights are allocated to local groups or communities. This includes managed sustainable extraction of natural resources.

By giving local communities rights to resources and making the protection of those resources beneficial to them, people may be more inclined to protect the resources and prevent them from coming to harm. In this way, biodiversity conservation is optimised while those with the natural rights to the resources can make a sustainable living from them.

An example of an extractive reserve is the Chico Mendez reserve in Brazil. The land is publicly owned. However, the people who live in the forests, and whose livelihoods and culture depend on access to its resources, have the right to sustainably extract resources from the forest (through hunting, fishing and harvesting wild plants for consumption or medicine). Access to the area requires a permit and is closely managed.

Figure 4.43: Extraction of rubber from a rubber tree in a reserve in Brazil. This practice, when correctly carried out, does not cause long-term harm to the tree.

Protection of habitats

Habitat conservation protects and restores habitats. This practice addresses species extinction, habitat fragmentation and species range reduction. It protects the whole habitat, and every individual species benefits.

Global wildlife occupies a diverse range of habitats, from tropical rainforests to frozen tundra. When these habitats are damaged or threatened, so is the biodiversity associated with them. By protecting habitats, biodiversity is protected. This protection can take different forms: national forests, nature reserves, conservation zones and national reserves. Each type of management encourages species protection and biodiversity conservation.

CAMBRIDGE INTERNATIONAL AS LEVEL ENVIRONMENTAL MANAGEMENT: COURSEBOOK

Nature reserves and national parks

Nature reserves and national parks are both conservation areas which protect nature. Nature reserves are designed to protect a particular habitat, while national parks protect landscapes, wildlife and the natural features of relatively large areas.

A nature reserve (also known as a nature preserve or wildlife sanctuary), is a legally protected area that is of importance for fauna, flora or even geology. It is legally designated by governments, private landowners or research institutions. Nature reserves are managed for the purpose of conservation and protection of natural resources. They are used as an opportunity for scientific research. Endangered species are often kept in nature reserves to protect them from being hunted.

The park can be natural or semi-natural or a combination of both, and it may have one or more undisturbed ecosystems. It can contain plant and animal species, geological sites and/or habitats that are of special conservational, educational or scientific interest. They can also be developed to protect areas of beauty. A national park is protected by national laws. However, visitors are allowed access to a national park under special conditions.

Marine conservation zones

Marine conservation zones (MCZ) are marine zones of national or international importance with rare or threatened species and habitats that require protection. MCZs do not usually have a fishing ban, however, there may be areas within

ACTIVITY 4.6

Explore your national parks

This activity is designed to raise your awareness of the number, location and features of national parks within a country of your choice.

As a class, carry out the following task for your country, or a country of your choice.

1. Print out a simple line map of the country. Only the outline of the country (and outlines of states or provinces if the country has these) is required.
2. Pin the map up on the wall or on the class notice board.
3. Research a list of national parks found within that country.
4. Allocate each of your classmates a national park.
5. Each classmate should then find ten facts about that park and write them on a card. They should consider facts like:
    a the location
    b the size of the park
    c the climate in the region
    d any special monuments or protected natural features.
    e the type of ecosystem or biome
    f plants and animals that are specific to the park
    g the number of visitors to the park each year.
6. Once everyone has completed the task, everyone should pin their national park cards next to their country maps. Everyone should draw a line connecting their card to the correct location for their national park on the map. This will show where the national parks are located.

Figure 4.44: Pin the notes to a map showing where your national park is located in the country that you have used for the group.

MCZs that are 'no-take' zones. MCZs form part of a greater network of protected zones that protect different marine factors. For example, in the UK, one MCZ may protect important marine species or habitats, while another MCZ may protect rare and vulnerable internationally important bird populations (Figure 4.45). In combination, these different sites form a network of ecologically important sites.

Figure 4.45: The sea around the Bass Rock Lighthouse is a marine reserve, helping to protect the birdlife that depend on the sea for food. Bass Rock has the largest colony of northern gannets in the world.

4.2 Questions

1. Explain the difference between a native and an invasive species.
2. Explain how a native species may become invasive.
3. What do you think is the most important reason for protecting biodiversity? Give reasons for your answer.
4. Explain how sustainable harvesting assists in protecting biodiversity.
5. Compare the roles that CITES and EDGE play in species and biodiversity conservation.

4.3 The impact of human activity on ecosystems

As human populations have grown, the demand they place on ecosystems has increased and exceeded the resources available. Natural habitats have disappeared under agricultural lands and cities. Ecosystems further from human settlements are being harvested for their resources at a rate that exceeds their ability to replace the resources that are lost.

Human impacts on tropical rainforests

Tropical rainforests around the globe are under pressure from human activities. These complex, stratified ecosystems with high biodiversity offer many resources that humans can use, including wood, medicinal plants, animal products, minerals and land for farming. Although rainforests cover only approximately 7% of Earth's dry land, they hold approximately 50% of its species. Many of those species are found in **microhabitats** within the forest, making them more vulnerable to extinction.

Human activities threaten the rainforest ecosystems. Global and local demand result in **deforestation** (Figure 4.45) for the collection of fuel wood and timber, agricultural expansion, urban expansion, mineral extraction and hydroelectric and reservoir projects. Deforestation causes the fragmentation and loss of tropical forests. Plants and animals in the **fragmentation** ecosystem remain increasingly vulnerable. The edges of the fragments dry out due to exposure to winds, with **marginal** vegetation often dying off. **Cascading changes** in the types of fauna and flora that can survive in the fragments cause a rapid decline in biodiversity. This results in the further extinction of species.

> **KEY TERMS**
>
> ***microhabitat:** a habitat that is small or limited in extent and that differs from the surrounding habitat
>
> **deforestation:** the action of clearing forested areas; the cutting down of trees

> KEY TERMS
>
> **fragmentation:** an ecosystem that has been broken up into patches that are too far apart for species to properly interact and reproduce
>
> ***marginal:** the trees found along the edges of a forest or cleared area
>
> ***cascading change:** a top-down process, where a change made at the top of a food web makes a change throughout the food web and the ecosystem

Figure 4.46: Deforestation in the Amazon Basin, Brazil. Natural vegetation has been burned to clear the forest for farming activities. The forest on the right shows what the area looked like.

In addition to the loss of species, there is also a loss of genetic material and a decline in genetic biodiversity. Those genes are not only potentially useful in medical treatments – they are also needed to ensure the ongoing stability and resilience of populations of species.

Although deforestation and harvesting from tropical rainforests meets some human needs, it also has devastating long-term consequences. These include social conflict, the extinction of plant and animal species, pollution of water resources, soil erosion and contribution to climate change. These challenges do not just affect the local ecosystems and societies – they have far-reaching global impacts.

Many equatorial countries are LICs and need resources in order to improve their standard of living. These resources often include timber or mineral deposits found in forests. However, there are indigenous territories where native tribes live off the forest, using subsistence farming. Where loggers, miner, colonisers and refugees move into forested areas, there is the risk of social conflict. In addition to this, negotiating the needs of indigenous tribes, the forest, and the economy can cause disagreement at a higher level.

Exploration for and mining of minerals (Figure 4.47) such as bauxite (for aluminium production), gold, and oil have resulted in large stretches of forest being felled for roads, infrastructure and mining. Mining also presents the problems of water, air, soil, noise and light pollution. Recovery of mined areas is slow due to the loss off topsoil and subsoils during the mining process.

Soils in tropical rainforests are thin and poor in nutrients. High rates of leaching due to year-round heavy rainfall has resulted in the development of an ecosystem where nutrients are stored in the biomass rather than the soils. When an area is completely deforested for farming, the soils collapse and rates of soil erosion are high. This results in farmlands becoming unable to support crop growth within two to three years. With the loss of topsoil and the nutrient store during this time, the repopulation of the area by the forest is slow.

Figure 4.47: Mining for gold in a rainforest in Venezuela, South America. Note the area of cleared vegetation and the mine scars.

Deforestation also contributes to climate change. Forests store vast quantities of carbon. Photosynthetic species absorb CO_2 and store the carbon in their branches, leaves, trunks, roots and soils. When trees are felled or burnt, the carbon is released back into the atmosphere. It is estimated that the global loss of tropical forests contributed 4.8 billion tonnes of CO_2 in the atmosphere annually between 2015 and 2017. This was estimated to be approximately 8–10% of annual human emissions during that time.

In every activity associated with deforestation, there is a potential for loss of biodiversity, with some species becoming extinct (Figure 4.48). It is believed that many rainforest species went extinct before there was an opportunity to identify them.

Figure 4.48: Spix's macaw is considered extinct in the Brazilian rainforests. Loss of habitat, poaching and the exotic bird trade mean that it is now only found in captivity.

It is not just large areas of forest that are under threat. Individual species are also targeted. Many of these individuals are taken illegally for the exotic pet trade where they are isolated from breeding populations. Examples include exotic birds (macaws), mammals (jaguars), reptiles (anacondas) and amphibians. Specific plants are also targeted. If too many of these individuals are removed from the habitat, it is likely that individuals who are left behind will be too small in numbers to support a viable population. In addition, the gene pool decreases and the remaining population becomes less resistant to the spread of disease.

Sustainable management of tropical rainforests

There are many reasons for preserving rainforests and managing them sustainably. The key reasons include:

- maintaining biodiversity
- protecting the ongoing production of resources (rubber, hardwood, medicines)
- managing climate change
- managing local water quality and protecting the local water cycle. Rainforests regulate the local water cycle and climate.

Sustainable management of forested areas can occur at a range of levels. These include local, regional, national and international levels. A variety of methods are employed to help manage forests sustainably.

Debt reduction

Debt reduction is an international approach to managing rainforests. Many tropical rainforests are found in LICs which often have high levels of debt. Rather than watching LICs continue to remove rainforest in order to generate the income they need to pay off debt, some HICs have agreed to write off debt. In return, the HIC asks for the protection of an area of rainforest. In 2010, the USA signed an agreement with Brazil where $21 million was written off in return for the protection of a certain area of tropical ecosystems. Instead of having to pay back the debt, the money was used to conserve portions of the Atlantic coastal rainforest.

International agreements and other protection measures

International agreements for the management of species and forested areas through sustainable harvesting (e.g. CITES, ITTO) are put in place to protect species and areas under threat from excessive harvesting.

High demand for forest resources such as tropical hardwoods have led to an increase in the rates of illegal deforestation. International agreements, although robust in theory, are difficult to implement on the ground. This is due to the scale of the area concerned and the limited resources on the ground. There are some ways of ensuring hat the end user knows materials are sustainably harvested (such as marking timber with the logo of a company that is registered as a sustainable harvester). However, international and local agreements need to include the education of those exploiting the resources, from the harvester to the end user in a remote country. If the consequences behind the actions and the use of a resource are understood, then the management of forested areas may become more sustainable.

Legislation and the formation of protected areas through the establishment of anti-deforestation public policies and private measures can significantly protect forested areas, biodiversity, and the ability of forests to absorb CO_2. From 2004 to 2011, Brazil reduced emissions by creating new protected lands. Satellite monitoring was introduced to monitor illegal activity, and to reduce credit extended to farmers who carried out high levels of deforestation. For this type of legislation to be of value, it needs to be rolled out across countries that have large areas of the tropical rainforest biome.

Other mechanisms that can be used to protect rainforest on a local scale is the implementation of selective logging, shift cultivation, reforestation and agroforestry.

Selective logging (Figure 4.49) is the process of identifying older or inferior trees of a specific species for extraction from the forest, without removing the forest around them. This protects the habitat, biodiversity and soils from erosion.

Shift cultivation involves subsistence farming where farmers keep fields small, and move to small new fields every 2–3 years, allowing the forest to re-establish in the old fields.

Figure 4.49: Selective logging in a commercially managed forest. Here the trees are being marked to indicate which ones can be felled.

Reforestation is the process of planting trees into areas that have been felled. A mix of indigenous trees are planted into an area that has been damaged to give it a better chance of recovering.

Agroforestry is a way of farming where trees and crops are grown alongside each other. The trees maintain some canopy cover, holding nutrients and protecting soils. The crops provide nutrients to the soil from dead organic matter, while supplying a harvest.

ACTIVITY 4.7

The Amazon rainforest

1. In small groups, investigate the threats to the Amazon rainforest and the steps being taken to minimise the damage being done. How effective are those preventative measures?

   Divide the task into the following categories, and allocate a heading to each team member:

   a   Threats to the Amazon

   b   Protective measures being put in place to slow or prevent loss of the Amazon rain forest

   c   The effectiveness of the steps being taken to protect the rain forest

   d   What could be done differently?

2. Once you have completed your individual research, discuss your findings as a group and present your research to the class.

Figure 4.50: A river in the Amazon in Venezuela, showing the thick Amazonian vegetation along each river bank.

4 Managing Ecosystems and Biodiversity

> **CASE STUDY**
>
> ### Biodiversity conservation in grasslands
>
> Biodiversity conservation often lies in the hands of official nature and game reserves. However, opportunities are increasingly arising for private companies to change their practices, setting aside sensitive or threatened ecosystems to protect biodiversity.
>
> The diversity of climate and topography gives rise to broad vegetation zones in South Africa. As a result, the country has one of the highest ranges of biodiversity in the world. The grassland biome in South Africa is mainly found on the high central plateau, the inland areas of Kwa-Zulu Natal and the Eastern Cape along the eastern side of South Africa (Figures 4.51 and 4.52).
>
> Figure 4.52: Grassland biome in South Africa.
>
> are common. Grasslands remain dominant due to the influence of frost, fire and grazing pressures which prevent the establishment of trees.
>
> This biome has an extremely high biodiversity with a number of rare or endangered species. However, urbanisation, overgrazing and timber production threaten this biome's existence.
>
> Figure 4.51: Distribution of the South African Grassland Biome.
>
> The grasslands consist mainly of flat, gently rolling plains, with altitudes varying from 2850 metres to near sea level. They fall within a summer rainfall region and have an average rainfall of between 450 mm and 1900 mm per year.
>
> The grasslands are dominated by a single layer of grasses, with cover varying depending on climate and grazing pressure. Trees are generally absent from this biome. However, bulbs (geophytes)
>
> ### Mt Gilboa and the Mondi Forestry Grassland Conservation Project
>
> In South Africa, a timber company known as Mondi Forestry has taken steps to conserve biodiversity in Natal. Mondi is a multinational packaging and paper company that operates in more than 30 countries around the world.
>
> In order to produce paper and packaging, Mondi plants fast-growing trees such as pine trees. The wood from these trees is harvested and processed to make paper. In South Africa, these forests are often planted in grassland areas, resulting in the loss of natural vegetation and a decrease in the biodiversity.
>
> Mondi has a Grasslands Programme team responsible for taking steps to protect grassland areas that are on land they own, and which have not yet been affected by forestry activities. The Grasslands Programme team work together to identify and implement the best methods for conserving grassland areas.

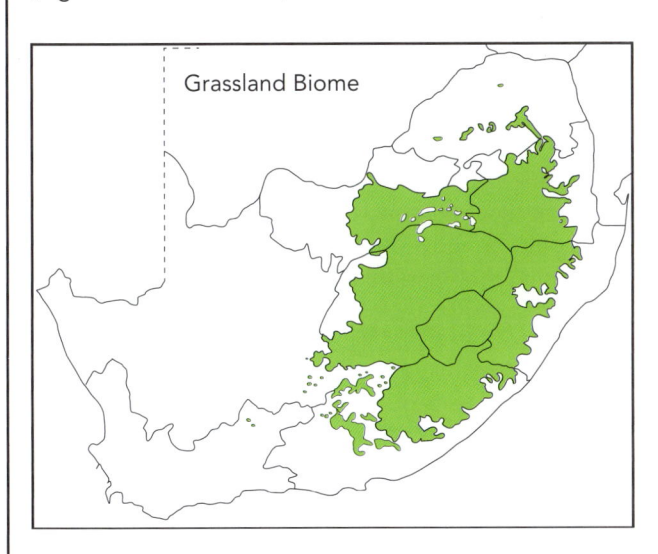

CONTINUED

The Mt Gilboa nature reserve is situated within a forestry estate, where non-native timber species are grown for wood. It was the first forestry property proclaimed as a private nature reserve in Natal using the Environmental Biodiversity Stewardship mechanism. The establishment of the grassland reserve took three years. The first step in this process was an environmental assessment to determine the best way forward.

Grasslands are high in biodiversity. They support a wide range of grass and shrub species, along with small and medium size mammals, birds and a wide array of insects. The grasslands in the Mt Gilboa nature reserve are the critically endangered Midland Mist-belt Grasslands. They are home to endangered species such as the wattled crane, the Samango monkey and the oribi antelope. By protecting the habitat, these species are also given refuge.

The Mt Gilboa reserve not only protects the sensitive grasslands, it also incorporates the source water of three of Natal's important river systems with associated peat wetlands which function as water purification systems and help with flood production. Within the reserve there are 238 hectares of critically endangered grassland. The selection of this site therefore protects biodiversity beyond its borders, as the headwaters of the longitudinal river ecosystems are also protected. This ensures clean fresh water in the upper reaches of the three rivers.

Unfortunately, grasslands are one of the most threatened biomes in the world as they are generally viewed as unnecessary. In South Africa only fragments of intact grassland remain, one of which is the Mt Gilboa reserve. The forestry industry in South Africa has more than 4000 km^2 of unplanted land in the grassland biome. This presents an opportunity for further biodiversity conservation.

The Mondi Grasslands Programme team is using lessons learned from the establishment of this reserve to develop tools to help landowners identify biodiversity hotspots, and how to manage their grasslands for fire events. The team has currently identified 37 key grassland areas that require conservation.

Figure 4.53: Oribi antelope standing in long grass in a grassland reserve in South Africa.

Benefits of establishing a grassland reserve include:

- The reserves act as islands for biodiversity conservation. They can be used to repopulate other grassland areas when carrying capacity has been reached. For example if the oribi populations grow too large for one reserve they could be moved to another grassland reserve within their natural range.

- The cost of the land to establish a private nature reserve within a forestry estate can be written off as a tax deduction over 10 years.

- Reserves offer opportunities for scientific studies or education of both school students and management and staff teams.

Case study questions

1. What could Mondi do to increase the protection of grasslands beyond their own projects?

2. What are the potential limitations for other companies wanting to undertake a similar project?

3. Explain the ecological benefits from the fact that the Mt Gilboa reserve incorporates the headwaters from three local rivers.

4 Managing Ecosystems and Biodiversity

Human impact on Antarctica

Antarctica is Earth's fifth largest continent (Figure 4.54), it is located almost entirely south of the Antarctic circle, surrounded by the Southern Ocean. Antarctica is the coldest place on Earth. It has neither trees nor bushes and the only plants that exist there are moss and algae. However, the oceans are rich in life and support a healthy food web with species such as seals and whales. The geographic South Pole is located on Antarctica.

Figure 4.55: Large tyres buried in the sand and in the water on a shoreline in Antarctica.

Growing threats to Antarctica include the following:

- Climate change is warming the oceans and resulting in the loss of both sea and land-based ice. This is the greatest threat to the region. Some ice slopes and glaciers have retreated and **ice shelves** are collapsing. Ocean acidification from excess CO_2 in the atmosphere has caused a loss of biodiversity, as some marine snails have disappeared. Some penguin species' breeding zones have been affected and their ranges are shrinking.

- Fishing, both legal and illegal, is a significant threat to Antarctic waters. Early in the 19th century, hunting for whales and seals resulted in heavy losses to the wildlife populations. Hunting was banned and wildlife sanctuaries were established (for example, in 1933, Macquarie Island became a wildlife sanctuary). With global oceans overfished, fisheries are looking further afield to areas where ice has retreated and fishing is possible. The significant decline of krill is a particular concern related to overfishing, as krill forms the base of many of the Antarctic food chains. With the disappearance or reduction of krill, entire food chains could collapse. Illegal fishing in Antarctic waters is difficult to control due to the remoteness of the location.

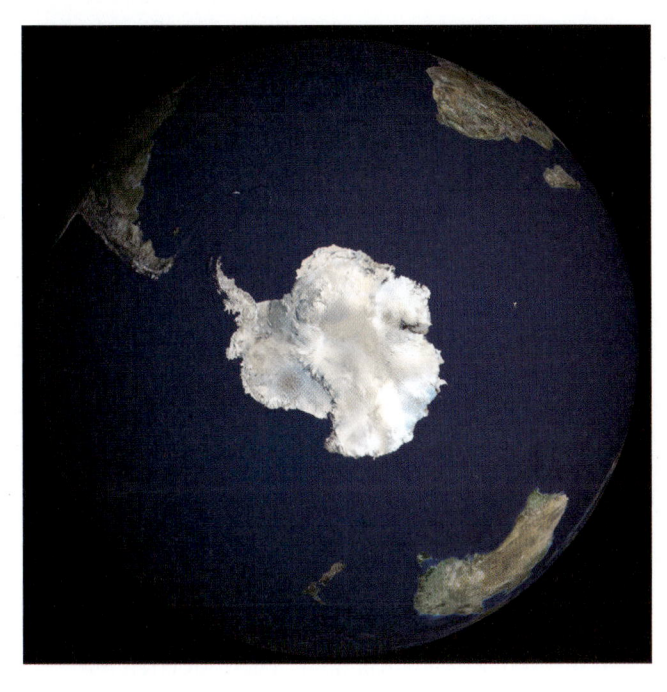

Figure 4.54: A satellite image of Earth, centred on Antarctica. The South Pole is at the centre.

Humans have been travelling to and exploring Antarctica for more than 100 years. Some species, such as fur seals, were exploited to the point of near extinction and many others have been disturbed. Non-biodegradable waste lingers in even the most remote parts of the continent (Figure 4.55). More recently, a greater understanding of the importance that Antarctica plays in the global ecosystem has developed.

Antarctica offers scientists the chance to study largely unexplored areas and species, and it also provides opportunities for tourism. Both scientists and tourists, even when carefully managed and limited, have the ability to damage Antarctica.

KEY TERM

*ice shelves: the place where sea ice and ice sheets meet. They are platforms of ice that extend over the edge of the land onto the oceans

- Tourism (Figure 4.56) brings more ships into the area, increasing the potential of oil and sewage spills into Antarctic waters. Other pollutants may be left in the area, impacting wildlife and the wider ecosystems. Tourists can also disturb colonies of Antarctic animals. The challenge is that no individual government has the power to set the regulations. This means that the number of tourists visiting the area could continue to increase, with different tour operators competing to reduce costs. Cheaper costs will mean the trips become accessible to more people. If the tourism carrying capacity of Antarctica is exceeded, this fragile ecosystem is at risk.

Figure 4.56: A cruise ship approaches Lemaire Channel on the Antarctic Peninsula.

- The emission of chlorofluorocarbons (CFCs) and other ozone depleting chemicals are responsible for the development of a hole in the ozone layer in the stratosphere above Antarctica. This has led to increasingly strong and frequent winds and storms. Winds are estimated to have increased by between 8% and 20% in the Southern Ocean over the last 30 years.

- Pollution is also an issue. In addition to the problems created by CFCs and tourism, chemicals that are produced thousands of kilometres from Antarctica have been found in the ice and in the bodies of wildlife. Plastic pollution and fishing nets and lines carried in by the oceans from other parts of the world are also causing harm to birds, fish and marine mammals.

- Exploration and exploitation of mineral reserves is currently not economically viable in Antarctica. The cost of accessing the resources is greater than the profits that can be made. However, as technology improves and resource reserves dwindle elsewhere, greater pressure will be put on accessing these resources in Antarctica. Currently, Antarctica has a total ban on all mining or mineral exploitation. However, this ban comes up for review in 2048, at which time mineral exploration and mining could threaten Antarctic ecosystems.

Figure 4.57: One of the USA's research stations in Antarctica.

Scientific research is vital for understanding the continent and the species and environmental processes that occur there. Scientists are also at the leading edge of observing and understanding how humans are impacting this environment. Their research has the positive impact of raising global awareness of the changes occurring in the region. The information collected in Antarctica's research stations is used to inform scientists and governments globally. It is also used in the formation of international laws regarding the environment. However, the research stations do have negative impacts on the environment in which they operate. There are 70 active research stations in Antarctica (up from 53 in 2012), and they represent 29 countries from every continent.

The presence and action of the research community can affect the fragile ecosystem, with problems ranging from pollution to disturbance of wildlife. It is important to realise that these impacts are cumulative. When you consider the different harmful factors together, their impact is greater, and the knock-on effects more harmful.

ACTIVITY 4.8

Graphing the decline of Antarctic sea ice coverage

Use the data in Table 4.5 to complete a line graph showing the change in Antarctic sea ice between 1980 and 2020.

Year	Antarctic summer sea ice size ($\times 10^6$ km^2)
1980	2.6
1985	2.5
1990	2.6
1995	3.4
2000	2.7
2005	2.5
2010	2.3
2015	2.1
2020	2.6

Table 4.5: Antarctic summer sea ice extent, 1980–2020.

1 Once you have completed the graph, add a trend line which shows how the ice coverage in Antarctica is changing over time.

2 The ice coverage in Antarctica for a number of years was recorded as follows:

Year	Antarctic summer sea ice size ($\times 10^6$ km^2)
2004	3.7
2013	3.6
2018	2.7
2022	1.9

Add these points to the graph you have drawn, showing the change in sea ice over time. How does this data impact the trend line that you added to your graph?

Discuss the following points in groups or as a class:

- Consider how this data could be used or misused to create an argument for or against a decline in Antarctic sea ice.
- What other data do you think should be included if you want to discuss the potential causes of changes to Antarctic ice coverage?

KEY TERM

*sea ice: the ice that floats on the surface of the oceans and seas

Strategies for managing the impacts of humans on Antarctica

The Antarctic Treaty (1959)

On 1 December 1959, 12 countries signed the first Antarctic Treaty. The treaty came into force in 1961 and has now been signed by 54 nations.

The key features of the Treaty are:

- Antarctica shall be used for peaceful purposes only. Military bases and weapons testing are expressly prohibited.
- Mining for minerals is banned.
- Freedom and co-operation in scientific investigation shall continue.
- Scientific observations and results from Antarctic research shall be exchanged and made freely available.
- In the original agreement, seven of the countries made claims to territories within Antarctica (some overlapping). However, the treaty sets aside the potential for sovereignty disputes between Treaty parties by providing that no activities enhance or diminish previous territorial claims, and that no new or enlarged claims can be made. It also made rules relating to the jurisdiction of each country. No one country can lay claim to Antarctica as their sole territory.
- No nuclear testing or disposal of nuclear waste is allowed.
- All areas of Antarctica, including all stations, installations and equipment within those areas shall be open to inspection at all times.
- The treaty prohibits the import of non-native fauna and flora species into the Antarctic without a permit. This is to protect Antarctica from potentially invasive species.
- The Antarctic Treaty system allows fishing in the Southern Ocean, but it has a strict focus on conservation and the precautionary management of fish stocks. Fishing is currently the only large-scale resource exploitation in Antarctica. Concerns about the overfishing of Antarctic waters resulted in the Antarctic Treaty, including The Convention

on the Conservation of Antarctic Marine Living Resources (CCAMLR) in 1982. This was mainly as a result of concerns over the increased catches of krill (Figure 4.58) in the Southern Ocean. Krill is a vital part of the Antarctic food web and needs to be protected from over fishing. The treaty targets sustainable fishing, rather than issuing total bans on it.

Figure 4.58: Krill (*Euphausia superba*) is an important food source for animals living in the Southern Ocean around Antarctica.

Figure 4.59: A young gentoo penguin explores the boundary sign warning people to keep out of the penguin study area on Goudier Island, Port Lockroy, Antarctica.

The Antarctic Treaty System grew out of the Treaty. It comprises the Treaty itself and a number of related agreements. It also includes a range of organisations that contribute to the work of the decision-making forums.

The Treaty System includes the measures, recommendations, decisions and resolutions of the Consultative Meetings relating to Antarctic matters. It is one of the first successful international agreements.

The Madrid Protocol was added to the Antarctic treaty in 1991. It requires national Antarctic programs to clean up abandoned worksites and waste sites. Sites must be cleaned up so long as the clean-up process does not cause more harm than leaving the site intact, or where the site is of historical importance.

Tourism has seen the government of the Netherlands host a meeting with experts, civil society groups and members of International Association of Antarctic Tour Operators (IAATO) and ASOC to identify key areas of concern regarding tourism in the Antarctica, which are:

- pressure on main tourist sites
- expansion of tourism into new areas
- the rise in tour operators who have not joined the IAATO and are not applying best practice.

Figure 4.60: Heli-skiing is when a helicopter takes skiers to remote locations. This allows skiers access to areas that are otherwise too difficult to get to.

The findings of this meeting were presented to the Antarctic Treaty Committee for Environmental Protection. Although stronger regulations have not yet been formed, they are in the process of being put together. These could include banning the use of heli-skiing or jet-skiing, strengthened visitor regulations, and the establishment of protected areas that are off limits to tourist vessels and tourists. Meeting attendees also want to determine how domestic laws can enable authorities to prosecute visitors who contravene these laws. New regulations could also require that any tour operator wanting to work in Antarctic waters has IAATO membership before being granted a permit, and that a limit is set on the number of visitors allowed each year.

4 Managing Ecosystems and Biodiversity

Scientific stations are now subject to environmental audits on both the land and the sea to determine the impact the base is having on the area around it. This means that disturbance is minimised. Any waste generated at the base is returned to the country of origin wherever possible. Scientific stations are also moving towards renewable energy, with wind powered generators, some solar energy for use in summer, and very well insulated buildings to minimise energy loss and energy consumption.

Protected areas have been and continue to be set up with different levels of protection. Some areas are no-go zones where no activity can take place, while others prohibit vehicles or place an annual limit on the number of people allowed into the area.

The use of heavy fuel oil in Antarctica was banned in 2011 under the MARPOL convention. Today's ships generally use less-polluting marine diesel, while others have taken this further and supplemented their traditional fuel with battery power to avoid the risk of oil spills in this fragile ecosystem.

4.3 Questions

1. Describe how deforestation impacts the soils and the nutrient cycle in a tropical rainforest.
2. Explain why it is so difficult to protect forested areas that are located in LICs.
3. Describe how debt reduction helps to protect tropical rainforests in LICs.
4. International agreements such as CITES are put in place to protect species. Explain why they are so difficult to implement.
5. Antarctica is protected from ownership by any single country, and from exploration for minerals. Explain why, despite these regulations, Antarctica is still being negatively impacted by human activities.

EXTENDED CASE STUDY

Rewilding Yellowstone National Park

Yellowstone National Park is an 8991 km² conservation area in the state of Wyoming, USA. The park is made up of high, forested volcanic plateaus (areas of flat high ground). Yellowstone has a large collection of volcanic **geysers**, and **hot springs** (Figure 4.61). The rewilding of Yellowstone National Park through the reintroduction of the North American Wolf to the park is considered to be one of the most successful rewilding projects to date.

Figure 4.61: Grand Prismatic Spring in Yellowstone National Park.

KEY TERMS

*geyser: a hot spring in which water boils, periodically sending up tall columns of water and steam

*hot spring: a source of ground water that is heated by underground volcanic activity

The causes

In 1872, Yellowstone was declared the world's first national park, protecting 20 000 km² of mountain wilderness. However, the westward movement of settlers and their livestock brought them into contact with wolves. The settlers thought it wasteful that the wolves killed other animals such as elk and deer. They thought that these animals could have been used to feed themselves. By the late 1800s, predator controls were implemented in the park, and this included poisoning the wolves. By the mid-1920s, more than 130 Yellowstone wolves had been killed. The last remaining wolf was reported to have been killed in 1926.

CONTINUED

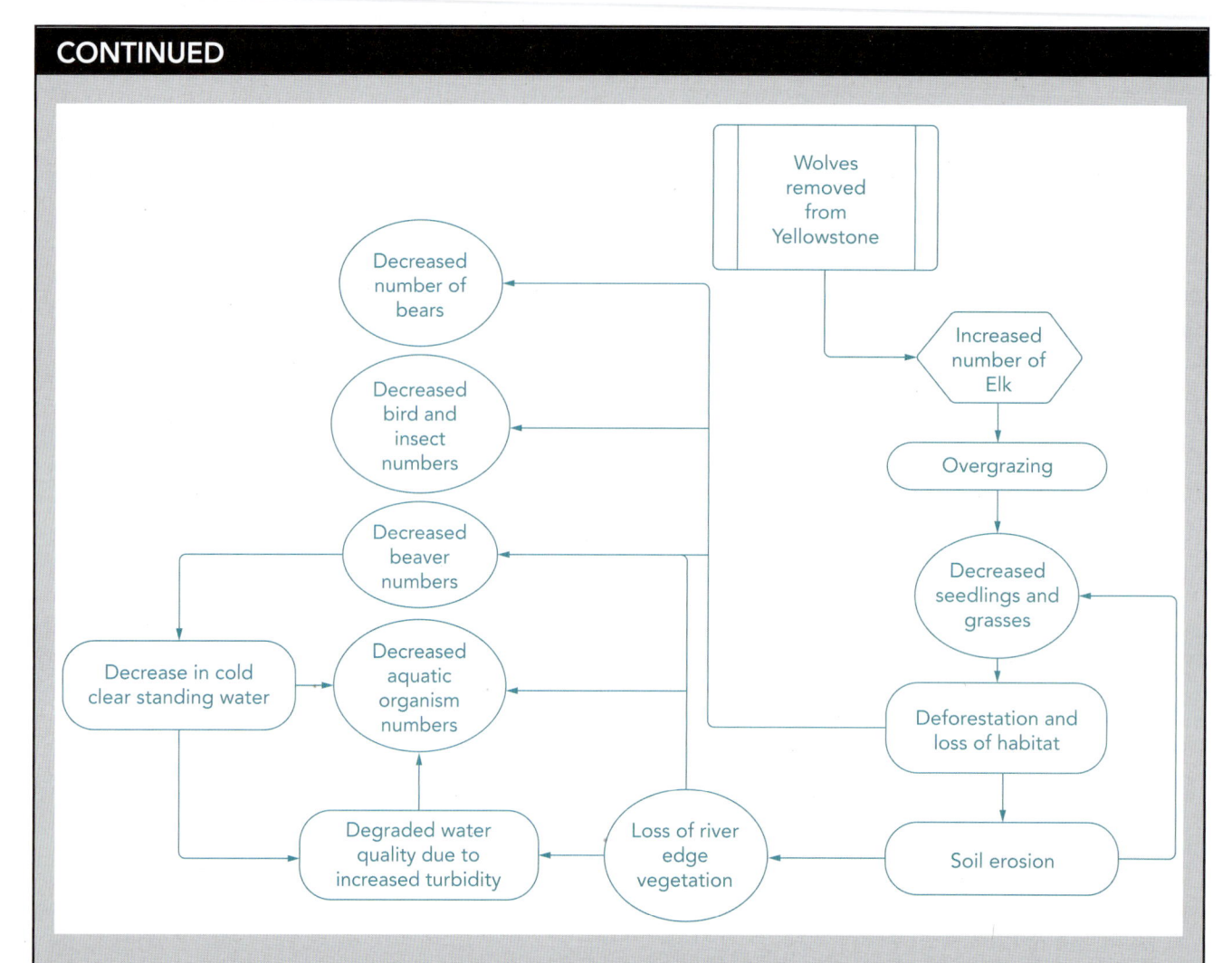

Figure 4.62: The knock-on effects of the loss of the wolves from Yellowstone national park.

These settlers did not understand the importance of an apex predator in the function of a healthy ecosystem. This resulted in the loss of a key species, and had a significant impact on the Yellowstone ecosystem as a whole.

The impact

The loss of the apex predator had a variety of knock-on effects (Figure 4.62). These included an increase in the number of elk and deer. These animals caused overgrazing of grasses, bushes and young trees. As a result, the area suffered loss of habitat as the forests and grasslands disappeared, and soil erosion where exposed soil was removed by wind and water. This meant a decrease in the number of other species such as bears, birds, insects, beavers and aquatic organisms. Soil erosion resulted in a lack of vegetative regrowth, and, it also caused the loss of the vegetation that grew along the edges of the rivers (marginal vegetation). The loss of vegetation meant that soil ran into the water and changed the clear water to muddy water, which caused the aquatic organisms reliant on clear water to disappear. Beaver numbers also declined. Beavers build dams which result in deeper, cooler, shaded water. Without beavers, the species that relied on that habitat also disappeared.

4 Managing Ecosystems and Biodiversity

CONTINUED

Figure 4.63: A North American beaver. This species increased in Yellowstone when wolves were returned to the area.

Figure 4.64: The North American wolf is the species that was reintroduced into Yellowstone Park.

The solution

Throughout the 1900s, wolf numbers in the USA fell, and in 1973 the Endangered Species Act was introduced. Wolves were on that list. This meant that by law, the US Fish and Wildlife Services had to restore wolves where possible. However, ongoing conflict continued between farmers and wolves, as wolves were killing livestock. As a result, the plan to reintroduce wolves into Yellowstone was delayed for a further 22 years.

It was in 1995, 69 years after the last known wolf pack had been eliminated from the area, that the first wolf pack of eight animals was reintroduced. More wolves were reintroduced as the program developed.

The successes

Within the first ten years, the impact of reintroducing wolves on the biodiversity of the Yellowstone Park was evident. When wolves were introduced there was only one beaver colony left in Yellowstone. Now there are nine colonies.

Reintroducing a small pack of wolves not only had an impact on deer and elk numbers, but also on their behaviour. The wolves kept the grazing animal numbers down, and kept the herds moving. This reduced the problem of overgrazing and allowed forests to regrow, as tree seedlings were no longer being destroyed.

With the return of the grasslands and the forests soils stabilised, bird, insect, mice and even bear numbers increased. However, coyote and rabbit numbers decreased due to presence of the wolves.

The riverbanks stabilised, which improved water quality. This, in combination with the return of the beavers, resulted in the recovery of aquatic ecosystems and the return of species to the area. The knock-on effect of reintroducing the wolves was felt across all levels of the ecosystem.

The rewilding at Yellowstone and in the USA faces challenges. Wolves that stray outside the safety of the park fall victim to hunting. Many people still see wolves as a pest, a danger, and as a trophy species. Keeping wolves protected inside the park is the main challenge.

Case study questions

1. In 1995, one pack of eight wolves was introduced. By 2019 there were 81 wolves in nine packs.

   Calculate the percentage increase in the number of wolves in the park between 1995 and 2019.

2. Create a flow diagram showing the changes to the Yellowstone ecosystem after the wolves were returned.

3. Bears are omnivores (they eat meat and plant material) and opportunistic feeders. Explain why bear numbers increased rather than decreased after the wolves returned.

> CAMBRIDGE INTERNATIONAL AS LEVEL ENVIRONMENTAL MANAGEMENT: COURSEBOOK

CONTINUED

4 Explain how the ecological cascade effect can result in a catastrophic change to an ecosystem.

Case study project

Research another rewilding project. Write a case study which introduces the area, explains the causes of the problem, the species involved, the rewilding process and the successes and/or failures of the project. How does it compare to the Yellowstone case study?

SUMMARY

A biome is a large zone characterised by similar soil, climate, vegetation and wildlife.
Soils develop in relation to the climate.
Deserts can be hot or cold. They have extreme climates with low precipitation (less than 250 mm).
Deserts have low biodiversity and organisms are adapted to the extreme climates.
Tropical forests are found in equatorial zones. They have rainfall all year around with high average annual precipitation (2000 mm or more), low seasonal temperature fluctuations and low diurnal temperature ranges.
Forest soils vary greatly, depending on the climate.
Grassland biomes are found in central parts of the continents between the equator and the tropics. They have distinct wet and dry seasons. Precipitation is unpredictable and droughts are frequent.
The Tundra biome is one of the coldest, harshest biomes. It has very cold temperatures, strong polar winds and long dark winters. It has low biodiversity with low rates of precipitation.
Native and invasive species: native species originate in a region or ecosystem, invasive species are able to outcompete the native species.
Benefits of conservation biodiversity: food security, economic growth, poverty reduction, combatting climate change, medical and other resources, ecological services and cultural and recreational value.
Conservation legislation and protocols involve the protection of species, ecosystems and biodiversity using enforceable laws.
Humans impact the ecosystem through the harvesting and use of the resources that ecosystems provide.
Sustainable management of ecosystems involves managing biodiversity, protecting resources, managing climate change and managing local water quality and use.
Human impact on Antarctica is due to climate change, pollution, legal and illegal fishing, tourism, ozone depletion, exploration and exploitation of mineral reserves and scientific research. Antarctica is protected by the Antarctic treaty.

REFLECTION

- Did the end of unit questions help to reinforce your learning? Or would you have benefitted from creating your own chapter summary?

- Do you find that carrying out research and writing your own case study helps with your understanding of a concept? Could you have presented your case study so that the rest of the class could have benefitted from your learning?

4 Managing Ecosystems and Biodiversity

PRACTICE QUESTIONS

1 Figure 1 shows climate data for two locations in the southern hemisphere.

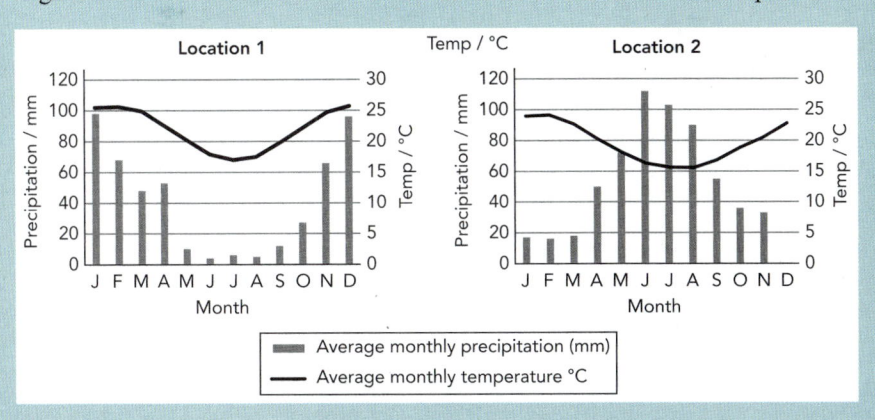

Figure 1

	Location 1 Climate data		Location 2 Climate data	
	Average monthly rainfall (mm)	Average monthly temperature (°C)	Average monthly rainfall (mm)	Average monthly temperature (°C)
J	98	25.5	17	20.0
F	68	25.6	16	20.1
M	48	24.9	18	18.9
A	53	22.5	50	16.9
M	10	20.2	72	15.1
J	4	17.9	112	13.6
J	6	17.0	103	13
A	5	17.5	90	13
S	12	19.7	55	14
O	27	22.1	36	15.7
N	66	24.6	32	17.1
D	96	25.7	**20**	19

a Complete the graph for location 2 in Figure 1 by plotting the precipitation for the month of December. [1]
b Calculate the annual rainfall for location 1. [1]
c Calculate the temperature range for location 2. [1]
d With reference to Figure 1, state the name of the major climatic region (biome) in which location 1 is situated. [1]
e With reference to the data in Figure 1, compare the climates of the two locations. [4]

2 The passage below gives information on sustainable coffee farming in South America. Coffee is consumed by people worldwide. There are environmental costs associated with the production of coffee to satisfy the ever increasing global demand.

CONTINUED

> **Sustainable coffee farming in South America**
>
> Commercial coffee farming can cause deforestation, chemical build up in soils, runoff of toxic chemicals into water ways, degradation of aquatic ecosystems, soil erosion and land degradation.
>
> Extreme temperature fluctuations, heavy rainfall and drought pose serious risks to coffee crops. Sustainable coffee farming includes forest, water and soil management. Agroforestry is combined with the planting of shade trees that are native to the area, resulting in the parts of the coffee industry planting 4.5 million trees.
>
> There are two types of coffee plants, those that grow in the sun and those that grow in the shade. The one that grows in the sun produces approximately three times the amount of coffee as the one that grows in the shade.

a Describe the relationship between the commercial farming of the sun-loving coffee plant and the loss of biodiversity. [3]

b Describe the process of agroforestry. Explain how it can help to prevent deforestation and make coffee farming more sustainable. [3]

c Explain how agroforestry can result in the reforestation of areas that have previously been deforested. [2]

d Describe and explain other methods that could be implemented to protect the forested areas from commercial farming. [6]

3 Figure 2 is a map showing two major biomes.

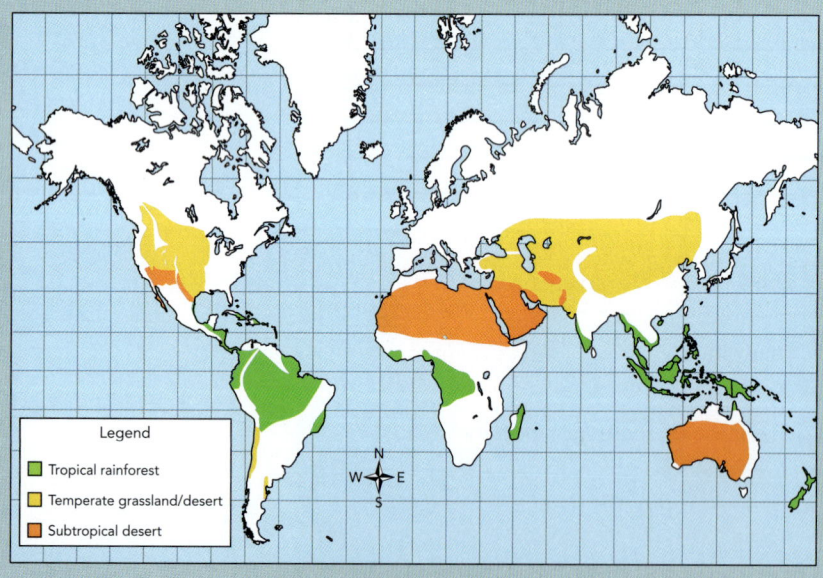

Legend
- Tropical rainforest
- Temperate grassland/desert
- Subtropical desert

Figure 2

4 Managing Ecosystems and Biodiversity

CONTINUED

a With reference to Figure 2, compare and explain the difference in distribution between hot desert and tropical rainforests. [6]

b Describe how the soils in tropical rain forests differ from those found in the tundra. [4]

4 Figure 3 shows Yellowstone National Park (Wyoming) which is one of the largest and best-known national parks in the USA. It is a biosphere reserve that covers 8991 km², and is surrounded by national forests on all sides. It is situated in a volcanically and seismically active region. Since 1950 average temperatures have risen by 2.3 °C and it has lost a quarter of its annual snowfall. Snowfall is forecast to decrease further in the future.

COMMAND WORD

identify: name/select/recognise

Figure 3

a Suggest the impacts of increasing temperatures on the water supply and species in the Yellowstone National Park. [4]

b Explain how biodiversity in a National Park like Yellowstone benefits from organisations such as CITES. [2]

c Describe the impact that an invasive fish species could have on the biodiversity of the aquatic systems in Yellowstone National Park. [3]

d Identify the renewable energy source that is derived from the heat of the Earth which can be harvested in Yellowstone National Park [1]

5 'Captive breeding is the most effective way of protecting biodiversity'. To what extent do you agree with this statement?

Give reasons and include information from relevant examples to support your answer. [20]

SELF-EVALUATION CHECKLIST

After studying this chapter, think about how confident you are with the different topics.
This will help you to see any gaps in your knowledge and help you to learn more effectively.

I can	Needs more work	Getting there	Confident to move on	See Section
Describe the world's major terrestrial biomes (vegetation, soil and climate).				4.1
Outline the characteristics of primary and secondary succession.				4.1
Define gross primary and net primary productivity.				4.1
Define ecosystem productivity.				4.1
Discuss energy transfer between trophic levels.				4.1
Explain the shape of ecological pyramids.				4.1
Define the terms native and invasive species.				4.2
Explain the impact of invasive species on biodiversity.				4.2
Explain the benefits of conserving biodiversity.				4.2
Describe and evaluate legislation and protocols as methods of conserving biodiversity.				4.2
Describe and explain the role of EDGE.				4.2
Describe and evaluate captive breeding and habitat conservation as methods of conserving biodiversity.				4.2
Describe the impacts of humans on tropical rainforests.				4.3
Explain and evaluate the strategies for managing the impacts of humans on tropical rainforests.				4.3
Describe the impacts of humans on tropical Antarctica.				4.3
Explain and evaluate the strategies for managing the impacts of humans on Antarctica.				4.3

Chapter 5
Managing Resources

LEARNING INTENTIONS

In this chapter you will:

- learn about the causes of food and energy insecurity
- explain the impacts of food and energy insecurity
- understand strategies for managing food and energy security
- identify methods of waste disposal and treatment
- describe the impacts of waste disposal
- evaluate strategies which reduce the impacts of waste disposal.

CAMBRIDGE INTERNATIONAL AS LEVEL ENVIRONMENTAL MANAGEMENT: COURSEBOOK

GETTING STARTED

Discuss the following questions with your class or in small groups.

1. Give two examples of non-renewable energy sources. Briefly explain why their use is problematic.

2. How do toxic spills affect living organisms?

3. How do you think food insecurity may affect people and their environment?

4. In pairs, make a list of the types of pollution a landfill site could cause. Compare your list with your classmates'. Do they differ?

ENVIRONMENTAL MANAGEMENT IN CONTEXT

The power of waste

Oahu in Hawaii generates more than 2 million tonnes of residential, commercial and industrial waste every year. Much of the waste is recyclable (glass, paper, aluminium and cardboard) or is compostable food waste. Recyclable waste is separated from the non-recyclable waste, which goes to landfills. Despite this, there is still a significant amount of waste that has to be managed.

In Oahu, a waste-to-energy plant reduces non-recyclable waste by approximately 90%. It produces roughly 10% of the electricity required by the city. This is enough to power 60 000 homes daily. In addition to this, Oahu also has a receiving station for sewage sludge, which is solid waste left after waste water has been treated, that is the first of its kind. This station burns sludge to produce energy. This energy powers 6500 homes.

Kaka'ako Waterfront Park was opened in 1992 on the site of a former city landfill (Figure 5.1). A well-managed and sealed landfill can be reused for recreational purposes when it has been permanently shut down.

Both residents and businesses in Oahu are encouraged to recycle their waste. Employees in the city recycle more than 120 000 kg of paper per year, and all yard trimmings and grass from the city parks are composted. This minimises the amount of waste going to the landfill and reduces the work that goes into separating and managing the waste.

Figure 5.1: Kaka'ako Waterfront Park on the site of a former city landfill.

Discussion questions

1. Discuss why careful waste management is important for an island population.

2. Discuss the potential problems caused by burning waste.

3. Why is it important to get individuals and commercial businesses involved in waste management?

4. 'Waste management offers an opportunity to support food production.' As a group discuss this statement. How far do you agree?

5.1 Food security

World food security (Figure 5.2) looks at **food security** in different countries. It ranks countries according to how secure their food supply is. LICs tend to have lower food security than HICs. HICs such as North America, Australia, and many European countries, have the highest levels of food security.

There are three main considerations when discussing food security:

1. food availability (if there is a sufficient quantity of food)
2. food access (how possible it is to obtain food)
3. food use (the correct balance of food for a nutritious diet, and correct handling of food to avoid the spread of disease).

According to the United Nations Environmental Programme (UNEP), the world produces sufficient food for the world's population. However, this does not mean that everybody has food security. In some parts of the world there is excess food production, while in other parts there is a shortage. Many countries with insufficient food do not have the financial means to either import food or to increase food production. Although there is sufficient food globally, it is estimated that approximately 11% of the world's population is undernourished.

Climate change and global shifts in rainfall patterns are having an impact on food production. Already arid areas are becoming drier and less capable of supporting food production. Areas already struggling with food production may find it harder to produce food in the future.

Inefficient farming methods and urbanisation also impact food security. In some instances, agricultural lands have been lost through soil degradation or residential development. Inefficient farming can lead to poor irrigation methods, resulting in water wastage and **salinisation**. This is when soils become too salty and infertile, causing crops to fail. Salinisation can also lead to water insecurity (Chapter 6, topic 2).

> **KEY TERMS**
>
> **food security:** when all people, at all times, have physical, social and economic access to sufficient, safe and nutritious food that meets their dietary needs and food preferences for an active and healthy life
>
> **salinisation:** an increase in salt content, usually of agricultural soils, irrigation water, or drinking water

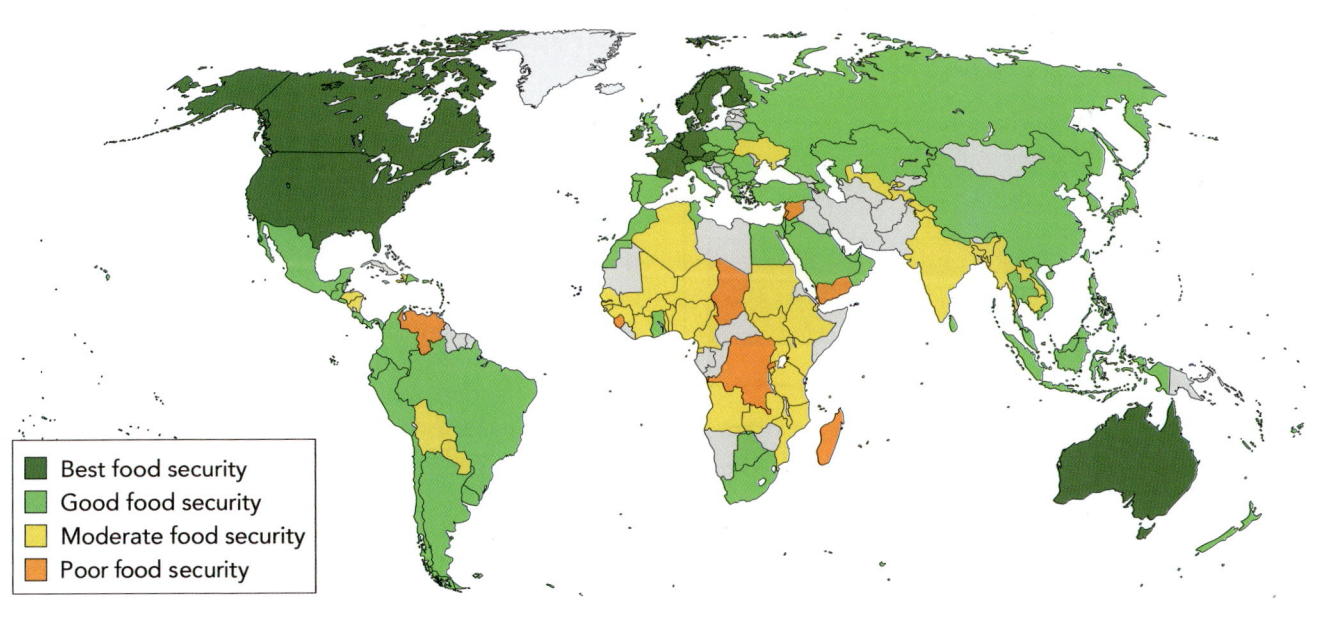

Figure 5.2: World food security map.

Figure 5.3: Excess bananas dumped behind a banana plantation in Panama, Central America.

Fast-rising food prices, combined with a decline in food production, have caused a significant geographical imbalance between food production and food consumption.

> ### ACTIVITY 5.1
>
> **Disparities in global food costs**
>
> Plot a bar chart to show how the cost of a plate of food (basic bean stew), shown as a percentage of an individual's daily income, varies around the world.
>
> | Country/City | % of an individual's daily income spent on a plate of basic bean stew |
> |---|---|
> | New York | 0.6 |
> | Iran | 3.9 |
> | Nepal | 13 |
> | Yemen (urban) | 22 |
> | Haiti | 35 |
> | Malawi | 45 |
> | North-east Nigeria | 121 |
> | South Sudan | 155 |

> ### CONTINUED
>
> As a class discuss what you observe.
>
> 1. What does this information tell you about food security in different countries?
> 2. Will the percentage of daily income spent on a plate of basic bean stew differ in rural and urban Yemen? Give reasons for your answer.
> 3. In pairs, find out why the South Sudanese spend so much of their daily wage on a plate of bean stew. Discuss your findings in small groups or as a class.

> ### SELF ASSESSMENT
>
> - Did your graph make the data more easy to understand than when it was presented in the table?
> - Would further analysis of the data would give you more information on the differences between countries? Why/why not?
> - Has this activity helped you to understand how the cost of food differs between countries, and how this impacts food security? Explain.

Causes of food insecurity

Food insecurity is caused by a variety of both natural and human factors.

Water shortages

A lack of water leads to drought, which results in the death of crops and livestock, and famine. Low rainfall can be seasonal or long term. Long-term drought can result in a community having to move to survive. Even where water is not scarce, water management is vital for successful food production. In areas with low water availability, incorrect use of water can result in depleted water sources and farms failing.

Land degradation

Misuse of land leads to land degradation, soil exhaustion and finally low crop yields as plants fail to thrive. Land degradation is caused by **over-cropping** (continuously growing crops on land without giving soil time to rest) and **monoculture cropping** (repeatedly growing a single crop).

Long-term depletion of nutrients leads to desertification and soil erosion, due to the combination of overgrazing and drought conditions (Figure 5.5). Top soil may be washed away, leaving behind infertile soils unable to support plant life.

Figure 5.4: Severe soil erosion, clearly showing the resulting land degradation.

Agricultural pests and diseases

Pests such as locusts (Figure 5.5) can consume hundreds of thousands of tonnes of food a day. Farmers can lose an entire year's crop to locust swarms as they move across fields. In addition, fungal diseases destroy both crops and stored food, putting pressure on food supplies.

Figure 5.5: Large locust swarms can consume entire crops, causing both food and economic loss for a region.

Population growth

Human population numbers are increasing at a faster rate than the increase in food production. This results in shortages of food for populations. Where **subsistence farming** is the main type of farming and rapid population growth occurs, food insecurity is even more of a challenge. Globally increased population numbers put pressure on food resources.

Diverting crops for biofuels

Increased production of food crops for **biofuels** by commercial agricultural companies puts pressure on food sources. Agricultural land once used for food crops is now used to grow energy source crops. This has≈led to a decrease in the amount of land available to grow food.

Figure 5.6: Mechanical harvesting of sugar cane. Sugar cane is high in energy and is often used as a biofuel crop.

Poverty

Poverty and food security are closely linked. When people have no money to buy food, or if they are undernourished and ill, they cannot work. For subsistence farmers this means that they cannot farm and produce food, so they remain undernourished. This leads to populations relying on aid to survive. Without donations, a country dependent on **food aid** cannot meet the needs of its population.

> **KEY TERMS**
>
> **overcropping:** Nutrient deficient soils due to crops being continuously grown on them
>
> **monoculture crops:** The cultivation of a single crop in an area
>
> **subsistence farmer:** When a farmer grows food for their family and not to sell at the market.
>
> **biofuel:** A fuel derived from biomass (plant or algal material, or animal waste).
>
> **food aid:** help given to a country or region suffering from food insecurity

Price-setting

Price setting is how the market determines the price of a food product. As demand increases or supply decreases, food product prices increase. When the price of food goes up, people struggling to buy their food have increased food insecurity, and are unable to feed their family. Price setting occurs along with drought or war in food-producing regions. It results in decreased supply and pressure on food supplies globally.

Climate change

Climate change is identified as a threat to food security and food supply systems. Increased temperatures, floods, droughts and storm size all impact the growth of crops and the stability of food supply. Changes in climate also affect the frequency of insect and fungal outbreaks that impact crops. Climate change has wider impacts, too, as food insecurity affects trade flows, food markets and price stability.

Unsustainable production

Global food supplies have become more **homogeneous** over the last 50 years. Human diets around world are now more similar. The list of available of foods is becoming shorter (wheat, maize and soya, with meat and dairy products, account for most of our foods). This increasing lack of food diversity reduces the nutritional quality of food as consumption of other grains and vegetables decreases. This also makes agriculture vulnerable to threats such as drought, insects and disease. When large areas have the same type of crop, there is a higher risk of crop failure and reduced food security.

> **KEY TERM**
>
> **homogeneous:** describing things of the same kind, e.g. the crops produced by a farmer may be all of the same kind

Impacts of food insecurity

Food insecurity has a wide range of impacts on countries. However, there are four main threats it presents:

1. increases the number of people living in poverty
2. slows or reverses economic development
3. causes political destabilisation, as food insecurity leads to food-related unrest.
4. causes population decline due to migration or death.

The impacts of food insecurity are felt most intensely in LICs. However, food insecurity can occur in HICs, as a result of natural disasters, global supply chain disruption or war. The 2022 war in Ukraine is an example of how the food supply can be disrupted regionally and globally due to conflict. Ukraine is a significant exporter of wheat, sunflower oil and seed. Russia also exports many key goods, but during the war it has not been exporting as much to countries who are not direct allies. The war has interrupted farming and transport of food products. This has caused a decline in exports and an increase in food insecurity in other parts of the world.

Food insecurity can cause **malnutrition** (Figure 5.7). This condition is a result of insufficient nutrients and it can result in **starvation**. Once malnutrition sets in, a person is less resistant to the spread of disease and may be unable to work. This further decreases their ability to farm and produce food.

Poverty worsens the problem. Hunger and malnutrition become serious issues when a population has insufficient finances to purchase food in the event of illness of crops failing. In cases of prolonged **famine**, people may die due to the lack of food.

Figure 5.7: A father in a refugee camp holds his daughter, who is suffering from malnutrition due to a prolonged lack of access to food.

> **KEY TERMS**
>
> **malnutrition:** lack of adequate nutrition, caused by not having a balanced diet, or enough to eat
>
> ***starvation:** suffering or death caused by lack of food
>
> **famine:** the extreme scarcity of food

5 Managing Resources

Figure 5.8: Volunteers in Scotland pack food parcels in 2022: significant increases in the cost of living resulted in food banks like this one having to provide more than 5100 food parcels daily.

ACTIVITY 5.2

The impacts of climate change on food security

1. Working in pairs, research the impact of climate change on food security in a region of your choice.

   Consider changes in temperature, precipitation, floods or drought, storm frequency and strength, diseases and insect swarms, and how they impact plant productivity.

2. Using this information create a flow chart. Show how climate change is impacting food security in the region you studied.

Strategies for managing food insecurity

With changes to farming, the introduction of technology and innovation, food insecurity can be reduced. Food science and the ongoing development of technology have made it possible to meet the growing demand for food and improve the nutrition supplied by the crops grown.

Agricultural technology and innovation have developed a wide range of techniques and equipment including hydroponics and aquaponics, the protection of pollinators, food supply chain intervention, reduction of food waste and food redistribution.

Improved agricultural techniques and efficiency

Improved agricultural techniques and efficiency have been achieved through the use of fertilisers, pesticides, herbicides, irrigation schemes and high-yield variety seeds. This is known as the green revolution, and it has resulted in an increase in food production globally.

Use of genetically modified crops and selective breeding:

- **Genetically modified (GM) crops** are crops that have been scientifically developed through changes to their genes. These crops can grow under more challenging conditions and provide higher levels of nutrition. GM crops can be designed to grow in drier, saltier or wetter conditions.

- **Selective breeding** is when humans breed plants and animals to achieve particular characteristics. For example, a wheat plant that needs less water to grow than its counterparts will be used to develop a crop that is drought resistant. Or a sheep that produces good volumes of wool will be used to breed sheep with good wool production.

Controlling limiting factors:

- Lack of water can be a limiting factor. Better **irrigation** systems allow the irrigation of larger areas of land and the development of commercial farming. Systems include pivot irrigation and drip irrigation systems (Figure 5.9). An efficient irrigation system also minimises water use and saves water.

- Lack of nutrients in the soil limits plant growth. Increased use of **fertilisers** can boost plant growth rates and productivity.

KEY TERMS

genetically modified (GM) crops: foods derived from organisms in which DNA has been changed by humans

selective breeding: when humans grow plants and animals for specific characteristics, e.g. high yields or drought tolerance

irrigation: the supply of water to land or crops to help plants grow

fertiliser: a chemical or natural product that can be added to soils to increase the nutrients available for plants

193

Figure 5.9: A water-saving drip irrigation system installed to irrigate grape vines in Napa Valley, California, USA.

Increasing productivity:

- Weeds compete with plants for space, water, light and nutrients. To reduce this competition, weeds need to be controlled. They can be reduced manually, or through the use of **herbicides**.

- Pests and fungi can harm the crops, and reduce their harvest. **Fungicides** can be used to limit their impact.

- Pest species can also be controlled through biological controls. This is when a natural predator is used to control the pest, for example, cats are used to control rodent populations in wheat fields.

Improved transportation of food:

- Improvements to transport infrastructure and the addition of freezing and refrigerated vehicles allow for the transport of products which results in better food distribution.

Hydroponics and aquaponics:

Newer technologies such as **hydroponics** (Figure 5.10) and **aquaponics** (Figure 5.11) have helped food production in areas where space or fertile soil is limited, or in urban areas where urban farming is being introduced.

Hydroponics (Figure 5.10) is the farming of crops without soil. Instead, mineral nutrient solutions in water are used. Water provides nutrients, water and oxygen to help the plant grow. This technique uses less space and up to 90% less water than in traditional farming, and the crops take a shorter time to grow to harvesting size. The need for pesticides is also reduced.

Aquaponics (Figure 5.11) is when fish are included in the hydroponic cycle. The fish eat the plant waste and then produce waste that plants can use as fertiliser. Aquaponics mimics a natural system, where waste and by-products from one organism are food for bacteria. The bacteria then coverts the waste and by-products into food for plants.

Both hydroponics and aquaponics are forms of **intensive farming**. They require a relatively small area but high financial investment. They achieve high productivity.

Figure 5.10: A hydroponic farming system growing lettuce.

Figure 5.11: An aquaponics farming system.

The improvement of agricultural techniques has allowed for the development of both intensive and **extensive farming** areas. In intensive farming, farmers produce more food in smaller areas using farming techniques like aquaponics. In extensive farming, farmers use large areas of land with relatively low financial investment and labour to produce their food products.

5 Managing Resources

KEY TERMS

herbicides and **fungicides:** chemicals used to control insects, unwanted plants and fungi in commercial food crops

hydroponics: the growth of plants without soil. Instead, plants are grown in nutrient-rich water

*aquaponics: a soil-free farming system that uses the waste produced by aquatic organisms (fish) to supply nutrients to plants being grown hydroponically

*intensive farming: a system of farming that uses large amounts of investment and labour relative to the area of land being farmed

*extensive farming: a system of farming that uses a small amount of labour and capital investment relative to the area of land being farmed

ACTIVITY 5.3

Disadvantages of improved agricultural techniques and efficiency

In pairs, research the main problems that improved agricultural techniques have caused. Create a table or spidergram to show your findings. Use the following factors to guide you:

- the impact of the increased use of herbicides and pesticides
- the impact of the increased use of fertiliser
- the development of new farmlands in natural areas
- the impact of machinery on farming jobs
- mismanagement of irrigation and salinisation of the soils.

INVESTIGATIVE SKILLS 5.1

Testing the impact of fertiliser on plant growth

The aim of this investigation is to understand the impact of inorganic fertilisers on the rate of plant growth.

You will need:

- ten seedlings of equal size per student group/class
- ten individual pots of equal size (empty yoghurt containers with holes in the bottom work well)
- inorganic fertiliser
- potting soil
- measuring cup/syringe for watering
- a ruler and a table to record growth rates.

Figure 5.12: Spinach seedlings prior to the start of the investigation. The seedlings should all be the same age and a similar size.

195

CONTINUED

Safety

Always take care when working with inorganic fertilisers.

- Do not expose them to an open flame.
- Work carefully to prevent dust forming and entering your eyes or lungs (wear a mask and goggles to prevent harm).
- Wear gloves or avoid handling fertilisers with your hands; wash your hands after handling.

Getting started

1. Do you think varying the levels of fertiliser will result in different growth rates in the plants?
2. Set a hypothesis for what you expect the outcome to be.

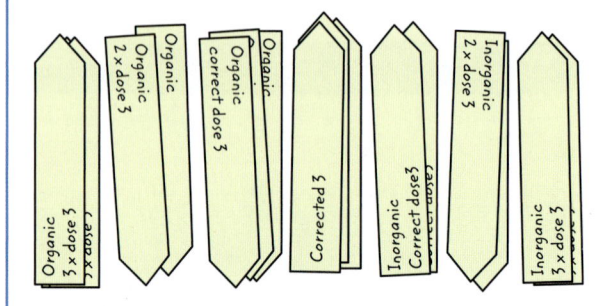

Figure 5.13: Labels made out of old yoghurt containers are used to ensure that each pot is clearly identified throughout the investigation.

Method

- Clearly mark the pots 1–5 (no fertiliser) and 6–10 (fertiliser). Old yoghurt cups can be cut up to make pot markers (Figure 5.13).
- Divide the potting soil into two separate containers.
- Add the recommended amount of fertiliser (follow instructions on the fertiliser packet) to one of the containers of potting soil.
- Fill each of the pots with an equal amount of potting soil (this is done by weight), adding potting soil with no fertiliser to pots 1–5 and potting soil with fertiliser to pots 6–10.
- Plant a seedling in each pot (Figure 5.14).

Figure 5.14: Seedlings being planted into individual pots.

- Measure and record the height of each plant on day one. Do not pull the leaves up; measure the tallest leaf without touching it. Use a ruler placed behind the leaf to measure the height.
- Water the seedlings with an equal amount of water when required. Position them so they get an equal amount of light.

Figure 5.15: Regularly water each pot with the same amount of water to ensure the plants have sufficient water to grow.

5 Managing Resources

> **CONTINUED**
>
> - To obtain results, measure the change in plant growth daily for up to two weeks. Calculate the average growth each day for the five plants in the potting soil and the five plants with fertiliser. Copy the table below to record your results.
>
> ### Questions
>
> 1. Plot the results on a line graph. Compare the growth of plants with fertiliser to plants without fertiliser.
> 2. Calculate the difference in growth between the plants that got fertiliser and those that did not.
> 3. What did you observe?
> 4. How do fertilisers support food security?
> 5. What are the limitations of using fertilisers?
> 6. Why do you think inorganic fertiliser was used?
>
> | Day | 1 | 2 | 3 | ... | (up to 14 days) |
> |---|---|---|---|---|---|
> | Pot 1 | | | | | |
> | Pot 2 | | | | | |
> | Pot 3 | | | | | |
> | Pot 4 | | | | | |
> | Pot 5 | | | | | |
> | Average | | | | | |
>
> | Day | 1 | 2 | 3 | ... | (up to 14 days) |
> |---|---|---|---|---|---|
> | Pot 6 | | | | | |
> | Pot 7 | | | | | |
> | Pot 8 | | | | | |
> | Pot 9 | | | | | |
> | Pot 10 | | | | | |
> | Average | | | | | |

> **SELF ASSESSMENT**
>
> - In this investigation, was your hypothesis correct, partially correct or incorrect?
> - Did you discover other variables that needed to be considered?
> - How confident were you in applying the scientific method to this investigation?

Other strategies for managing food security

In addition to improving agricultural techniques and increasing efficiency, there are other strategies for managing food security, as outlined below.

Protecting pollinating insects

Pollinating insects, such as bees and butterflies, are insects that move pollen from one plant to another, helping to fertilise the plants and allowing them to produce seeds and fruit. Protection of pollinating insects is important. Insect pollinators are disappearing from ecosystems due to pesticide misuse and disease among bee populations. Insects are an essential component of the world's terrestrial ecosystems, and form a vital part of food production.

Farmers can do the following to protect pollinators.

- Reduce herbicide and pesticide use by adopting organic farming methods.
- Manually clearing weeds while allowing them to grow and bloom during pollination season to attract pollinators to the area.
- Planting crops that flower at different times of the year.
- Allowing portions of the farm to return to natural vegetation to form natural habitat patches that attract pollinators to the farm.
- Farming bees as pollinators alongside crops (Figure 5.16).

Subsistence agriculture

Subsistence farming can be encouraged in areas where people have limited access to food. The health and economic stability of the local population are improved when they can feed themselves.

Figure 5.16: Beehives next to fields of sunflowers and lavender in Provence, France. The bees help pollinate the crops.

Subsistence farming can take place in both rural or urban areas. Rural areas are likely to have space readily available for small-scale farming. In urban areas, roof-top farming (Figure 5.17) is becoming more popular.

Figure 5.17: Roof-top farming in an urban setting, allowing crops to be produced in an area with limited space.

Large-scale food stockpiling

Countries prone to natural disasters such as earthquakes or periods of drought often practise **stockpiling** of food resources. Food supplies that can be stored for periods of time (such as grain) in silos (Figure 5.18) are kept to meet food demands when there are food shortages caused by an interruption in production. This interruption can be due to seasonal changes, as there is lower production of food crops in winter, or can be after a natural disaster. Stockpiling also offers protection against interruptions in the international food supply. This can be due to factors such as pandemics, regional wars, or government export polices that may interrupt the food supply chain both within and between countries.

Figure 5.18: Large food storage silos on a farm.

The World Food Programme and food aid

The World Food Programme and food aid assist people experiencing food insecurity (Figure 5.19). The World Food Programme is a humanitarian organisation that provides hungry children and families with nutritious meals. The programme is available in over 80 countries worldwide, many of them in Africa. The programme responds to food supply emergencies through using technology which identifies areas of developing concern.

Food aid is when one country provides food to another. This food is often free of charge, or provided under fair terms to assist the country in need. Food aid may be organised by non-governmental organisations. Recently food aid has been organised by countries for their own citizens. During the global COVID-19 pandemic, many people lost their jobs and struggled to purchase food. Food supply chains were disrupted as transport drivers had to stay at home and transporting food posed infection risks, resulting in shops running out of food. Many people also stockpiled food, meaning supply for others was limited.

Figure 5.19: A man collects food aid that has been supplied by the World Food Programme (WFP) to aid the Afghan people.

Rationing

Rationing is another way of ensuring the fair distribution of food during periods of scarcity. By rationing food, governments can work towards ensuring that people get an equal amount of food, as each person is only allowed access to a certain amount. Rationing first occurred after the Second World War when food production was below demand. People were given ration books and had to register to buy their food from different shops.

In more recent times, the outbreak of COVID-19 saw the rationing of food products around the globe as panic buying saw the shelves in shops being emptied (Figure 5.20). Those who could afford to purchased food to keep at home in case of shortage. In contrast, people with less savings, or those who were working long shifts were sometimes unable to obtain food. By rationing the number of a specific item that any one person was able to purchase, the supply chain was protected, so everybody could have access to food.

Figure 5.20: Empty shelves in a supermarket in Sydney, Australia during the COVID-19 pandemic.

> ### KEY TERMS
>
> **stockpiling:** to store large amounts of goods or materials, in this instance food stores
>
> **rationing:** to limit the amount of food each person or family is allowed to purchase

Reduction of food waste and food redistribution

Ways of reducing food waste (Figure 5.21) are being developed. As much as 24% of all food currently produced goes uneaten. Reducing this waste would help to tackle global food insecurity.

The first step is to reduce loss close to the farm. Low-cost storage methods can cut food loss before it gets to the market. If the amount of food lost to pest, spoilage and transportation damage could be minimised, more food would be available for the population.

Food spoils unless it is properly processed and stored. Preserving foods in jars and bottles, or drying food sources can make them last longer before they perish. In shops, food products nearing the end of their shelf life could be redistributed to those in need.

Figure 5.21: Waste fruit and vegetables in a bin.

Reduction in livestock and increase in growing crops

The farming of livestock (Figure 5.22) plays an important role in food insecurity. Crops that could be grown directly for human consumption are being grown to feed livestock. Grain-based feeding of livestock has significantly increased the production and consumption of meat in recent decades. However, intensive livestock farming requires a significant amount of resources, from the land needed to water used. It is estimated that three-quarters of all agricultural land is used for farming livestock. Moving the majority of this land over to crops would increase the amount of crops grown for human consumption.

Figure 5.22: An intensive cattle farm in Drucat, France.

5.1 Questions

1. **Define** the terms *food security* and *food rationing*.
2. **Give** the main impacts of food insecurity.
3. **Describe** the impact that the green revolution had on agriculture.
4. **Explain** how hydroponics differs from traditional farming.

5.2 Energy resources

Energy is essential for modern life, as we need power for most of our daily tasks. Whether we are switching on the lights, cooking a meal, or driving a vehicle, we need an energy source. In this section, we will look at two types of energy sources: renewable and non-renewable.

Renewable resources

Renewable resources use nature in a sustainable way. Currently, only a small quantity of the world's energy comes from renewable sources, yet there are many forms of sustainable energy sources. One of the main advantages of this type of energy is that it does not release greenhouse gases into the atmosphere. However, this does not mean that no environmental damage occurs when developing these energy sources. Mining for materials, clearing sites for construction and decommissioning old equipment all have an impact on the environment. For example, old solar panels that are no longer useful need to be recycled, otherwise they go to landfill, contributing to the problem of waste.

Examples of renewable energy

Hydroelectric dams

Hydroelectricity is produced using moving water. Dams are the most common type of **hydroelectric power (HEP)**. A dam is built across a river to block the flow of water. The flowing water is then channelled through a pipe so that it can turn the turbines which produce electricity. The greater the flow of water, the greater the potential energy produced. Hydroelectric power is currently the most important source of renewable energy in the world.

Figure 5.23: Clyde Dam Power Station in New Zealand.

Solar energy

Solar panels (Figure 5.24) generate electricity when exposed to sunlight. They can be used on either a small or large scale. They are ideal for use in remote locations as they do not require an electrical grid system to supply energy to the end user. **Solar power** is relatively safe and efficient. It is also limitless. It has good potential for LICs in the equatorial regions and regions with high rates of insolation. However, solar power's usefulness is limited in colder regions where direct sunlight is less common.

> **KEY TERMS**
>
> **renewable resources:** any source of energy that can be naturally and quickly replenished, e.g. wind and solar power
>
> **hydroelectric power (HEP):** electricity that is generated using the energy of flowing water
>
> **solar power:** electricity that is generated by utilising the energy of the sun

5 Managing Resources

Figure 5.24: An engineer carrying out maintenance on a solar panel at a solar power plant.

Wind energy

Wind energy harnesses the movement of air to turn the blades of wind turbines (Figure 5.25). The turbine drives a generator, creating electricity. Installing wind turbines at sea limits the loss of land area for other uses. It also reduces the impact of the noise caused by the turbines on surrounding towns. There are limitations with this energy type. Firstly, for this renewable resources to be reliable a location with regular wind is required. In addition, although prices are coming down, wind turbines are still expensive to build.

Figure 5.25: Offshore wind turbines.

Wave and tidal energy

Wave and tidal energy (Figure 5.26) can potentially supply vast amounts of electricity. However, it is not yet as developed as other renewable resources. Power from the movement of waves and the tides of the ocean and river mouths is used to turn turbines. It is then converted into electricity. There are some small, non-commercial wave power installations. However, harnessing wave power is extremely complex and expensive. Building off-shore comes with challenges. There is also the issue of corrosive salt water and the potential damage that waves can cause.

Figure 5.26: Workers maintaining an off-shore electricity generator based on wave power off of Portugal's Atlantic coast, 2008.

Biomass

Biomass is obtained from organic material, such as plants, and animal waste. Sources of energy using biomass can be divided into wood, **bioethanol** and **biogas**.

- Wood can be burned to generate heat and light and to cook with. It relies on the availability of wood from trees and has resulted in the deforestation of large areas globally. Wood is used when people do not have an alternate energy source. Its use has both environmental and health implications.

> **KEY TERMS**
>
> **wind energy:** electricity that is generated using the power of wind
>
> **wave and tidal energy:** electricity that is generated using the energy of waves or the tides
>
> **bioethanol:** an alcohol produced from plant matter such a sugar cane or maize which can be used as an alternative to petrol
>
> **biogas:** a gas such as methane that can be used as a fuel and is produced by fermenting organic matter

- Bioethanol is a flammable liquid that can be used as a motor fuel or as an additive to petrol. All existing petrol engines in vehicles can operate with blends of up to 10% ethanol with petrol. To produce bioethanol, feed-stocks such as corn, sugar cane, algae or other glucose-rich biomass are fermented using a yeast to digest the sugars.
- Plant or animal waste can be fermented to create biogas in the form of methane that can be used for cooking. Biogas digesters can be small (for a single household) or large commercial systems (for supplying energy to a town) (Figure 5.27).

Figure 5.27: A small-scale biogas digester on a farm that uses cattle manure to generate methane for cooking.

Figure 5.28: Geothermal electricity generation in Iceland.

Geothermal energy

Geothermal energy (Figure 5.28) uses heat from within the Earth to heat water or generate electricity. To create geothermal energy, cold water is pumped down into boreholes where it is heated by the Earth. The water then returns to the surface as steam. This steam is used to turn turbines and generate electricity. This energy is sustainable and does not add to carbon emissions in the atmosphere. The disadvantage is that it can only work in areas where the heat of the Earth is near the surface, in places like Iceland, New Zealand and Yellowstone National Park in the USA.

Renewable energy does have some limitations. For example, if the energy source is unreliable: wind energy does not work if there is no wind. Renewable energy may not supply enough energy and some renewable resources take up a lot of space. It is also still relatively expensive to install. This means that those on limited budgets may find it hard to move over to renewable resources.

Non-renewable energy sources

Non-renewable energy sources are ones that will run out and not be replaced for thousands or even millions of years. They include coal, oil, natural gas (fossil fuels) and uranium.

Fossil fuels include oil, gas and coal. Fossil fuels become usable by humans when their energy is released through combustion in industrial processes. These are formed over millions of years from the remains of dead organisms.

Examples of non-renewable energy

Oil and gas

Oil and gas (Figure 5.29) are formed from the remains of dead marine organisms over millions of years. These remains fall to the bottom of the ocean where they are covered by sediments. Over time, as more sediments settle, the increased load changes the pressure and temperature. The heat and pressure turn the remains of the dead organisms into oil and gas, trapped in the sedimentary rocks.

> **KEY TERMS**
>
> **geothermal energy:** energy generated from the heat under the surface of Earth
>
> **non-renewable resources:** resources that will run out and not be replenished for millions of years, for example oil, gas and coal

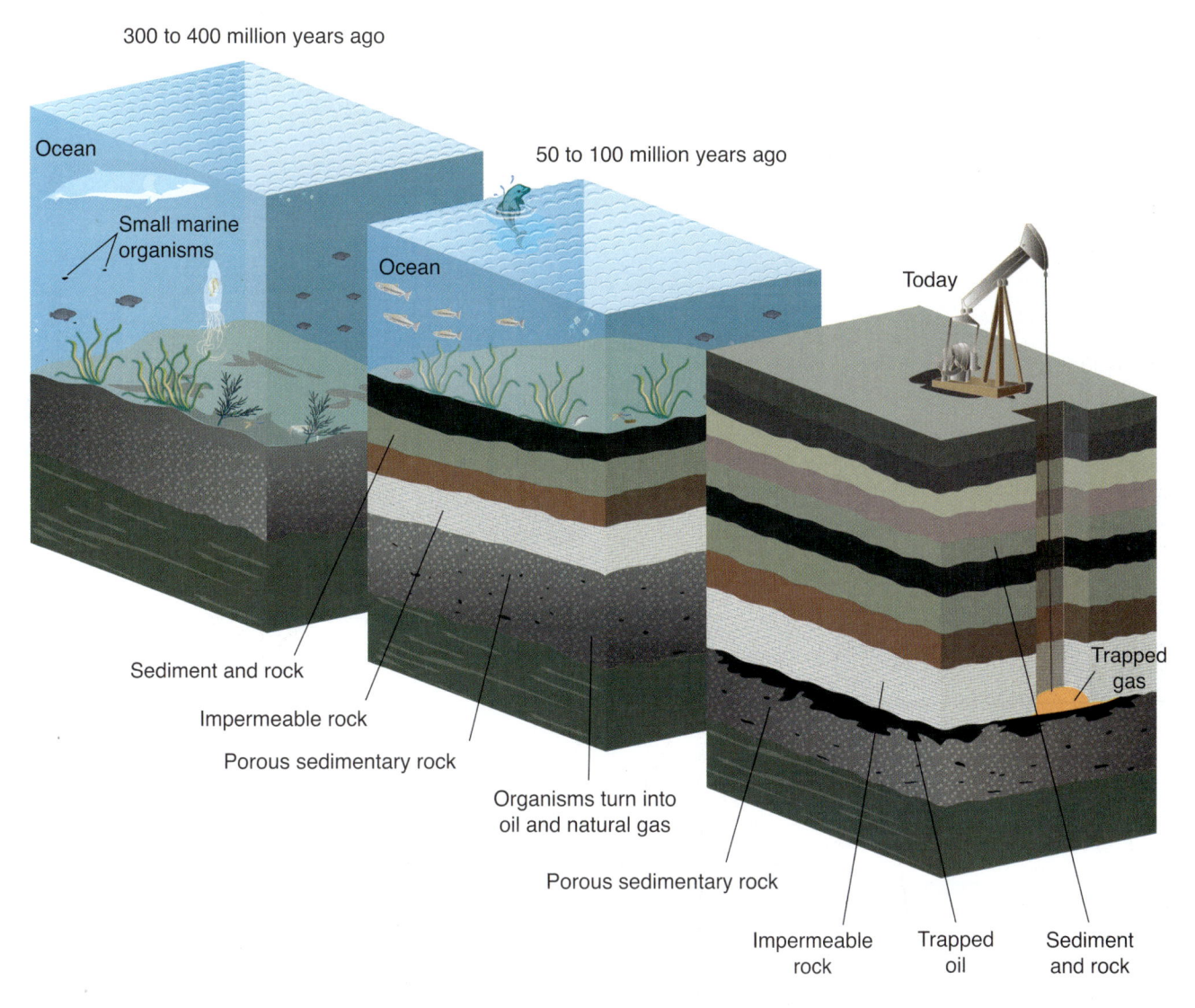

Figure 5.29: Oil and gas formation at the bottom of the ocean.

Coal

Coal (Figure 5.30) is formed from dead trees and other plant materials lying on the surface of the land, when rates of decomposition are low. Over time, the dead material gets buried under sediments on the surface. The change to coal is the result of increasing heat and pressure over a long time.

Nuclear power

Nuclear power is a non-renewable energy resource because it uses uranium. When depleted, spent uranium fuel is replaced with more uranium mined from the Earth. Estimates by the Nuclear Energy Agency suggest Earth's accessible uranium supply will last around 230 years at our current rate of consumption. It is estimated that between six and ten percent of commercial energy is nuclear. However, it uses very small amounts of uranium and does not emit greenhouse gases when producing energy. This makes it more sustainable than fossil fuels.

> **KEY TERM**
>
> **nuclear power:** nuclear power uses radioactive materials such as uranium or plutonium. These materials undergo reactions and power is produced from the energy released

One of the main disadvantages of using nuclear power is the risk of an explosion contaminating an area with radiation. New technology is working to make it safer.

Huge forests grew millions of years ago covering most of Earth

The vegetation dies and forms peat

The peat is compressed between sediment layers to form lignite

Further compression forms coal

Figure 5.30: Coal formation takes millions of years and requires large amounts of organic material.

> ### ACTIVITY 5.4
>
> **The impacts of using different energy sources**
>
> 1. In groups of three or four, research the negative environmental impacts of using different types of energy sources.
> 2. Think about the impacts involved in getting the materials needed to build the equipment, or during the construction of the energy sites (dams, mines or wind turbines), as well as the disadvantages occurring when they are operational.
> 3. Copy the table below and expand it to include the energy sources that you have investigated.
>
> | Energy source | Impacts |
> |---|---|
> | e.g. Hydroelectric power | Flooding of ecosystems upstream of the dam |

Consider both the construction and the operational phases. Where do you see a difference in the impacts of the energy sources? What differs between renewable and non-renewable resources?

Figure 5.31: This nuclear power station under construction in Bridgewater, England will generate energy for approximately 6 million homes.

Energy security

Energy security is defined by the International Energy Agency as 'the uninterrupted availability of energy sources at an affordable price'. When a country suffers from energy insecurity, it means that either the energy supply is either interrupted, unaffordable, or both.

Long-term energy security is the supply of energy in line with the proposed economic development of an area. This requires planning and foresight on behalf of those investing in the development of the energy supply in a country. It can include long term planning for systems such as hydroelectric or nuclear plants, and upgrading the electricity grid to ensure supply for future needs.

Short-term energy security is the supply of energy systems that are able to react to sudden changes in energy supply. This can be achieved through the use of systems that store energy for use when energy demand increases. It could include oil or gas stores to be utilised during periods of international price instability. It could also include stores to be used during periods of high pressure on the electricity supply, for example, during seasonal temperature changes that result in the increased use of heating or cooling equipment.

5 Managing Resources

> **KEY TERMS**
>
> **energy security:** the reliable availability of energy sources at an affordable price with a consideration of the environmental impacts
>
> **long-term energy security:** the supply of energy that is in line with economic developments and environmental needs
>
> **short-term energy security:** systems that react promptly to sudden changes in the supply-demand balance

Causes of energy insecurity

There are various causes of energy insecurity. These include population growth, location of energy resources, fossil fuel depletion, supply disruption and the differing energy needs of countries experiencing different levels of development and climate change.

Figure 5.32: Cities across the world, like Cairo, are expanding as the population increases. These cities require lots of energy to function and remain lit up at night time.

Population growth

In many countries, population growth is occurring at a faster rate than the supply and development of new energy resources. Increasing population size puts pressure on systems that are already strained or struggling to supply enough electricity. This results in energy insecurity. On a global scale, increasing populations and industrialisation have led to a higher standard of living and greater personal wealth. Increased use of cars, electrical white goods (such as washing machines) and personal electronic equipment has resulted in more demand on available energy resources.

Figure 5.33: The increased standard of living has put pressure on energy resources, as individuals own more electronic goods such as laptops and mobile phones.

Global energy resource distribution

Geographical distribution of resources plays a role in the availability of energy in different regions. Some parts of the world have easy access to resources because those resources are abundant in that region. For example, the Middle East has abundant oil reserves. Other parts of the world have limited known reserves and rely on importing energy resources.

Figure 5.34: An oil well in Bahrain in the Middle East. These wells access oil deep underground. The region is rich in oil reserves, ensuring a good energy supply.

205

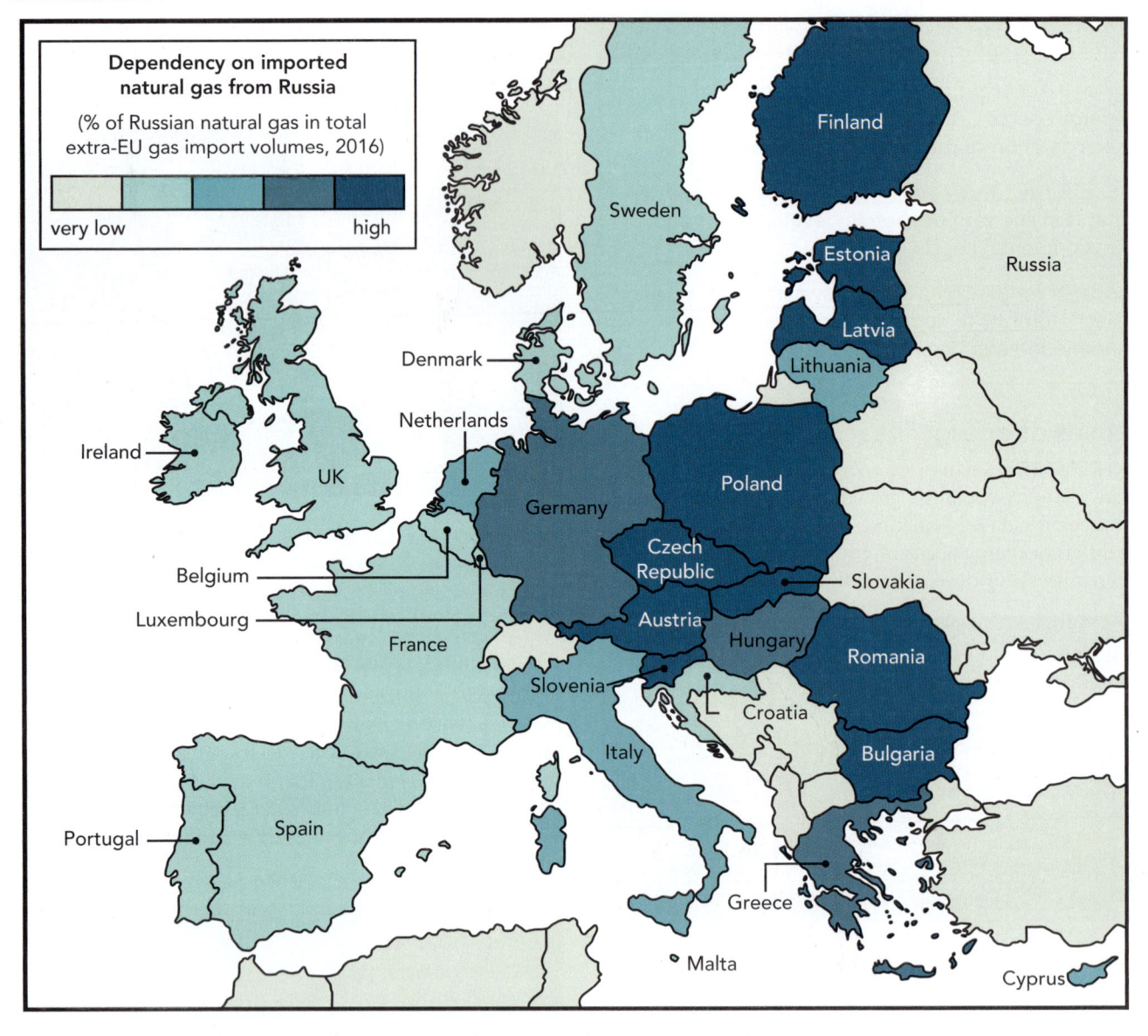

Figure 5.35: A map of Europe showing the level of gas supply from Russia to each country.

Energy needs of countries in different income groups

Countries with different levels of development have different energy demands. HICs have a much higher energy demand than MICs and LICs. This is due to their level of economic development. HICs are more industrialised and have a higher standard of living. They therefore use more energy-hungry technology. Fewer than 25% of the world's population lives in HICs, but HICs use 66% of the world's energy.

Climate change

Climate change considerations play a role in energy security. Climate change has put pressure on countries, especially HICs, to combat increasing levels of CO_2 in the atmosphere by reducing their fossil-fuel consumption. The need to change from one energy source to another puts energy security at risk in many countries, as changes will be expensive and time consuming. Adapting to climate change also has the potential to limit economic growth and impact job availability.

Supply disruption

Supply disruption through natural disasters, war, political instability, piracy or terrorism is a significant cause of energy insecurity. Disruption to the supply of fossil fuel resources from one part of the world to another puts significant strain on the energy security of a country or region. Disruption to the supply of oil from a source area to the rest of the world during times of conflict not only limits supply but also increases prices.

The 2022 war in Ukraine is an example of supply disruption due to conflict. The countries that border Russia are more reliant on the country's natural gas than those further west. However, Europe depended to a great extent on Russian gas supplies (Figure 5.35).

This meant that supply disruption caused by the war resulted in significant energy insecurity within Europe. Gas and oil supplies coming from Ukraine or through Ukraine from other places were reduced. Energy supplies to Europe and the UK were also reduced and prices increased across the region and globally, triggering energy supply challenges all over the world. For example, the cost of petrol in South Africa increased by 35% between February and July 2022.

ACTIVITY 5.5

Energy detectives

An energy audit is a tool that helps to investigate and record information on energy consumption. In this task you will be carrying out an audit that investigates the energy being consumed in your school.

1. As a class, create a basic energy audit for the school.
2. In small groups, carry out the audit for different parts of the school.

   This audit should consider:

   a What type of energy is being consumed (electricity, gas, oil or renewable)? Where energy is being consumed (lighting, heating, other technology).

   b Where energy is being lost. Heating or cooling both require energy. If either heated or cooled air is being lost, energy is being lost. In this case, consider windows and insulation.

   c Where and how can energy use in your study area be reduced?

   d If there are air conditioners or heaters in classes, what temperature are they set to? Can they be adjusted to save energy?

   e Create the audit as a checklist that can be used to check the same points in each space audited.

Energy audit checklists are also available online. Do some research to find out what other factors you can consider. Use search terms which include 'simple', 'school', 'energy audit', 'for students'. Some energy audits are extremely complex, so look for one that is easy to carry out.

If your school has equipment such as power meters, light meters and flicker meters, or classroom temperature records, include them in the audit process to get more precise energy consumption records.

Collect your data and create a short report that you can share and compare with the rest of the class. Are there easy solutions to reducing power consumption in your school?

Area	Low energy bulbs y/n	Light off if class not used y/n	Air conditioners (y/n)	Air conditioners off if class not in use y/n
1				
2				
3				

Impacts of energy insecurity

Energy insecurity causes social, economic and environmental problems. The disruption of energy supply to homes and industry directly impacts a population's ability to work and thrive. Without a consistent, predictable supply of energy, it is difficult for both individuals and industries to plan their work day, which limits productivity.

Poverty and low standards of living:

- The need to ensure a secure energy supply can lead to people using fuels that are harmful to their health. Burning of biomass in homes for cooking or staying warm can result in long term lung damage, for example. Illness caused by smoke from fires makes it impossible for people to work, trapping them in the poverty cycle.

- Energy poverty has a negative impact on the life expectancy and literacy levels of a population. Education is negatively impacted as learners are unable to study at night time or access internet resources. This could potentially limit their future employment opportunities.

Figure 5.36: In Zambia, a group of young women carry their bundles of firewood back to their village.

Civil unrest:

- Energy insecurity can result in civil unrest and conflict.
- Fossil fuel resources in a region can cause conflict, as groups may compete to control the energy-rich mines.
- Conflict can come about as a result of some areas being rich in energy sources while others experience energy scarcity.
- War can break out when one country wants to control the energy resources of another.

Increasing energy prices and economic recession:

- People experiencing energy insecurity are vulnerable to changes in energy prices. An increase in the cost of electricity or gas can cause energy poverty. Where people have to choose between buying food and buying energy, their standard of living declines.

- When industry has to increase prices in response to rising energy costs, economic instability occurs as costs become too high for customers.

- Industries unable to meet the higher cost of energy are forced to shut down. This causes job losses and economic recession.

- When jobs are lost, communities become poorer.

Figure 5.37: A store in Austin, Texas closed due to power outages during a winter storm that put pressure on the state's energy supplies.

Reliance on imported energy sources:

- When a country does not have its own energy source, it needs to import energy to meet demand. This often results in significant amounts of international debt.

- The country may be unable to pay off its debt, pay for energy and also invest in its own infrastructure like roads, hospitals and schools. This can result in the development of the country being slowed down.

- The country is also far more vulnerable to any changes in global energy prices. If energy prices increase, the country has to pay more to import it.

5 Managing Resources

ACTIVITY 5.6

Energy sourcing: mining oil sands and tar sands

Read the following information on oil sands mining. Then, carry out the task below.

Oil sands/tar sands are a deposit of sticky black oil sands, sand, clay and water. They are found beneath boreal forest areas in Canada and cover a region of approximately 170 000 km², equivalent to the size of Florida (USA).

There are two methods for processing oil sands: either in situ (in place) or mining them when the deposits are deeper than 70 metres below the surface.

Large equipment scoops the oil sand into trucks that transport it to crushers for breaking down. The sands are then mixed together with hot water and pumped through to a plant where the oil (in the form of bitumen) is separated from the sand, clay and water.

Extracting oil sand and converting it into usable fuel is expensive and requires large volumes of water. The waste water and materials are toxic and cause water, air and soil pollution. However, in the 1990s, a rapid rise in oil prices made it financially viable to mine this source of oil. Canada produces approximately one million barrels of oil a day through this method of oil production.

Figure 5.38: Oil sands mining operation near Fort McMurray in Alberta, Canada. The bitumen lies in the soils near the surface.

Task:

Research the impact of oil sand/tar sand extraction on the environment. Write a short summary to highlight your findings.

Strategies for managing energy security

A number of strategies can be employed to improve the levels of energy security both globally and within a population.

- Energy usage can be managed through increasing energy efficiency or reducing energy waste. Improving housing insulation is one example of how this can be achieved. Insulation results in less energy being used for heating or cooling the interior temperature of the house.

- Energy usage can also be managed through improved technology. Technological developments in cars, electrical equipment and other machinery can reduce the energy used.

- Another way of managing energy security is increasing energy production to meet demand. This can be done on an individual level, for example, by installing solar panels on a house. At a more strategic level, it can be undertaken by a country. This involves long term planning and considerations as to which type of energy would best fit the location and the conditions under which the energy supply needs to perform.

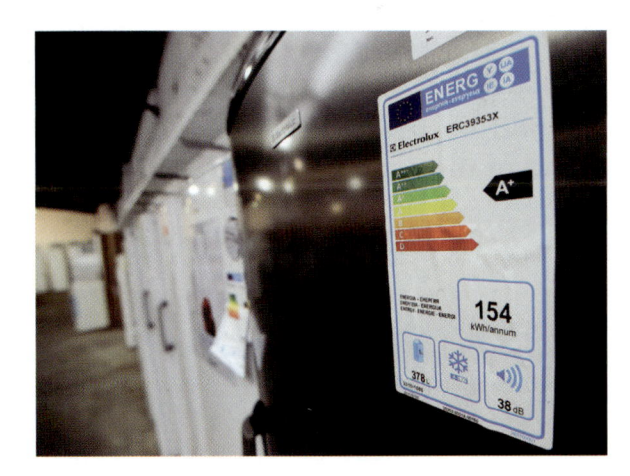

Figure 5.39: An energy rating label displayed on a home appliance at a store, 2022. These labels inform buyers about energy efficiency. The highest energy efficiency rating on this label is an A+++.

209

- Greater investment in renewable resources and carbon neutral sources means a reduction in our reliance on fossil fuels. However, countries that have an inexpensive supply of local fossil fuels (for example, coal in India, China and South Africa) or are reliant on importing fossil fuels as an energy source find it harder to move to renewable energy.

- Governments can support local energy projects as this will encourage the development of their own power supply. European countries encourage individuals to install renewable energy supplies such as wind or solar energy in their homes and to feed the excess back into the grid. That way, a country benefits from a diverse and widespread supply system. This results in greater security than when a country depends on one energy source.

- Where energy supply is limited, the energy supply can be rationed. This results in power supply to parts of the energy grid being either reduced or stopped for periods of time. Rolling blackouts allow the supply system to continue functioning under pressure, and attempt to manage the damage caused to industry and the economy.

Figure 5.40: Large-scale solar panel installation on water. Ongoing investment in new technologies has resulted in the development of solar farms that are installed on water rather than land.

5.2 Questions

1. Define the terms *energy security* and *geothermal energy*.
2. Compare renewable and non-renewable energy resources. Include examples of each type of resource in your answer.
3. Explain the meaning of the concepts of 'long-term' and 'short-term' energy security.
4. Give the main causes of energy insecurity.
5. Explain why energy insecurity is a problem for a country.
6. Describe the strategies that a country can employ to achieve energy security.

CASE STUDY

Worlds apart: energy security at either end of the economic scale

Countries at different levels of development (HICs and LICs) face different challenges when determining the requirements for future energy security.

HICs have the finance, skills and long-term planning abilities to ensure that they put appropriate mechanisms in place to meet future demands. However, their high level of energy demand means that they may struggle when the supply of imported energy resources is interrupted. HICs are also making sigificant changes to their energy supply in order to meet carbon emission

Figure 5.41: The hydroelectric power dam across the Zambezi river with Kariba lake behind it in Zimbabwe, southern Africa. The dam is currently undergoing major repairs.

CONTINUED

commitments to combat climate change. If an HIC identifies a threat to energy security (such as a change to supply), it will typically investigate and then activate programs to minimise these threats in the long term.

LICs are often short on power even before threats to supply or future demands are considered. Lack of funds, skills and, in many cases, political inertia, results in the energy security problem becoming greater over time instead of being managed. The cost of changing from existing power supplies to renewable ones are often too high for an LIC to manage. As a result, little or no change occurs.

Examples of countries at either end of this developmental scale are Germany (HIC) and Zimbabwe (LIC). Their challenges and resulting risks are significantly different. A comparison of these challenges can be seen in Table 5.5.

Factors affecting energy security	LIC (Zimbabwe)	HIC (Germany)
Supply diversity (energy mix)	Limited energy options, declining HEP facilities and coal generated power. 80–90% of the population rely on wood fuel and kerosene for lighting and cooking. 100% of their oil supply is imported and 44% of their electricity is imported from Mozambique or South Africa.	Germany has a diverse energy supply mix, including fossil fuels, wind, nuclear power, solar power, biomass, biofuels and hydroelectric power. Diverse mix gives greater security: if one fails then there are other energy options to rely on.
Economic and political stability	Economically and politically unstable with limited foreign investment.	Politically and financially stable.
Population growth rates	High fertility rates with rapid population growth rates, partially offset by high emigration rates.	Low fertility rates with population decline.
Education	Decline in level of education within population due to high cost vs income ratio. Poverty limits access to secondary and tertiary education so skills required for further development limited.	High level of education, high level of skills available for ongoing development of energy supply and management.
Drought	Zimbabwe relies heavily on HEP power from Kariba Dam. In the past, drought has reduced the flow of water into the dam and significantly reduced power output from the Kariba power station (Figure 5.41).	Drought has not played a role in energy security issues in Germany. Diverse supply prevents such situations arising.
Rural/urban	Rural areas still have limited access to modern energy resources, with only 5.8% of rural households having access to modern energy sources.	All homes in Germany have access to modern energy sources.

CONTINUED

Factors affecting energy security	LIC (Zimbabwe)	HIC (Germany)
Power supply	In June 2012, shortages occurred of up to 30% of demand. 2020 saw power outages for up to 17 hours a day and significant increases in the price of electricty to the consumer.	Power supply constant. However, in times of uncertainty in imported energy supplies (like natural gas from Ukraine), Germany may have to resort to burning coal to meet demand. Coal power stations still exist in Germany but are no longer operating due to acid rain and associated carbon emission problems.
Price of energy	The government has increased the price of electricity to consumers to discourage usage and reduce pressure on power grid.	Energy prices in Germany are among the highest in Europe. Increased pressure on energy resources caused by Ukrainian conflict will continue to drive prices up. High prices also required to meet need to develop renewable resources. However, German household disposable income means that the population is able to afford it, generally.
Education	Lack of energy makes studying at night challenging for students and limits their learning time. Also, there is limited access to the internet, technology and therefore potential for remote learning.	Students are able to study at any time of day or night. Access to the internet and remote learning available across the population.
Farming	Lack of energy for sustained commercial farming. Food security is therefore also impacted.	Extensive energy available for food production.
Industry	Industrial development in decline as industry is not sustainable.	Ongoing industrial development.
Medical care	Hospitals and medical facilities running without reliable power are unable to offer more technical services that rely on energy supply.	High level of medical care available.
Poverty	Poverty increasing with more of population struggling to put food on table.	Low levels of poverty.
Conflict within the population	High levels of crime as survival becomes primary goal.	Low levels of crime.
Emigration rates	Large numbers leaving the country for South Africa in hope of finding work and better standards of living.	Immigration is high from countries with lower levels of development as people look to improve their standard of living.

5 Managing Resources

CONTINUED

Factors affecting energy security	LIC (Zimbabwe)	HIC (Germany)
Political stability	Political instability and high rates of corruption contribute to the decline of the economy and lack of access to secure energy supplies.	Political stability and long term planning well defined.

Table 5.1: Energy security in an LIC versus an HIC (Zimbabwe and Germany).

Table 5.1 shows a clear comparison between the two countries. It also demonstrates how energy insecurity hampers the development of an LIC on many fronts: it can education, medical care, farming, and industry. In contrast, a country with a robust power supply system can continue developing. It is more likely to have a skilled populace and a strong economy. However, even an HIC may have to resort to using energy supplies that are not optimal (such as coal) when imported energy streams (like natural gas) are unpredictable.

Case study question

Using the information given in Table 5.1, write a 500 word comparison of the causes and effects of energy insecurity in Germany and Zimbabwe.

5.3 Waste management

Waste management is the collection, transport, treatment and disposal waste.

Waste can be categorised based on the material it is made of, for example, paper, glass, metal, plastic, organic and hazardous waste. **Hazardous waste** includes radioactive, infectious, toxic, flammable and non-toxic waste.

Historically, not a lot of waste has been recycled. However, there has recently been a move towards changing how waste is managed. The main goal of modern waste management is to reduce, reuse and **recycle** in order to prevent waste going to landfill and creating potential environmental and health hazards.

The main advantage of effective waste management is a cleaner, healthier environment. Effective waste management lowers the risk to the environment. It also limits the spread of disease and the risk of poisoning the environment from toxic **waste streams**. However, as many of the natural resources that humans utilise are limited, an additional advantage of effective waste management is the conservation of resources through reusing, recycling and **upcycling**.

> **KEY TERMS**
>
> ***hazardous waste:** waste that has properties which make it dangerous or capable of harming the environment or human health
>
> ***waste stream:** the flow of specific types of waste from their source through to recovery, recycling or disposal
>
> **recycling:** the action or process of converting waste into reusable material. For example, glass is melted down and reused to form a new product

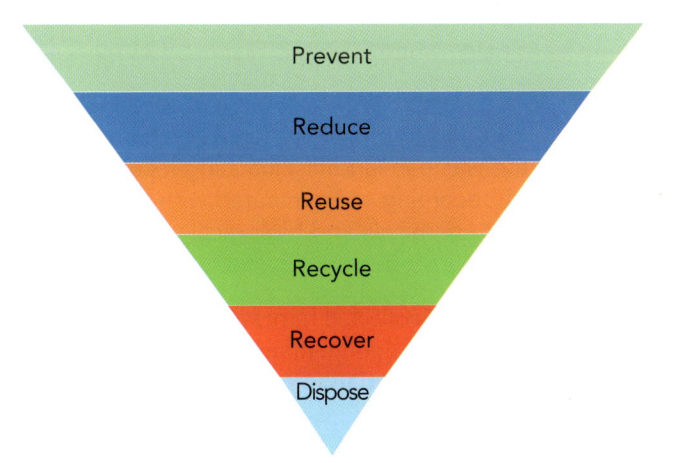

Figure 5.42: The waste management hierarchy. The most important aim, prevention of waste, is at the top.

CAMBRIDGE INTERNATIONAL AS LEVEL ENVIRONMENTAL MANAGEMENT: COURSEBOOK

> **KEY TERM**
>
> ***upcycling:** reusing a discarded item in such a way as to create a product of a higher quality than the original. For example, using materials from discarded plastic bottles to make new shoes

Methods of waste disposal and treatment

Various methods of waste disposal management are employed around the world. The method of disposal usually depends on the category of waste. Toxic waste needs to be treated very differently to non-toxic waste, while medical waste needs to be treated in a very particular way to ensure that it is safe. Some examples of different methods of waste disposal are outlined below.

Storage

Non-hazardous solid waste can be stored at the point of waste generation. For example, in large waste bins (or skips, Figure 5.43) or smaller storage bins/recycling bins for smaller volumes.

For the short- and long-term storage of hazardous waste (Figure 5.44), all storage containers must meet specified legal requirements for the type of waste they store. Adequate measures need to be taken to prevent accidental spills or leaks, or waste being blown away. Hazardous waste needs to be managed very carefully – stored in a container that does not react to the waste and which can be well sealed to fully contain it.

Figure 5.43: Short-term waste storage in a city street.

Figure 5.44: Hazardous waste storage with clear labelling and instructions.

Landfill

Landfill sites are among the most common waste-disposal methods currently employed around world. The process involves burying waste underground, then covering it up to minimise the health, visual and odour risks that unburied waste presents.

Landfill sites need to be designed and constructed to prevent ground and surface water pollution, to control the spread of diseases, and to prevent waste being blown around. They must be sealed to prevent **leachate** from polluting ground and surface water.

To prevent pollution, landfill sites must be divided into separate cells, which are sealed as they fill up. When a landfill cell is full, it is covered with layer of soil so that grass can be grown over the top to stabilise area and prevent erosion.

> **KEY TERMS**
>
> **landfill:** a place where waste is disposed of by burying it
>
> **leachate:** a typically acidic fluid that has filtered through the waste in landfills; leaching results in the fluid becoming contaminated with heavy metals, toxic chemicals and biological waste

Waste water must be collected to prevent water pollution, and the CO_2 and methane gas given off by the landfill should be captured to prevent explosions and atmospheric pollution.

5 Managing Resources

Figure 5.45: A waste compactor working on an landfill site.

Figure 5.46: An engineered and well-managed landfill site on the island of Hawaii along the Kona Kohala Coast.

Recycling

An important part of recycling is the separation of waste, which involves different waste categories being separated into different storage containers. Recycling bins should be clearly labelled to indicate the waste that should go in each bin (Figure 5.47).

Figure 5.47: Recycling bins at an event centre, clearly labelled to show the type of waste that should be placed in them.

Incineration

Waste **incineration** (Figure 5.48) involves the controlled burning of waste. It turns waste into heat, ash and gases. This process uses high temperatures that destroy many pathogens and some toxic materials. Used in the production of electricity, it reduces the mass of waste by up to 90%. In addition, it limits the volume of material needing to go to landfill. Smaller countries with limited space for landfill may be more inclined to utilise this method of waste management.

Figure 5.48: The Belvedere Riverside Resource recovery plant, which recycles waste from various parts of London in the UK, and generates electricity for the city.

> **KEY TERM**
>
> **incineration:** the process of burning materials

Disposal at sea

Historically, oceans have been used for waste disposal, including radioactive waste, rubbish, septic waste, industrial waste and contaminated soil. Today, most material intentionally disposed of at sea are uncontaminated sediments, human remains and fish waste. However, it is internationally illegal to dispose of radioactive waste, chemicals, biological warfare agents, synthetic materials (e.g. plastics), sewage sludge, medical waste, industrial waste, hydrocarbons or materials with toxic heavy metals at sea.

Figure 5.49: Disposal of waste via a pipeline into the ocean.

Exporting waste

The global trade in waste is the export of waste from one country to another, which is driven by economics and financial profit. The exporting country pays the importing country to take its waste. The argument in support of global waste trade rests on the idea that LICs need the money to develop economically. Exported waste includes:

- hazardous materials
- household waste
- recyclable waste
- e-waste (electronic waste)

Figure 5.50: A worker inspects household rubbish in an export container that was returned to Britain from Brazil.

ACTIVITY 5.7

Learner awareness of solid waste management

1. In small groups, create a questionnaire to find out how much students at your school know about waste recycling. Your questionnaire should aim to find out:

   a   If students know what solid waste management is

   b   If students understand what recycling is

   c   If students can correctly identify types of waste that are recyclable

   d   If students recycle at home and, if so, what type of waste they recycle

2. Look at the questionnaires your classmates have written, and create a single questionnaire using the best questions. Get between 20 and 40 learners to complete the class questionnaire. Analyse your answers to calculate the percentage of students that know, partially know, or do not know about solid waste management.

3. Use this data to create graphs that represent the answers to the questions you set.

4. Discuss your findings in small groups or as a class.

5. Is there a need for students in your school to be educated on how to manage waste? Think about how this could be done.

KEY TERM

e-waste: electronic waste

The impact of waste disposal

The rate at which humans are generating waste is unsustainable. Even when waste is correctly disposed of, it is taking up valuable space and costs money. Waste that is not biodegradable or easy to recycle takes a long time to break down and is harmful the environment if it is not managed safely.

5 Managing Resources

The impacts of waste on human and environmental health vary depending on the type, scale and length of time involved. Some waste will rot and biodegrade, eventually returning to the nutrient cycle. Some waste streams are more harmful, remaining in the environment for long periods of time, causing significant damage. Waste disposal methods can result a variety of impacts.

Soil contamination

Mismanaged waste on land can result in contaminated soils. Liquid waste, such as oils, toxic chemicals and heavy metals can infiltrate soil and be washed deeper into the soil profile through leaching, contaminating both soils and ground water (Figure 5.51). Contaminated soils can kill vegetation and result in the loss of habitat of species that live in the area. Contaminated groundwater can surface, either through springs or wells, contaminating surface waters and killing aquatic organisms and other species that rely on the water source.

Figure 5.52: A landfill site in Chennai, India, burns after hot weather and gas build up resulted in the waste catching on fire.

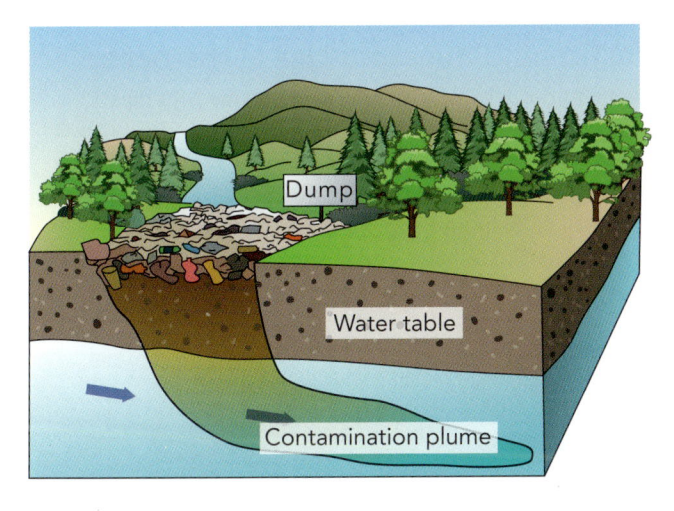

Figure 5.51: Groundwater pollution. Contaminants in the ground form a plume.

Dangerous gases

During the degradation of biological waste, greenhouse gases such as methane (CH_4) and CO_2 are released into the environment. These gases are potentially explosive and can lead to increased carbon-based greenhouse gases in the atmosphere and, in turn, global warming (see Chapter 8 for a discussion on climate change).

Visual, noise, smell and spread of diseases

In addition to the release of gases, waste creates visual, noise and smell pollution. Solid waste left unmanaged becomes unpleasant. If it contains organic or chemical materials, it smells and can attract vermin. This can result in the spread of parasitic, bacterial and other infectious diseases. Toxic waste can cause illnesses including cancer, lung conditions and skin conditions. Where septic waste is included in the waste stream, diseases such as cholera can spread rapidly among the populations.

Figure 5.53: Surface water pollution, with piles of plastic waste on the banks of a river.

Release of toxic substances

Toxic waste such as heavy metals or radioactive materials can impact the environment in different ways:

- Leachate that has percolated through waste in landfills can contain toxic chemicals and heavy metals. If not managed appropriately, this leachate can enter soil and then ground water, causing contamination. Living organisms exposed to this leachate can become ill and die.

- Incineration of waste causes problems, as many of the materials burned (such as plastics) produce toxic substances. Gases that are released may also cause air pollution and contribute to global warming and acid rain.

Figure 5.54: An oil spill washing up on beaches in southern California, poisoning both water and the surrounding land.

Bioaccumulation and biomagnification

Bioaccumulation refers to the build-up of toxins in the body of an organism, while **biomagnification** is the build-up of a toxin in an ecosystem's food-chain.

An example of bioaccumulation is the build-up of mercury in the tissues of marine organisms such as fish. The release of heavy metals through activities such as mining results in mercury entering the food chain. Krill absorb small amounts of mercury.

Krill are eaten by salmon, in which mercury continues to build up. Over time, tuna consume smaller fish with mercury in their systems, and accumulate the metal in their own muscles (Figure 5.55). This continues along the food chain. Mercury also bioaccumulates in humans, and humans eat tuna. In this way, pollution could ultimately contaminate the humans.

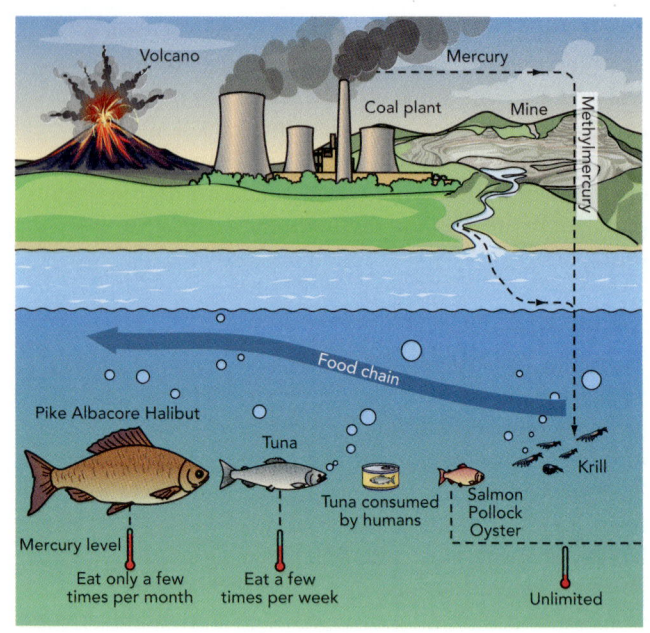

Figure 5.55: Bioaccumulation of toxins in the food chain.

Plastics and microplastics in the ocean

Plastic and microplastics all harm marine life. Plastic pollution has a direct impact on marine wildlife, with seabirds, turtles, seals, whales and other marine animals regularly impacted. Sea turtles mistake floating plastic bags for jellyfish, and can choke or sustain internal injuries, die or starve from consuming plastic. Seabirds are also often found dead with plastic in their stomachs.

Figure 5.56: Plastic waste and other waste floating on the surface of the ocean.

Over time, plastic in the oceans breaks down as a result of moving water and exposure to sunlight. This creates smaller pieces of plastic that are known as **microplastics** and can be extremely dangerous to marine wildlife because they are small enough to swallow. These plastic particles contain harmful chemicals that can increase the risk of disease and decrease fertility. After ingesting microplastics (Figure 5.57), marine organisms such as seals and dolphins may suffer ill-health for a long time prior to death.

> ### KEY TERMS
>
> **bioaccumulation:** the buildup of a toxin in the body of an organism
>
> **biomagnification:** the buildup of a toxin in a food chain, e.g. the concentration of mercury increases up the food chain as each consumer eats organisms that have mercury in their tissues
>
> **microplastics:** extremely small pieces of plastic waste in the environment. This results from discarded plastic breaking down into very small fragments

Humans consume fish as part of a healthy diet. However, the build-up of microplastics in fish stocks has resulted in people consuming increasing amounts of microplastics.

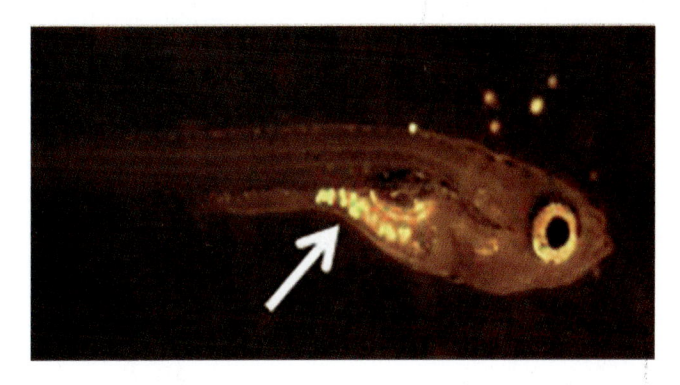

Figure 5.57: Microplastics in fish fry (baby fish). The fish fry ingest microplastics. They then die from starvation as they are unable to eat or digest food.

INVESTIGATIVE SKILLS 5.2

Primary micro-plastic investigation

The aim of this investigation is to find out what microplastics exist in our day-to-day products. In order to do this, you will need to isolate microplastics from cosmetic products using filter paper and hot water.

You will need:

- An assortment of cosmetics (that do not claim to be micro-plastic free). Facial scrubs and makeup containing glitter are products which are likely to show visible microplastics.
- Filter paper.
- A funnel.
- A beaker.
- A clamp and stand to hold the funnel.
- Hot water.

Safety

- Take care working with glassware.
- Boiling water can burn your skin, so be careful when pouring it over the samples.

Figure 5.58: The setup for the beaker, funnel and stand for this investigation.

Getting started

1. Look at the samples you are going to test. Which ones do you think you will find microplastics in?
2. Why do you think microplastics are used in the products that you are testing?

> CAMBRIDGE INTERNATIONAL AS LEVEL ENVIRONMENTAL MANAGEMENT: COURSEBOOK

CONTINUED

Method

- Place the funnel into the clamp. Place this over a beaker large enough to catch the water as it filters through.
- Place the filter paper into the funnel and add a sample of some of the cosmetic being tested.
- Slowly pour hot water over the cosmetic, giving it time to dissolve. Take care with the hot water.
- Check what has been left behind in the filter. Are there microplastics present? Samples can be placed under a microscope to get a closer look at what has been left behind.
- Different teams can test different cosmetic products to observe how microplastics can vary between brands.

Questions

1. What did you observe?
2. Did different cosmetics have different quantities of microplastics? Did the microplastics vary in size?
3. These cosmetics are washed down the drain every day. What implications does this have for the environment?

Strategies for reducing the impact of waste disposal

Rapid industrialisation, economic progress and globalisation have resulted in a consumer society with a lot of throwaway or one-use products. Products are wrapped in plastic to maximise shelf life. The latest model of phone, car or computer is designed to be passed on or discarded for a newer model when it has outlived its usefulness. To combat the impacts of waste creation, waste management strategies need to be employed at town, city, government and global level.

The waste management hierarchy (minimisation, recovery, transformation and disposal) is the strategy adopted by most developed nations to manage their waste and reduce the impact of waste streams.

Some countries have developed waste management to the point where the public is fully involved and active in the management of their waste streams. For example, in Singapore littering in the street carries a fine and/or community service (the perpetrator will have to clean a public place). This strategy is driven by two factors: national pride and also the very limited space available for landfill. Singapore is an example of how well the impact of waste can be managed.

Reduce, Reuse, Recycle

Waste minimisation is the implementation of practices designed to reduce the amount of waste produced. Waste minimisation should be at the top of the solid waste management hierarchy (Figure 5.42). It consists of two main parts: source reduction and recycling.

In source reduction, the amount of waste that needs to be disposed of is reduced. A waste reduction programme will vary from source to source. In a school, the main waste stream may be paper, while the main waste streams from a restaurant are more likely to be food, packaging and glass waste.

An example of waste minimisation is the 'naked produce' for sale in some shops (Figure 5.59). Naked produce is produce without packaging or labels. This limits the amount of non-recyclable waste created when selling a product.

The three Rs – reduce, reuse, recycle – all fall within the waste-minimisation category. These practices are designed to limit the volume of waste being produced by any one source. The first steps in the RRR cycle are the establishment of education programs to ensure that the population understands how to manage their waste. At the same time, the country needs to develop an infrastructure that can manage the recycled waste. Unless the country has the means to transport and turn that waste into a new product, this waste stream will remain a problem.

5 Managing Resources

Figure 5.59: Naked produce: fruit and vegetable products that have not been wrapped in plastic or other packaging.

Figure 5.60: An example of biodegradable plastic.

Biodegradable plastics

Biodegradable plastics are made from materials that will break down when exposed to micro-organisms. Eco-friendly biodegradable plastics break down faster than other plastics, and the components they break down into (CO_2, water vapour, organic material) are not toxic to the environment. However, many bio-plastics need aerobic conditions to break down effectively. This means it takes them longer to break down in oceans and in landfill.

Food waste for animal feed

Animal food waste ranges from crop residue to waste from restaurant kitchens. When farmers have harvested their crops, they are able to use the leftovers to feed their livestock. However, to effectively use fresh food-waste for animal feed, the waste needs to be processed. This means converting it into animal food that has a shelf life and is free of pathogens. This waste management option minimises the waste going to landfill, limits the risk of spread of disease due to rotting food, and ensures that valuable waste is being effectively used. Here, waste is being viewed as a product stream rather than a waste stream.

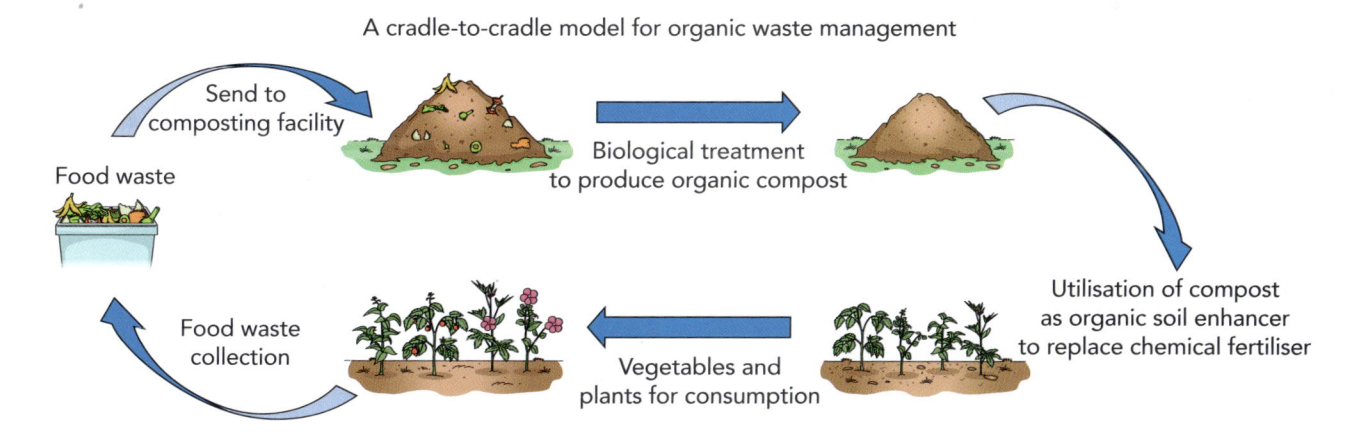

A cradle-to-cradle model for organic waste management

Figure 5.61: Food waste, composting and food production form a continuous sustainable system to reduce waste and maximise food production.

By converting food waste into safe and nutritious animal feed, the issues of waste management, food security, resource conservation, pollution and climate-change are all being addressed. This ultimately makes livestock farming more sustainable.

Composting

Composting is a natural process which involves the decomposition of organic materials into a nutrient rich product that is used to feed soils and improve farming productivity. Vegetable food waste and garden waste are two waste streams that can be used in the composting process.

Composting can be carried out at on a large or small scale. Material can be composted in an urban garden's compost pit. Organic waste can also be composted through a city-wide garden and vegetable waste collection system that composts on a commercial scale.

Fermentation

Waste **fermentation** is an anaerobic (oxygen free) process. It allows for the conversion of biological waste into sources of energy such as ethanol and biogas. The process is carried out using a bio-digester. Bio-digesters can be small scale and used to manage biological waste on a small farm. They can also be used on a large scale, for example, to manage biological waste from a town.

The biogas produced in the waste fermentation process has energy potential as it can be used for generating electricity, heating and refrigeration. The remains of the fermented materials can be recycled and used in composting. The equipment required for fermentation takes up relatively little space. It is also a closed-cycle system which significantly reduces odours.

Figure 5.62: A biogas production plant near Warton, Lancashire, UK. The plant uses excess grass from surrounding farms and produces methane.

> **KEY TERMS**
>
> **composting:** decomposition of biotic/organic material that can be used as a fertiliser for plant growth
>
> **fermentation:** the chemical breakdown of substances by yeast or bacteria anaerobically to create an alcohol and biogas

Education

For waste management to be effective, a population first needs to understand the benefits of effective waste management procedures. Education gives them the tools to understand the decisions they are making.

The outreach and marketing departments of waste management groups encourage the population to change their behaviour on a specific issue. With waste management (recycling, preventing dumping, waste minimisation), this education process needs to be ongoing to be effective. The education of consumers helps to increase awareness. It also encourages the use of zero-waste shops, where customers can reuse glass bottles and other containers to minimise waste (Figure 5.63).

Town management can play a vital role in educating a population about waste. Signage, public education campaigns and making waste recycling and safe disposal as easy as possible for those who do not have direct access to waste management systems are all good examples of this.

Figure 5.63: A customer refilling shower gel into a reusable glass bottle in a zero-waste store.

Financial incentives

Financial incentives are programs where people earn money, or gain money back, by selling their waste. In this system, waste has value. This is already common practice globally for waste streams such as aluminium and other metals that have a relatively high value. For example, in some countries, a small amount of money will be provided in exchange for empty cans or tins. The practice means that people will recycle these metals. Once plastic and glass bottles hold a value, people are less likely to throw them away.

Norway is a world leader in recycling. The country offers a 97% recovery rate on all plastic bottle waste. Norway has implemented a loan scheme. When the customer purchases the product, the bottle does not belong to them: they are borrowing it for a small fee (US13–30 cents). The fee can be recovered by taking the plastic bottle to a reverse vending machine that gives the money back when the bottle is put in. They can also be returned to gas stations and small shops for credit or cash. Up to 92% of the plastic recycled can be reused up to 50 times.

ACTIVITY 5.8

Event waste minimisation

1 Work in teams of 2–4. You are the event manager for a golf tournament. You want the tournament to achieve zero waste, making sure no waste from the event goes to a landfill. Your team has put together a waste separation system. How can you manage the following items of separated waste?

   a   Organic waste from the kitchens (scraps)

   b   Unused food (safe for consumption)

   c   Signage advertising the event

   d   Building materials

   e   Used golf balls

   f   Glass, tin, paper, cardboard and plastic waste

   g   Grey water (not sewage water) from the kitchens

   h   Non-recyclable waste

2 Present your decisions to the class, giving reasons for your choices

3 As a class, discuss how your plans differ. Is one idea easier to implement than another?

Figure 5.64: A commuter trades plastic bottles for transit credit, saving money on bus and train travel costs.

Legislation

Laws regarding waste management can be complex. They can begin with the local town and the by-laws that apply to a local area. For example, Seattle in the USA has specific composting laws, while New York has different ones. Laws need to comply with the local provincial/state laws for waste management as a minimum. These local laws are governed by the laws of the government. In some instances, there are international laws that regulate the management of specified waste streams.

Globalisation has resulted in waste being moved around the planet. As a result, laws have been put in place to limit or manage the harm being done to environments far from where waste is produced. Such laws make waste producers responsible for the full, correct disposal of their waste. However, there is no single global law that governs waste management, reduction and transborder flow.

Marine legislation

The Marine Protection, Research and Sanctuaries Act (**MPRSA**), 1972, is a US act which regulates the dumping of any materials which have a negative impact on human health and welfare or the marine environment. The MPRSA led to the implementation of one of the first international agreements for the protection of the marine environment from human activities, the London Convention (1972).

The London Convention is a globally applied international agreement which protects the marine environment from pollution through dumping of waste and other matter into the oceans. The signatories agreed to control waste dumping in the oceans, promote pollution prevention and prevent the dumping of specific hazardous materials. The agreement also promoted better waste management strategies.

The London Protocol (2006) was developed from the London Convention and was designed to be more protective of marine ecosystems. It specifically bans the incineration of waste at sea, or the export of waste for the purpose of ocean dumping. By 2019, 53 countries had signed the London Protocol.

Hazardous waste legislation

The Basel Convention (1989) was designed to control the international movement of hazardous waste and its disposal following the discovery that HICs were offloading their hazardous waste in LICs. The Basel Convention aims to protect human health and the environment from the harmful effects of hazardous waste. Its key aims are to reduce hazardous waste generation, promote environmentally sound management of the waste, restrict transboundary movement of hazardous waste (except where it is being moved to be treated in an environmentally sound manner) and regulation of the transboundary movement of the waste.

However, the convention does not precisely define 'hazardous waste'. This means that e-waste, which causes significant environmental harm, may or may not be considered under the convention. This limits the effectiveness of the convention.

The Stockholm Convention on Persistent Organic Pollutants (2001) is an international treaty. Its purpose is to eliminate or restrict the production and use of organic pollutants that do not break down quickly in the environment, or which result in bioaccumulation in an ecosystem. The convention outlawed nine of the most harmful persistent organic pollutants (POPs), one of which is DDT (which was widely used in the control of mosquitoes that carry malaria). However, the list of banned chemicals has grown over the years. The treaty also aims to prevent the production of dioxins and furans and ensures that the toxic waste of POPs is disposed of in an environmentally responsible manner.

Figure 5.65: Workers wearing full body protective clothing while working with the asbestos roof tiles, which are a hazardous material.

5.3 Questions

1. Explain the meaning of the terms *waste management* and *leachate*.
2. List five waste management strategies.
3. Explain how the waste management hierarchy helps to minimise waste.
4. Briefly describe the challenges posed by landfill sites.
5. Energy is a by-product of an incineration plant. Explain how an incineration plant produces electricity.

5 Managing Resources

EXTENDED CASE STUDY

Global food security during the COVID-19 pandemic

SARS-CoV-2, or COVID-19, a coronavirus that causes an infection of the throat, nose and lungs, broke out in late 2019. Due to the nature of modern day travel, the virus spread quickly. This particular strain of coronavirus caused severe symptoms and was transmitted person to person. The rapid spread and the mode of transmission resulted in global lockdowns, with people having to stay home and isolate from each other. This was done in an attempt to stop the spread of the disease.

The impacts of this lockdown were far-reaching. Many industries closed, including those associated with food production, manufacturing, refrigeration and transport. However, it was not just the shortage of supplies that was a problem. With many people out of work, there was also a shortage of money to purchase food; the financial impact on many was severe. COVID-19 increased global food insecurity in almost every country.

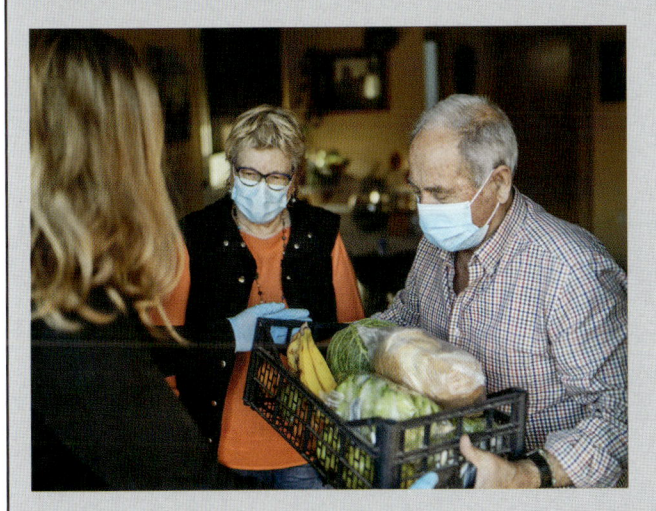

Figure 5.66: Food deliveries to people unable to leave their home during the COVID-19 pandemic.

The causes of food insecurity during the COVID-19 pandemic

Lockdowns around the world had a significant impact on food systems. Major disruptions to food supply chains triggered a global health crisis. The following list contains examples of how the food supply chains were impacted:

- A fear of food insecurity resulted in the stockpiling of food products by those who had the financial means to do so. This limited the amount of food available to the general population.

- The production of food on farms in some instances declined as workers were unable to get to work due to illness or travel bans.

- Where food was being produced on farms, there was difficulty getting it to market. Transport systems were no longer functioning as truck drivers had to stay at home.

- Factories where food is normally processed closed as staff stayed home. This resulted in a decrease in the production of food goods normally found on supermarket shelves. Products like flour and canned goods were in high demand.

- There was a decrease in crop harvest in farms which employed migrant farm workers. This was because workers were unable to travel across borders due to the lockdown regulations that were in place.

- Grocery store staff were unable to get into work due to enforced isolation, and with supply chains failing, the shelves were often empty. Goods that were available increased in value.

- Restaurants had to close due to the risk of COVID-19 spreading. This led to a sharp

CONTINUED

decline in the demand for perishable foods such as milk, potatoes and fresh fruit. This added to farmers' challenges, as the demand for their products declined.

- The movement of food between countries and continents slowed down. Borders were closed and shipping lines limited. This hit farmers, who rely on export, especially hard. It also meant that food products normally found on the shelves around the world were not available.

- Some exporters of staple goods (main food types, like rice and wheat) also imposed restrictions on the export of those products to increase their own stores in the event of a shortage. This resulted in a global shortage of these products and an increase in the prices of crops like wheat and rice.

- The global economic slowdown also resulted in families having lower incomes, while the cost of food increased. Some foods became too expensive for many to buy. The poorest and most vulnerable people faced the greatest risk of starvation.

The impact of food insecurity during the pandemic

Global food insecurity impacted the world's population in a variety of ways:

- Poverty increased and social inequality widened. The poorest people struggled to access basic needs such as food, water, healthcare and access to jobs.

- Malnutrition and starvation (Figure 5.67). There was a steady decline in instances of malnutrition and starvation between 2005 and 2017, but between 2017 and 2019 there was a slight increase. During the pandemic, the numbers of undernourished people increased significantly.

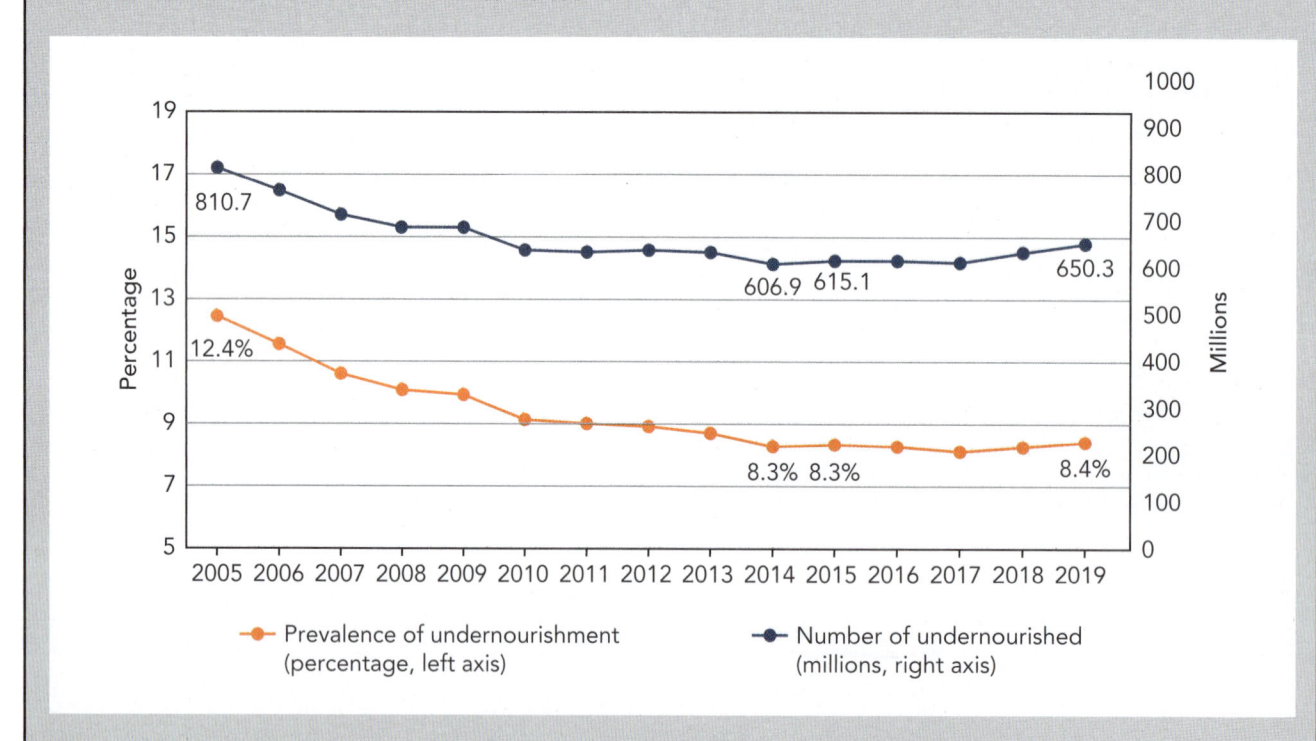

Figure 5.67: The number of undernourished people in the world 2005–2019.

CONTINUED

- As food demand has decreased due to the increase in cost, there is a smaller market for farmers to sell their produce to. It is estimated that 35% of the global workforce (an estimated 450 million people) that worked in the food supply chain lost their jobs due to lower demand.
- Many feeding programmes, such as school lunch schemes, shut down during COVID-19. Where feeding programmes sent food supplies to schools and impoverished communities, many lost access to the food they relied on. With schools shut, the most vulnerable children no longer had access to this vital source of nutrition.

Solutions to food insecurity during the pandemic

- Governments that responded quickly to the emerging food crisis, by identifying problems in production, transport and delivery, were able to develop alternative plans and move food along different routes. Their actions lessened the impact of the problem.
- People stopped relying on restaurants and started producing meals for themselves. This reduced both cost and waste.
- Countries reduced the export of food, keeping those resources for their own populations. This had an impact on a global consumers as supplies decreased. However, it helped to meet local demand.
- In some countries, governments put systems in place to ensure that social protection programmes got food to the people who most needed it.
- Crops were no longer exported for biofuel. Instead, they became a local food supply.
- Staple crops were grown in place of luxury crops to meet the most basic food demands.
- Food systems were diversified. What was grown, how it was distributed and whether it met the basic needs for both local and export markets were all questions that were considered.

Despite these measures, some governments failed to effectively manage the overwhelming crisis that COVID-19 caused. Their actions resulted in an increase in the number of people undernourished globallies. In the years following the pandemic, these effects are likely to continue to be felt as food industies and economies recover.

However, lessons have been learnt. Governments, communities and individuals are more aware of the challenges being faced. There has been a realisation that local solutions need to be used, rather than over-relying on international help. Partnerships between the private sector, academics, governments and banks will develop effective and flexible responses to challenges of this scale.

Extended questions

1. Consider the combined impact of COVID-19 and a country at war. What impact would this have on the food security of the population?
2. 'Subsistence farmers were less likely to be impacted by food insecurity during the COVID-19 pandemic.' How far do you agree? Give reasons for your answer.
3. Figure 5.68 is a pie chart showing the number of people around the world who were hungry in 2010.

    a. Copy and complete the pie chart using the data in the table.

Region	% of people hungry	Number of people hungry
HICs	2.05	19 million
North East and North Africa	4.00	37 million
Latin America	5.72	53 million
Sub-Saharan Africa	25.81	
Asia	62.42	578 million
Total	100.00	

CONTINUED

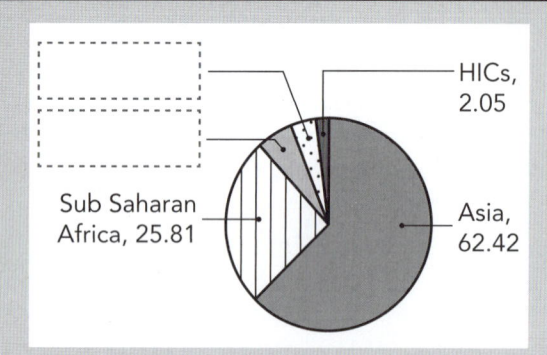

Figure 5.68: The percentage of people hungry in 2010 by region.

(Pie chart showing: HICs, 2.05; Sub Saharan Africa, 25.81; Asia, 62.42)

b Complete the labels in the pie chart you have copied to show the regions and number of people that were going hungry in each section.

c Calculate the number of people who were going hungry in sub-Saharan Africa.

d Calculate the total number of people who were going hungry in 2010.

e With reference to Figure 5.68, describe the distribution of people going hungry by region.

Extended case study project

In pairs, investigate the impact of the COVID-19 pandemic on food security in both an LIC and an HIC country of your choice. Compare the impacts of food security on the two countries. Consider factors such as:

- Why food security declined. Include ideas around production, transportation, retail shutdown and loss of jobs.
- Who was most affected? Consider both financial challenges and illegal or undocumented immigrants in your discussion.
- What solutions were put in place to fight the problems faced. Were these solutions effective?

Produce a large, poster-sized table that compares your findings for the two countries.

SUMMARY

Food insecurity is linked to climate, soils, population growth, economic limitations, incorrect farming methods, biofuels, political unrest, war and natural disasters.
Food insecurity results in malnutrition, poor health, higher risk of the spread of disease, low levels of productivity and the increased risk of poverty.
Food insecurity can be tackled by using technology, employing good farming techniques and reducing wastage.
Energy security is the uninterrupted availability of affordable energy. There are both long term and short term energy requirements for energy security.
Energy insecurity is caused by population growth, location of resources, fossil fuel depletion, supply disruption, climate change and increased industrial energy demand.
The impact of energy insecurity includes poverty, health issues, economic limitations, civil unrest and conflict, environmental decline.
Waste management is the collection, transport, and disposal of waste streams. It includes hazardous, non-hazardous and biodegradable waste.

5 Managing Resources

CONTINUED

Waste can be disposed of through storage, landfill, incineration, disposal at sea, and exporting waste.
The potential impact of waste disposal includes loss of habitat and biodiversity, pollution and the spread of disease.
The Basel Convention (1989) dealt with hazardous waste management requirements, while the Stockholm Convention (2001) considered the management of persistent organic pollutants.

REFLECTION

Think about what you have learned in this chapter. What was of more significance to you: gaining knowledge or how you learned that knowledge?

Did you enjoy one topic more than another? Consider the reasons for this. Was it the content or was it the style of learning that you adopted? Could a change of approach increase your enjoyment across all topics?

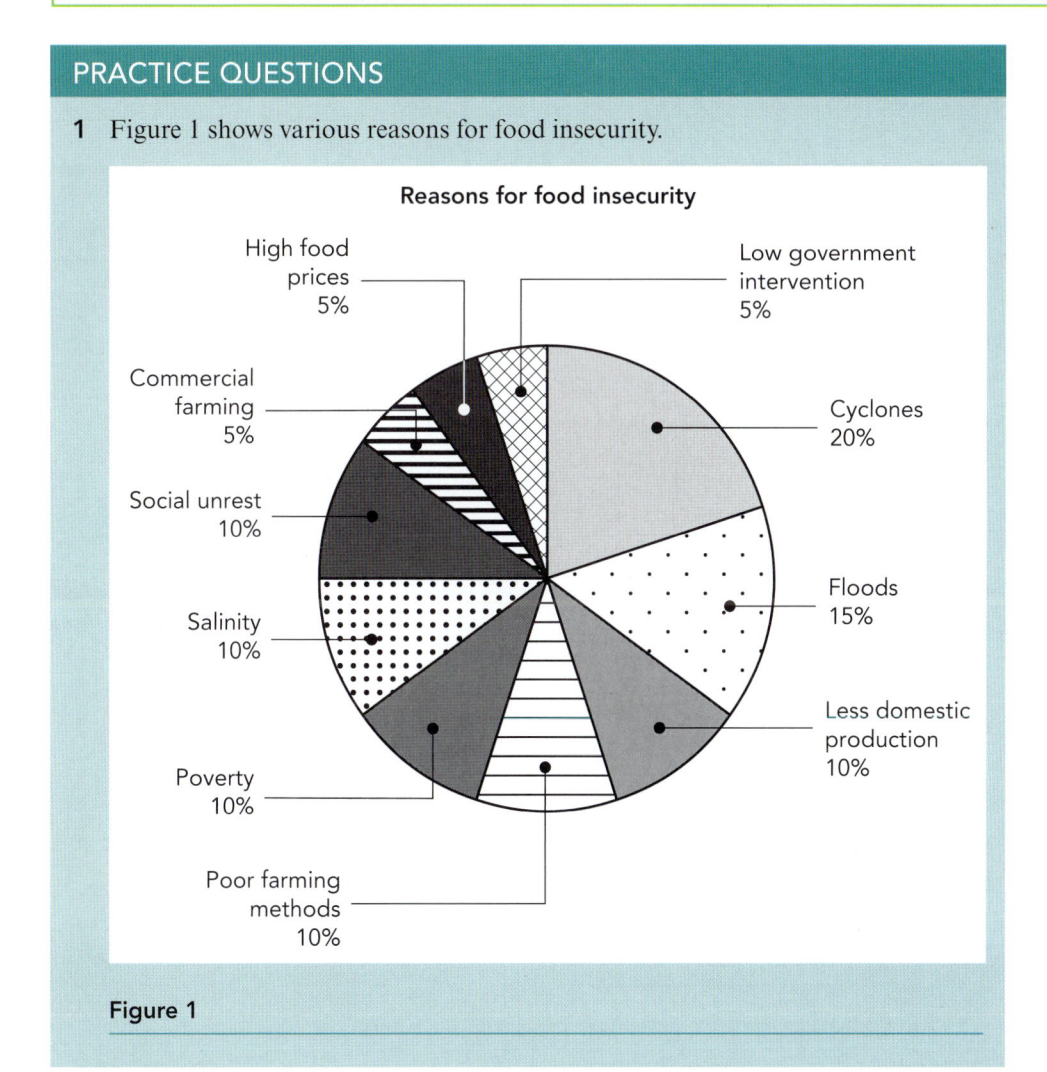

PRACTICE QUESTIONS

1 Figure 1 shows various reasons for food insecurity.

Reasons for food insecurity

- High food prices 5%
- Low government intervention 5%
- Commercial farming 5%
- Cyclones 20%
- Social unrest 10%
- Floods 15%
- Salinity 10%
- Less domestic production 10%
- Poverty 10%
- Poor farming methods 10%

Figure 1

> **CONTINUED**
>
> One reason for food insecurity is the high price of food. This makes it difficult for less financially stable people to purchase food.
>
> Explain **two** other reasons for food insecurity. [4]
>
> 2 Explain how the production of biofuels impacts food security in LICs. [4]
>
> 3 Many subsistence farmers in sub-Saharan Africa live in poverty. Climate change and the loss of pollinators have resulted in a decline in food production. Read the report about the methods which farmers in Uganda are using to increase their food output and productivity.
>
> > ## Beekeeping to alleviate poverty in poor communities
> >
> > Beekeeping has been promoted as a potential way to reduce poverty in rural sub-Saharan Africa. In Uganda it was found that beekeepers have tended to take up beekeeping after coming into contact with non-government organisations.
> >
> > Bees help the farmer to pollinate their crops. However, honey is also a source of nutrition and income for poor families, and has medicinal value too. To be successful, the beekeepers need training, protective equipment such as suits, gloves and smokers, honey spinners and containers for the honey. These are critical tools as they allow the farmer to work safely with the bees and harvest the honey.
> >
> > Uganda is licensed to export honey to the European Union. Despite this, it has failed to meet either the export quota or the domestic demand for honey due to a weak uptake of beekeeping by farmers.
>
> a Describe the role that bees play in increasing crop production for a farmer. [2]
>
> b Suggest how diversifying and farming bees helps a farmer become more sustainable. [2]
>
> c Explain why farmers in rural Africa might be unwilling to start beekeeping. [3]
>
> 4 Energy insecurity impacts low-income countries (LICs).
>
> Outline **three** causes of energy insecurity. [6]

CONTINUED

5 Figure 2 shows the main export/import flows of liquid natural gas around the planet.

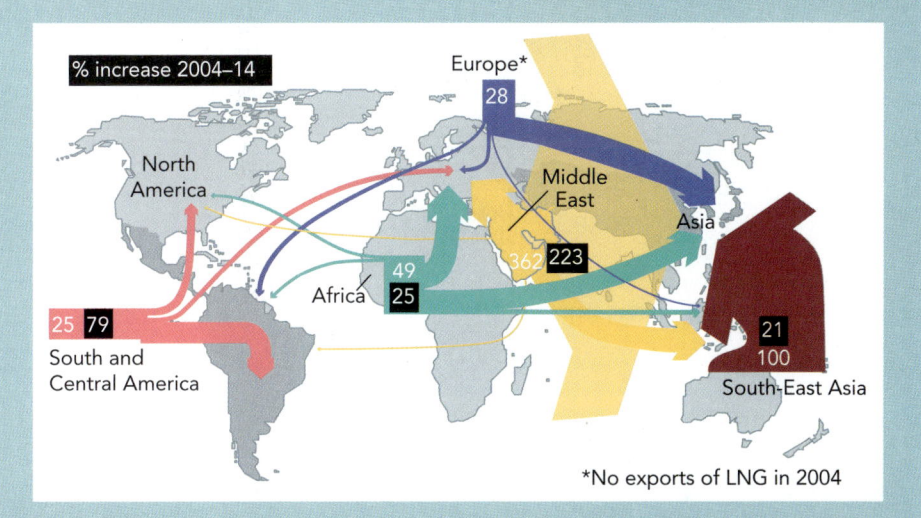

Figure 2

a State which region showed the largest percentage increase in the export of liquid natural gas between 2004 and 2014. [1]

b State which region imports the most natural gas. [1]

c Explain how a decrease in the availability of natural gas from the Middle East, Europe and Southeast Asia would impact the energy security of Asia. [3]

d Describe some strategies that Asia could employ to reduce its reliance on liquid natural gas from international sources and ensure better energy security. [6]

6 Pollution can be categorised as point source and non-point source.

Compare the terms: Point source and non-point source pollution. [2]

CONTINUED

7 Figure 3 shows the flow of pollution underground from a ruptured underground storage tank at a factory.

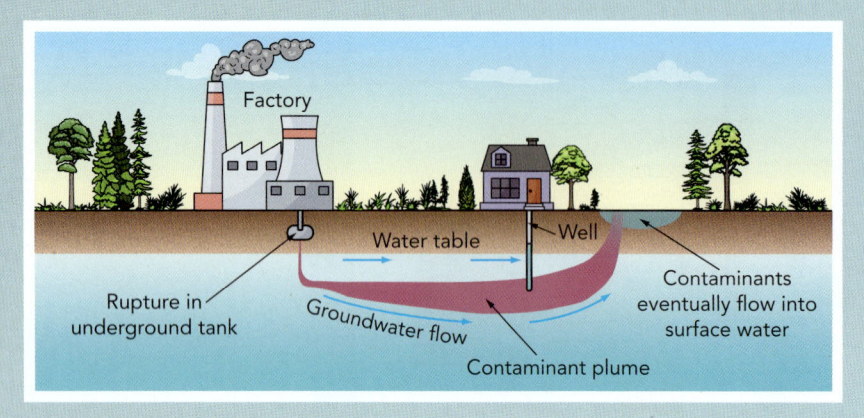

Figure 3

a Discuss the impact of the flow of pollution within the ground water on downstream users. [6]

b Explain what steps could have been taken to prevent an event like this happening. [4]

8 'Human population growth is the main reason for food insecurity globally' [20]

To what extent do you agree with this statement?

Give reasons and include information from relevant examples to support your answer.

9 Evaluate the success of strategies employed to manage energy security in a location of your choice. [20]

Give reasons and include information from relevant examples to support your answer.

10 'Toxic waste is the most environmentally damaging pollutant'. [20]

To what extent do you agree with this statement?

Give reasons and include information from relevant examples to support your answer.

5 Managing Resources

SELF-EVALUATION CHECKLIST

After studying this chapter, think about how confident you are with the different topics.
This will help you to see any gaps in your knowledge and help you to learn more effectively.

I can	Needs more work	Getting there	Confident to move on	See Section
Define food security.				5.1
Describe and explain causes of food insecurity.				5.1
Describe and explain threats to food security.				5.1
Outline the impacts of food insecurity.				5.1
Describe and evaluate strategies for managing food security.				5.1
Classify energy resources as renewable and non-renewable.				5.2
Define energy security.				5.2
Describe and explain the causes of energy insecurity.				5.2
Outline the impacts of energy insecurity.				5.2
Describe and evaluate strategies for managing energy security.				5.2
Describe methods of waste disposal and treatment.				5.3
Explain the impact of waste disposal methods.				5.3
Describe and evaluate strategies to reduce the impact of waste disposal.				5.3
Compare and contrast the impacts of energy insecurity in HIC and LIC.				5.3

> Chapter 6
Managing Water Supplies

LEARNING INTENTIONS

In this chapter you will:

- find out about the distribution of the water on Earth
- explore what is meant by the term 'water security'
- identify the causes and impacts of water insecurity
- evaluate the strategies that can be used for managing water security.

6 Managing Water Supplies

> **GETTING STARTED**
>
> In pairs, discuss the following:
>
> 1 What impact does a lack of water have on a region?
>
> 2 How might a pollution event upstream in a river affect people or organisms downstream?
>
>   Consider the scale and duration of the event in your discussion.
>
> 3 Why is groundwater such an important water resource?
>
> 4 What water saving strategies do you use, or could you use, to reduce water consumption in your house?
>
> As a class, discuss the distribution of water around the planet. How does that impact people/organisms in different regions?

ENVIRONMENTAL MANAGEMENT IN CONTEXT

Water in crisis: Cape Town, South Africa, 2018

The water we drink today has been on Earth in one form or another for millions of years. The water cycle ensures that freshwater stores remain relatively constant over time. However, in recent years, human population increases, industrialisation, pollution and climate change have caused water supplies to come under pressure.

In 2017–18, Cape Town, South Africa, faced extreme drought conditions due to three years of unusually low rainfall. It was a rare and severe climate event, with studies suggesting that a drought like that occurs only once in 300 years. Winter rainfall is the main supply of water to Cape Town. The rainfall is captured in six large dams in the mountains. These dams run low during the summer and refill in the winter.

Figure 6.1: One of Cape Town's main reservoirs – Threewaterskoof Dam – during the 2017–2018 drought. The dry edges show how far the dam had receded.

Figure 6.2: Residents collect their water rations from a spring in Newlands, Cape Town, South Africa in February 2018. Confronted by the worst drought on record after years of disastrously low rainfall, city authorities put tight water controls in place to stop the city running dry.

The drought of 2017–2018 was severe. Main water supply reservoirs dropped so low that it seemed that water would run out. This would have made Cape Town the first major city in the world to 'run dry'. Reaching 'day zero' – the day the water ran out – would have forced the city to turn off the water supply.

Reservoir levels had declined from 2015. By 2017, they had decreased to approximately 15% of their capacity. To prevent water from running out, water rationing had to be implemented.

Water usage was successfully halved to save the city from running dry. This was achieved by water usage being limited to 87 litres per person per day. When rain returned, the supply dams filled up. Day Zero restrictions, which would have limited individuals to

> CAMBRIDGE INTERNATIONAL AS LEVEL ENVIRONMENTAL MANAGEMENT: COURSEBOOK

CONTINUED

South Africa during drought: 87 litres per person per day

USA: 574 litres per person per day

UK: 334 litres per person per day

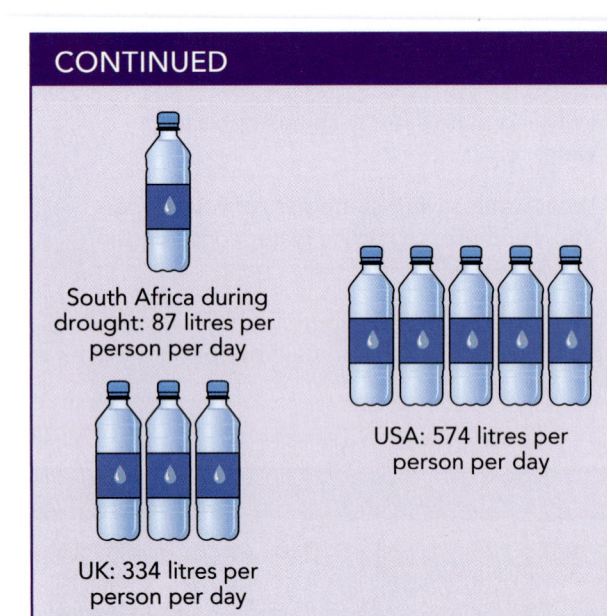

Figure 6.3: The average amount of water available to the people in Cape Town during the drought, in comparison to the water consumption per person in the UK and the USA.

25 litres per day, did not have to be implemented. Figure 6.3 compares the availability of water during the drought to estimated 'normal' usage in the USA and UK.

Discussion questions

As a class, discuss the following questions.

1. Considering the severity of the drought, do you think city planners should have had a strategy in place for it? Explain your answer.
2. Apart from rainfall, what other water stores could the city have used to try and supply water to the population?
3. Do you think that it was the responsibility of the people or the city to manage the water crisis?
4. How does education play an important role in helping to manage water supply?

6.1 Global water distribution

Distribution of Earth's water

The distribution and availability of water varies across the globe and even within regions and countries. Of the water available on Earth, only 2.5% is fresh (drinkable). The rest (97.5%) is found in the oceans and is saline, meaning it contains salt. Approximately 71% of the Earth's surface is covered water.

The distribution and availability of water is further affected by the movement of water across the globe in the hydrological cycle. This cycle is influenced by local geology, climate and rivers.

Geology determines where water can be stored as ground water, while climate influences the availability of precipitation and rates of evaporation. Climate fluctuates over the short and long term, changing the amount of water available by region.

Rivers move water out of one drainage basin into another, sending water from areas of high rainfall to areas of water scarcity. One example of this is the Nile River in Africa.

Salt water in oceans makes up the biggest water store globally. This water contains dissolved salts and is not suitable for human consumption.

6 Managing Water Supplies

(Freshwater)

(Saltwater) Oceans 97.5%

Groundwater 20%
Ice caps and glaciers 79%

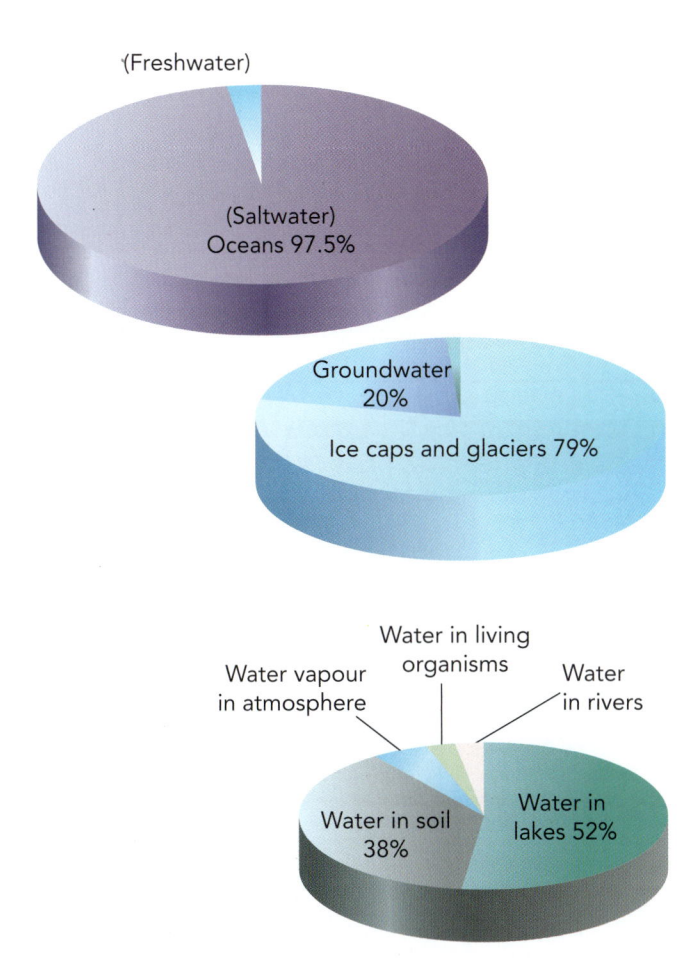

Water vapour in atmosphere
Water in living organisms
Water in rivers
Water in soil 38%
Water in lakes 52%

Figure 6.4: Distribution of Earth's fresh and salt water supplies.

Surface fresh water stores

Fresh water stores on land are essential to life. Surface water includes lakes, rivers, swamps, marshes, reservoirs, ponds, streams, canals and freshwater **wetlands**. These water resources are considered renewable, but they depend on the water cycle and need protecting from overuse and pollution. The amount of water in surface stores is constantly changing due to inflows and outflows. Inflows are from precipitation, overland runoff, **groundwater seepage** and glacial or snow melt. Outflows include evaporation, infiltration into ground stores and abstraction by humans.

The amount, location and quality of surface water changes over time, due to both climatic changes and mismanagement by humans.

Surface water that has been polluted is no longer viable for use by either humans or nature unless it is processed. Areas where water is over abstracted may dry out and suffer from human-induced drought conditions.

Figure 6.5: Cave Point County Park, which includes the fresh water store found in Lake Michigan, USA.

KEY TERMS

*wetlands: areas where water covers the soil, or is present either at or near the surface of the soil all year or during periods of the year

*groundwater seepage: when excessive amounts of groundwater push to the surface as a spring or an area of saturated soil

Sub-surface fresh water

Sub-surface water is either ground water held in saturated rock strata below the surface, or **soil moisture** held in unsaturated surface layers of Earth. Ground water includes permafrost which is water frozen in the soil.

In many parts of the world, ground water exists in quantities and at depths that wells or boreholes can access, making it a usable fresh water resource, even in arid regions. It is one of the most important global water sources.

KEY TERM

soil moisture: the amount of water found in the soil

237

Groundwater and aquifers

Large amounts of water are stored in the ground. This water, although slow-moving, forms part of the water cycle. Most of the water found in the ground comes from precipitation that has filtered into the ground. Infiltration of water into the **permeable** soil and rocks occurs until it reaches a layer of impermeable rock or clay. The upper layer is the unsaturated zone, where the amout of water varies over time. The layer below it is the saturated zone, where all the pores and cracks are filled with water.

Aquifer is another term used to describe this saturated zone of rock. The top of the saturated zone is known as the **water table**. The water table rises and falls depending on the rates of input versus output of water. When more water is lost than is going in, the water table will drop as the aquifer drains. For the water table to replenish (refill), there has to be a point of recharge. This could be seepage from rivers and lakes, infiltration from a rainfall event, or even the pumping of water back into the aquifer during periods of high rainfall at the surface.

Aquifers are filled with moving water. Amounts of water vary from year to year or season to season depending on precipitation. The speed of flow depends on how permeable the rock is, and every aquifer has a **discharge zone**, where water leaves the aquifer (Figure 6.6), and a recharge zone where water enters the aquifer. Aquifers are replenished through **natural recharge** during periods of precipitation or through the sideward movement of water from rivers, streams or other aquifers.

The rock in an aquifer must be both permeable and porous. Beneath the aquifer, there needs to be an impermeable layer that enables the aquifer to retain the water. Rock types typically found as aquifers are sandstone, conglomerate, fractured limestone and unconsolidated sand and gravel.

Rubble zones between volcanic flows make excellent aquifers, as they are able to store large quantities of water in the spaces and pores between the rocks.

> **KEY TERMS**
>
> ***permeable:** a material that allows liquid to pass through it (sandstone is a permeable rock type)
>
> ***water table:** an underground boundary between the soil surface and the area where ground water saturates the in the rocks

> **KEY TERMS**
>
> ***discharge zone:** the zone where water originating from an aquifer flows out into water courses such as lakes, rivers and wetlands
>
> ***natural recharge:** water that moves from the land surface or unsaturated zone into an aquifer. Where porous rock at the surface allows water to seep into an aquifer

Figure 6.6: A stream flowing through oak woodland in Dartmoor National Park, Devon, UK. Streams like this may start as a spring where ground water comes to the surface and flows over the land.

6 Managing Water Supplies

Figure 6.7: Grotta Giusti in Italy, an underground cave that catches the groundwater filtering through the rocks above.

Aquifers and artesian wells

All aquifers are unique (Figure 6.8). However, an aquifer's boundaries generally blend into other aquifers, making them interlinked and part of a larger system. Therefore, abstraction from one aquifer can impact another aquifer, as the water will flow from an area of high pressure to an area of low pressure.

There are different types of aquifers. The first is a **confined aquifer** which has layers of impermeable material (an **aquitard**) both above and below the aquifer. When punctured by a well or borehole, the water in the aquifer will rise to the surface under pressure. This is known as an **artesian well**.

An **unconfined aquifer** is an aquifer whose upper surface (the water table) is at atmospheric pressure and therefore able to rise and fall. These types of aquifer are is normally closer to Earth's surface than unconfined aquifers. They are more susceptible to decline due to drought conditions.

A **perched aquifer** occurs above the regional water table in an area where there is a layer of impermeable rock above the main aquifer but below the surface of the land. It results in a water store that remains when the rest of the surrounding water table declines. If a well is bored into this aquifer, it is likely to dry out quickly as the surrounding water table declines.

Non-renewable aquifers get little or no recharge, so taking water from such an aquifer is similar to mining. The resource becomes depleted and is considered non-renewable.

> ### KEY TERMS
>
> ***confined aquifer:** an aquifer below the land surface that is found between layers of impermeable material
>
> ***aquitard:** a zone within Earth that restricts the flow of groundwater from one aquifer to another; comprised of either clay or layers of non-porous rock
>
> **artesian well:** underground water that is under pressure. When punctured by a well or borehole the water will rise to the surface
>
> ***unconfined aquifer:** an aquifer in which the water table is at atmospheric pressure. There is no impermeable layer between the water table and the ground surface
>
> ***perched aquifer:** an aquifer that occurs above a regional water-table. It is usually a relatively small body of water that lies above the large aquifer due to an impermeable rock layer blocking the downward flow of water

239

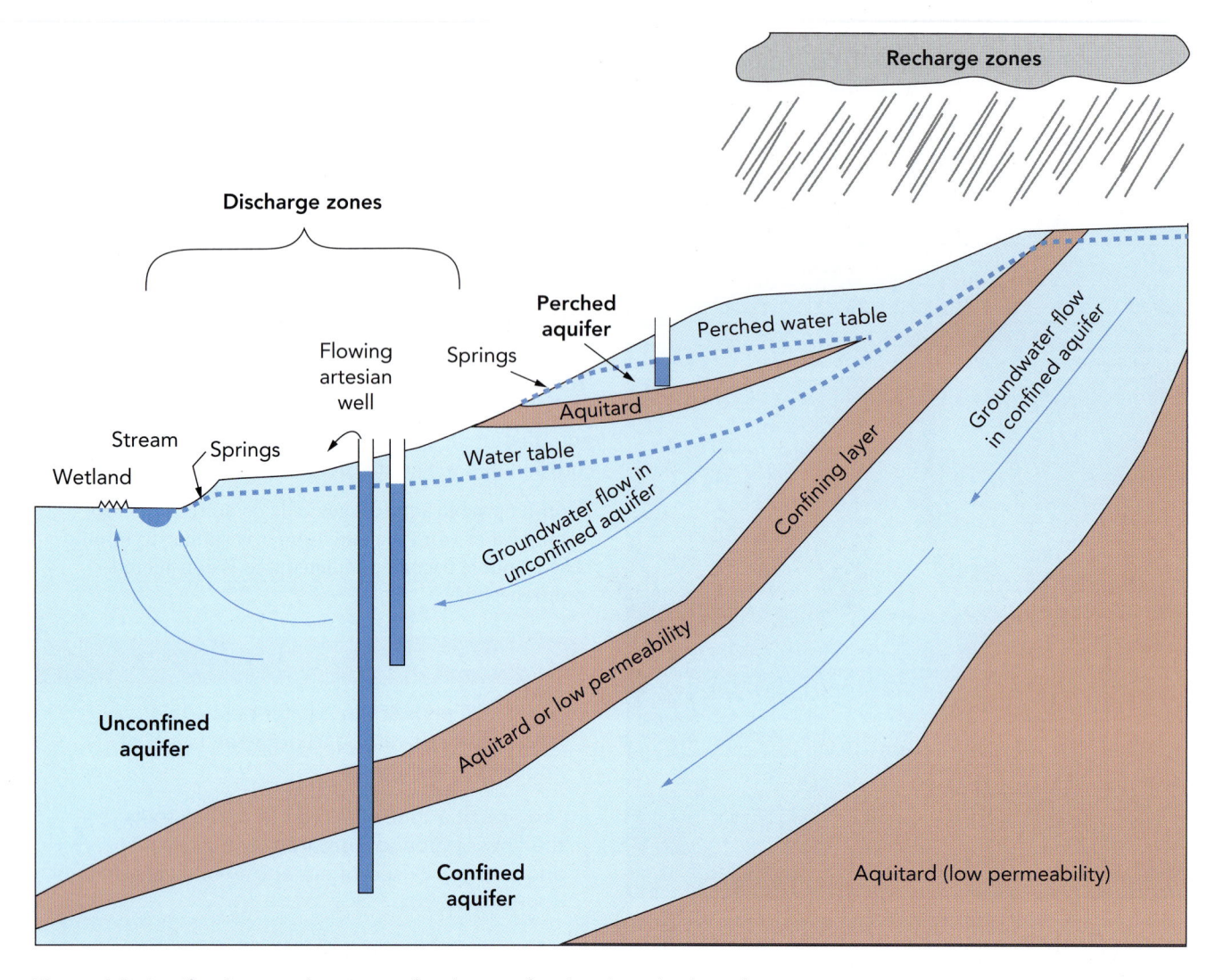

CAMBRIDGE INTERNATIONAL AS LEVEL ENVIRONMENTAL MANAGEMENT: COURSEBOOK

Figure 6.8: Aquifer diagram showing confined, unconfined and perched aquifer structures.

ACTIVITY 6.1

Water filtration

In this activity you will test different materials to see how effective they are at filtering water.

You will need:
- four funnels
- four beakers
- four clamps and stands to hold the funnels
- muddy water
- a cup measure
- a stop watch
- four different materials to test for their filtration capabilities. These could include sand, a coffee filter, paper, cotton wool, sponge, woodchips, paper towel, charcoal

Safety
- Always take care when working with glassware.
- When working with muddy water, wear gloves to protect yourself from potentially harmful chemicals or organisms.

CONTINUED

Getting started

- Consider the chemicals or harmful organisms that could be in the muddy water.
- Do you think the different materials will show different levels of filtration?
- Which material do you think will be the most effective at filtering the water?

Methods

1. Set up the funnels, clamps and beakers.

Figure 6.9: The setup. A funnel is held in a clamp over a beaker to allow for safe filtration of water.

2. Select four different filtration materials. Place a different one into each funnel.
3. Shake up the muddy water. Pour a cup of water into each funnel.
4. Record the time it takes for the cup of water to filter through each material.
5. Observe what happens to the water that collects in each of the beakers.
6. Which filtration material worked the best at cleaning the muddy water?

Next steps

Take time to consider your findings. Then answer the questions and complete the tasks below.

1. Was there a difference in the time it took for the water to filter through?
2. Was there a noticeable difference in the quality of the water after it was filtered? If so, why do think this was the case?
3. What happens if you combine two of the filtration materials in a funnel? Does it make a difference to the length of time it takes the water to filter through? What do you notice about the clarity of the water after it has passed through?
4. Is there a difference in the speed at which water moves through the filters if you compact them first? What do you observe and why do you think this has happened?
5. Identify your dependent and independent variables for this investigation.
6. Within this investigation there is opportunity for error. Can you identify where this could happen?
7. Do you think this water is now safe to drink?

Frozen water stores

Frozen water found on Earth includes glaciers, ice sheets, frozen parts of the ocean and frozen soils. Frozen soils include ice sheets, **ice caps** and permafrost.

Ice sheets

Ice sheets are the mass of glacial land ice that covers the polar regions (Figure 6.10). The Antarctic ice sheet is the largest single mass of ice on Earth, covering almost 14 million km². The Greenland ice sheet covers approximately 1.7 million km² and 80% of Greenland's surface. These two ice masses contain more than 99% of all the global freshwater ice. If these two ice masses were to melt, it is estimated that sea levels would rise by approximately 66 m.

> **KEY TERM**
>
> **ice sheets:** the mass of glacial land ice that covers the polar regions
>
> *****ice caps:** found in mountainous areas such as the Himalayas and Andes

Figure 6.10: A section of the West Antarctic Ice Sheet, viewed from the window of a NASA Operation Ice Bridge aeroplane.

Permafrost

Permafrost (Figure 6.11) is frozen soil, rock and sediment that has remained frozen for two or more consecutive years. In some parts of the world, this layer is just a few metres thick, while in the far north of Canada and Russia it is over a kilometre thick.

Figure 6.11: Permafrost, seen at the top of the cliff, melts into the Kolyma River outside of Zyryanka, Russia, 2019.

Atmospheric water

Water also exists in the atmosphere as water vapour, humidity, clouds or precipitation. It is estimated that atmospheric water makes up 8% of the easily accessible fresh water on Earth. Most of it is held as water vapour which is produced during the water cycle when liquid water is evaporated into gas.

The amount of heat available at any point on Earth impacts the amount of water vapour present. Warmer air is able to hold more water vapour than cool air. With each temperature increase of 1°C, water content can increase by up to 7%. This results in the highest volume of water vapour being found at the equator and the lowest at the poles.

Despite being one of the smallest water stores on Earth, this water source is vital because it has a number of functions:

a   It is a greenhouse gas that absorbs and reflects insolation and longwave radiation, helping to create our current climate.

b   It is an open system. This means that it allows for the redistribution of water around the planet.

c   It creates fresh water through the evaporation of water from the oceans.

d   It helps to clean the atmosphere as precipitation removes contaminates from the air when it falls.

Figure 6.12: Water heated by Earth emits steam into the atmosphere in Hvita River in Iceland, 2021. This steam is an example of atmospheric water.

6.1 Questions

1   Describe the meaning of the terms permafrost and water table.
2   Explain why so little of the water on Earth is available for human use.
3   Compare an ice sheet and sea ice.

4. Explain how the abstraction of water from one part of an aquifer can impact a user further away on the same aquifer.
5. Explain what is meant by atmospheric water. Why is it so important in the water cycle?

6.2 Water security

The term **water security** refers to the ability of a population to access a reliable source of clean water in sufficient quantities. Water supplies must allow the population to maintain an adequate standard of **sanitation**, sustainable health care, food production and manufacturing of goods.

> **KEY TERMS**
>
> **water security:** the ability to access sufficient quantities of clean water to maintain adequate standards of food and manufacturing of goods, adequate sanitation and sustainable health care
>
> **sanitation:** the provision of clean drinking water and sewage disposal

Due to the effects of climate change, urbanisation, natural disasters and population growth, challenges to water supplies are increasing globally.

Causes of water insecurity

Climate change

Climate change impacts the world's water. Warmer air holds more moisture, and warmer waters evaporate more rapidly. This results in the formation of larger and more frequent storm systems in areas where storms are normal weather events. Areas that are normally dry may suffer increased drought as the area heats up more.

More frequent and heavier precipitation leads to flooding. This results in significant erosion of topsoils, reducing the ability of the soils to support plant growth. Flooding also pollutes the water supply with sediments and contaminants such as fertilisers, herbicides, insecticides, oils and heavy metals. This makes the water unsafe to drink.

In areas that normally get winter snowfall, melted snow provides a sustainable supply of water through the summer. However, with climate change, many areas are experiencing a decline in snow fall. This means the supply of water through the drier months is threatened.

Climate change is making it harder to predict water availability. Globally, there is an increase in droughts and floods. Higher temperatures and less predictable, more extreme weather affect both the availability and distribution of precipitation, glacial melt, snowmelt, river flows and groundwater availability. This puts pressure on both the quantity and quality of water in areas that are already water stressed or becoming water stressed.

Natural disasters

Natural disasters such as floods and drought have a significant impact on water sustainability. Excess water during flooding results in the destruction of water points and sanitation facilities. Floods also contaminate drinking water sources with bacteria. This can cause diseases like cholera.

The shortage of water during periods of drought negatively impacts human health, productivity and the ability to produce food. This can lead to food shortages and potential dehydration and starvation.

Damage to infrastructure and water stores caused by earthquakes and volcanos can compromise the delivery of water to people in need. This results in water insecurity in those areas.

> **KEY TERM**
>
> **natural disaster:** an event such as a flood, earthquake or hurricane, that causes great damage or loss of life

Pollution events

The impact of a water pollution event depends of a number of factors: the scale and duration of the event and the type of spill. These factors determine how severe the pollution is and how difficult it is to clean up.

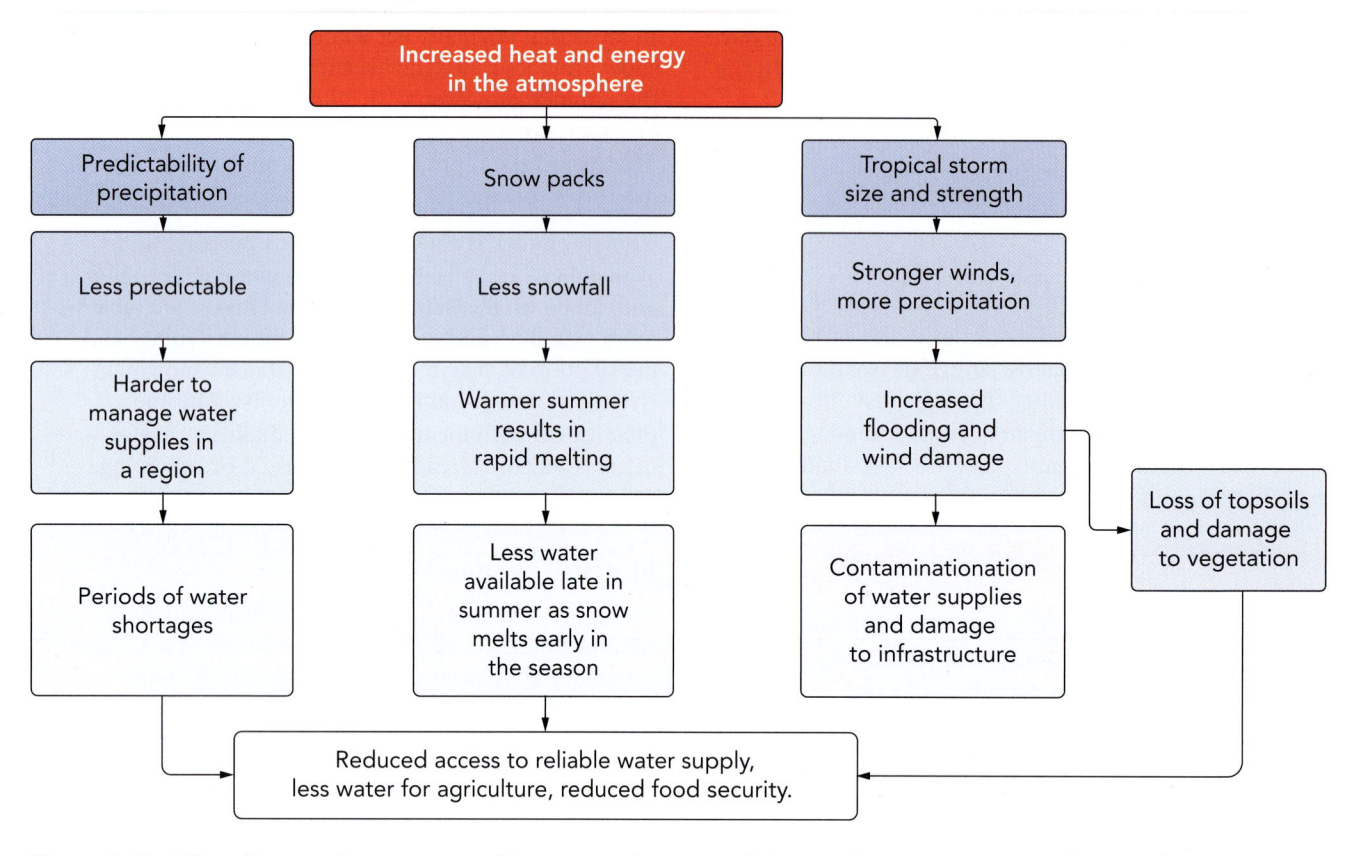

```
                    ┌─────────────────────────┐
                    │  Increased heat and energy │
                    │     in the atmosphere      │
                    └─────────────────────────┘
                       │         │         │
        ┌──────────────┘         │         └──────────────┐
        ▼                        ▼                        ▼
┌───────────────┐       ┌───────────────┐       ┌───────────────┐
│ Predictability│       │  Snow packs   │       │ Tropical storm│
│of precipitation│      │               │       │size and strength│
└───────────────┘       └───────────────┘       └───────────────┘
        ▼                        ▼                        ▼
┌───────────────┐       ┌───────────────┐       ┌───────────────┐
│Less predictable│      │ Less snowfall │       │Stronger winds,│
│               │       │               │       │more precipitation│
└───────────────┘       └───────────────┘       └───────────────┘
        ▼                        ▼                        ▼
┌───────────────┐       ┌───────────────┐       ┌───────────────┐
│  Harder to    │       │ Warmer summer │       │   Increased   │
│ manage water  │       │  results in   │       │  flooding and │
│ supplies in   │       │ rapid melting │       │  wind damage  │
│   a region    │       │               │       │               │
└───────────────┘       └───────────────┘       └───────────────┘
        ▼                        ▼                        ▼                ┌───────────────┐
┌───────────────┐       ┌───────────────┐       ┌───────────────┐         │Loss of topsoils│
│               │       │  Less water   │       │Contaminationation│      │  and damage    │
│ Periods of water│     │available late in│     │of water supplies │─────▶│  to vegetation │
│   shortages   │       │summer as snow │       │  and damage    │         │               │
│               │       │ melts early in│       │to infrastructure│        └───────────────┘
│               │       │  the season   │       │               │
└───────────────┘       └───────────────┘       └───────────────┘
        │                        │                        │                        │
        └────────────────────────┼────────────────────────┘                        │
                                 ▼                                                 │
                    ┌──────────────────────────────────────────┐                   │
                    │  Reduced access to reliable water supply,│◀──────────────────┘
                    │less water for agriculture, reduced food security.│
                    └──────────────────────────────────────────┘
```

Figure 6.13: A flow diagram showing some of the potential impacts of climate change on water supplies in different regions of the world.

An example of fresh water pollution took place in Woburn, Massachusetts. Twelve industrial solvent spills occurred into the Woburn river between 1969 and 1979. This resulted in an increase in the incidents of childhood leukaemia, liver, kidney, prostate and urinary cancers and birth defects. High chemical contents in the drinking water were believed to be the cause of all these increases in illnesses.

Water pollution can occur after a point-source event, where a spill into a water source results in pollution from a single point. It can also result from the long-term misuse of a water resource for the dumping of waste from a non-point source, as seen in Figure 6.14, where the source of the waste is hard to identify. Pollution can also be caused by unintentional runoff. This may be domestic waste (sewage and grey water), industrial waste (such as the waste water from tanneries which contain heavy metal and carcinogenic chemicals). It could even be agricultural chemical and fertiliser runoff, which causes eutrophication.

Figure 6.14: Litter and raw sewage in the Bagmati river in Nepal.

6   Managing Water Supplies

The clean-up of any spill depends on the type of spill. It also depends on the receiving waters and the ability of the ecosystem to absorb and filter the pollutants out of the water system. In some instances, it is possible to use bioremediation to clean up a river system. Normally, though, clean-up is a long and costly process. If pollution has entered into the groundwater it is impossible to clean (see next section). This effectively turns a renewable resource into a non-renewable one.

Pollution of water sources therefore results in the decline of access to safe, fresh water sources which puts pressure on water security in the area.

Aquifer pollution

Groundwater is clean because porous rock acts as a filter for all particles. However, in some cases, it contains a high concentration of elements such as iron, fluoride, uranium or arsenic. Such elements can result in the **contamination** of a groundwater source, effectively polluting it so that it is no longer a viable source of fresh water (see the Extended Case Study in this chapter).

Humans pollute water by storing toxic materials on Earth's surface in landfills, septic tanks, storm water drains and even injection wells, where toxic materials are pumped into the ground. The process of fracking for oil and gas also pushes toxic materials into the ground. This can result in contamination of the ground water source.

In coastal areas, over-abstraction of water from aquifers can lead to the contamination of ground water by salt water. An aquifer open to the sea contains a layer of fresh water above the salt water. Over abstraction reduces the pressure of the fresh water and this means that salt water can enter the aquifer. This is known as **salt water intrusion**. It is prevalent along coastlines where saltwater intrusion causes the ground water to become salty.

> **KEY TERMS**
>
> **contamination:** the action of making or being made impure by pollution
>
> ***salt water intrusion:** the movement of salt water into fresh water aquifers

Population growth

Figure 6.15: Steps leading down into a well in India. Groundwater resources like these are under pressure around the world due to increased population size.

Population growth has a significant impact on water security. As population numbers increase, the demand for domestic, agricultural and industrial water use also increases. In many areas, high demand is further increased by rising temperatures as well as the pollution of many surface and ground water resources.

Continued population growth and rising incomes lead to greater consumption and greater waste. Urban populations in many LICs are growing rapidly. This results in demand beyond the capacity of the already stretched water and sanitation infrastructure. Globally, the pressure on water supplies is expected to continue increasing. This will put severe pressure on water supply to all sectors.

Increased population also results in pressure on land use. Forested and wetland areas, which used to function as natural water reservoirs, have been replaced with hard, impermeable urban landscapes. These hard surfaces do not allow water to infiltrate into the soil and recharge groundwater supplies. Surface water in urban areas is also more likely to become contaminated with pollutants such as oils and septic waste. This will ultimately mean the loss of a natural supply of clean water. Urban populations are at risk of becoming reliant on treated water supplies.

Figure 6.16: Large areas of cities like Hong Kong have been deforested in favour of urbanisation. This reduces the ability of an ecosystem to fulfil its role in the water cycle, as it limits infiltration and evapotranspiration.

Population pressure on aquifers

Increased population sizes can result in the over abstraction of water from aquifers. When water is abstracted at a rate faster than it can recharge, the aquifer depletes. The water table then drops, making access to the groundwater more difficult. It is then necessary to drill deeper, more expensive wells and install bigger pumps, making it difficult for some people to access the water.

A fall in the level of the water table also has an impact on the flora and fauna in the area. This is because aquifers are a natural water source for surface water during periods of low rainfall. Springs and small streams that local wildlife rely on may dry out, causing wildlife to die off or migrate. Trees that access the water table with their deep roots will no longer be able to reach the water. In time, they will dehydrate and die.

Areas such as California in the USA suffer from extensive periods of drought. In such areas, the population uses groundwater as a significant water supply. Wild fires occur regularly. As the water table drops and vegetation dries out, these wild fires can get progressively worse.

ACTIVITY 6.2

Water pollution research task

In pairs investigate a region that has a problem with water pollution.

Create a poster:

- introducing the region/area/location
- describing the causes of the pollution
- describing the impacts of the pollution
- explaining what action is being taken to remedy the problem, if any.

Figure 6.17: An aerial photograph of the eutrophication that has developed in pond next to a waste dump.

Competing demands for water supplies

Water conflict is the term used to describe conflict between groups, regions or countries over access to water. A water dispute may result from the opposing interests of water users. As freshwater is a vital but unevenly distributed natural resource, its availability impacts the economic conditions of a country or region.

Agriculture, industry, the energy sector and domestic sectors all require access to water. Conflict can develop when demand exceeds supply. If a farmer has a privately owned dam on their property, and a town is running out of water, who has the right to access that water? The farmer has managed their water, and legally owns it, but is it right that the town goes without a fundamental, ever-moving resource? This is an ethical, social and political question.

Water conflict also occurs where rivers are dammed upstream for water storage and/or hydroelectric power. The rivers may also be diverted for use elsewhere. Downstream users are deprived of that water source and may no longer be able to live in the area due to water scarcity. This happened in the Aral Sea where water was diverted to farm in the desert. As a result, the Aral Sea was deprived of its water source. Over time, it dried out and the local population and ecosystem collapsed. This meant that fish disappeared, crops failed as the land dried out, and local people lost their livelihoods. In this case, the use of a water resource for agriculture resulted in water stress for another area.

International competition for water

International competition for water can happen when the source of a river starts in one country and runs through another country. The country that has the river source can dam the river, denying downstream users access to both the water and the fertile sediments that are flushed down the river in times of high flow. Rivers like the Mekong River in Asia, the Jordan River in Syria/Israel and the Nile River in Ethiopia/Egypt are examples of water resources that cross international boundaries.

Inequality of water availability

People in low income countries (LICs) are less likely to have access to clean drinking water than people in high income countries (HICs). HICs are typically able to pay for the infrastructure required to secure a safe water supply and to develop sanitation systems.

Urban versus rural water access

Urban populations usually have better access to clean water than rural populations. Due to the number of people in urban areas, there will be the combined financial resources to install treated, piped water supplied to houses. Less dense rural populations may not have the financial resources to have the same safe water supply. Rural areas often rely on access to groundwater or rainwater for their fresh water. These supplies are not always safe as they are untreated and may be contaminated with bacteria or heavy metals, especially in areas where surface water is used for bathing, washing and removal of septic waste.

However, rapidly growing urban areas can face significant water supply problems.

ACTIVITY 6.3

International water sharing

Figure 6.18 depicts a river crossing an international border. Write a 200–300 word evaluation of the cartoon explaining the message behind the image.

Consider the following:

1. what the message is
2. the impact of the international border on human populations and the environment downstream
3. what the political impact could be.

Figure 6.18: A cartoon depicting an international border on a river.

When the speed of urbanisation exceeds the rate of local infrastructure development, impoverished urban dwellers do not have access to clean water or sanitation. This results in the rapid spread of water-related diseases through the population.

Figure 6.19: Young girls cross the sand dunes carrying water from their local well, Thar Desert, Rajasthan, India. People in rural areas often have to walk great distances to access fresh water sources.

The development of water treatment facilities (Figure 6.20) and sanitation is important for both rural and urban areas. This protects the population from the spread of diseases such as cholera.

Figure 6.20: A water treatment centre in Thailand.

Where large cities have developed, the challenge is often not the supply of the water, but sourcing a sustainable water supply from surrounding rural areas to meet the population's demand. During periods of drought, rural areas may fare better, as they can closely manage their own resources. City dwellers may not understand the challenges of supplying water. As a result, they may be slower to conserve water. This means a city may run out of water while rural areas still have access to it.

On the other hand, cities may have the resources to use technology for processes such as desalination, which makes salt water drinkable. In this case, they can access groundwater to obtain a supply during dry periods. Rural areas may not have the financial means to do this.

Ultimately, the issue of rural versus urban water supply is not clear-cut. It is not always the case that urban areas have water while rural areas do not. Many complex and interacting factors need to be considered for each discussion of rural/urban water supply.

ACTIVITY 6.4

Urban vs Rural water security

1. Draw a spider diagram to summarise the reasons why urban areas may have access to safer water resources than rural areas. Include reasons why a rural area may have better water resources than an urban one.

2. Use your spider diagram to help you write a 300 word evaluation of how water security differs between rural and urban areas.

Mismanagement of irrigation and sanitation

Mismanagement of the use of water for irrigation, or failing to protect water supplies from poor sanitation, both have a direct impact on water security.

Irrigation mismanagement

If the irrigation of farm lands is not properly controlled, water supplies can run low or dry up. In areas of long-term mismanagement, human-driven drought conditions can occur. In these instances, overuse of both surface and ground water results in stores becoming so depleted

6 Managing Water Supplies

that drought conditions, normally only associated with low rainfall, develop. If vital surface and groundwater stores dry up, the people living in these areas may be driven away (Figure 6.21).

Figure 6.21: A water well for accessing groundwater supplies in an extremely arid part of Morocco, Africa. People rely on groundwater stores.

Mismanagement of irrigation (Figure 6.22) can also result in the pollution of surface water and ground water. Excessive irrigation results in runoff from farmlands. This runoff takes topsoil and any chemicals or fertilisers that have been applied to the crops. This can result in a number of issues for other water users, including, but not limited to:

- eutrophication
- increased turbidity
- toxic water pollution making water unsafe to drink
- harm to the natural ecosystems, biodiversity and food chains
- pollution of groundwater resources.

Figure 6.22: Water flows rapidly from irrigation pipes to supply crops used for cattle grazing. This farmer is irrigating the land during a period of drought in Colorado in 2021. However, this type of irrigation technique uses a lot of water. There are more efficient irrigation techniques.

Sanitation mismanagement

Lack of sanitation has a serious impact on water security. Changes in supply, demand and quality, combined with aging water infrastructure in many countries make it difficult for governments to manage the problem. When the amount of septic waste exceeds the capacity of a sewage system, raw septic waste is released into water stores. Once water stores are contaminated with septic waste there is an increased risk of disease such as hepatitis, cholera, intestinal worms and diarrhoea being transmitted.

Improved sanitation protects the available water resources. It helps to reduce the spread of disease, the impacts of malnutrition, and poverty.

Impacts of water insecurity

Global water systems are increasingly stressed, and some rivers, lakes and aquifers are drying up. Overuse or misuse of water supplies causes greater water stress for populations. This results in a range of issues.

Figure 6.23: A young girl washes her hands and drinks water from a modern borehole in Nsanje District, southern Malawi. The modern borehole protects the ground water from potential contamination. The supply of a clean water source to a population helps with managing sanitation.

Crop yields, livestock and food shortages

Water scarcity has a significant impact on food production. Without access to water, people cannot grow crops or water livestock. This results in food insecurity, malnutrition and famine. With the continued growth of the global population, pressure on fresh water availability is increasing. Consequently, associated food insecurity is becoming an ever increasing problem.

Illnesses related to contaminated water

Surface and ground water pollution comes from many sources including fertilisers, industrial waste and human waste. Polluted water resources can also cause water insecurity, as polluted water means that there is no water available to drink.

When harmful bacteria from human waste contaminates water, it becomes unsafe for drinking and bathing. Diarrhoea and cholera are examples of illnesses that can be caused by exposure to bacteria in septic waste.

Cholera can result in death if untreated. Areas that have experienced natural disasters or extreme water stress may have outbreaks of cholera because in these circumstances, septic waste can often mix with drinking water.

Figure 6.24: Severe pollution in a river. Visible waste can be seen trapped in and floating on the murky water.

Toxic chemicals from industrial processes (such as heavy metals such as mercury and cadmium) can also contaminate surface and groundwater sources. This contamination is a major cause of immune system suppression, reproductive failure and serious poisoning.

Poverty

Lack of access to water is often an overwhelming obstacle for people in poverty. They may be unable to grow food, build houses, work, attend school or stay healthy.

Good hygiene (for example, handwashing) is one of the best ways to prevent the spread of disease, but this is not possible without clean water. Poor sanitation and a lack of clean drinking water poses a threat to the health of over one third of the population worldwide. Disease also impacts a person's economic stability. A sick person may be unable to go to work or school, and this affects their ability to earn an income. They may then struggle with the costs of transport and doctor's fees and find it harder to recover. This vicious cycle is known as the poverty trap.

Access to water has a significant impact on food security. Effective poverty reduction is supported where stable agriculture is established. Water shortages therefore make it harder for people to escape the poverty trap. When drought occurs, low-income farmers often have to abandon their farms.

A person who lives in a country affected by military conflict is more likely to face poverty than a person in a country at peace. Water can cause conflict and it can also be harder to access during periods of conflict. It can even become a weapon during conflict as it can be used to control the population.

Figure 6.25: African girls from the Borana tribe carrying water to their village in Ethiopia. Women and children in Africa often walk long distances to collect water.

6.2 Questions

1. Explain the meaning of the terms water security and salt water intrusion.
2. Describe how climate change is impacting water security.
3. Give the main causes of water insecurity, with a brief explanation for each.
4. Explain how the pollution of an aquifer occurs, and the impact this has on water security.
5. How does fracking threaten fresh water resources?
6. Explain why people from LICs are more likely to suffer from water insecurity.
7. Write a short essay, explaining the impacts of water insecurity.

6.3 Water management strategies

Water management strategies vary in scale from individual actions to government policy designed to minimise water waste. On an individual level, waste management strategies include reducing the length of a shower, or switching off the tap while brushing teeth. On a larger scale, they include municipal water conservation strategies. These strategies start by identifying the demand for water. Once this had been established, demand can be compared to supply. From there, an investigation can be undertaken to determine the best way to meet the water shortages.

True water security relies means ensuring water security is resilient during times of drought. This relies on a range of strategies, such as expanding or improving current water supply (for example, increasing the number of reservoirs, or improving water transport and reducing losses), or using technology like desalination. Cost and time need to be taken into account when considering new methods.

A water management system needs to address the following key principles:

- Water should be treated as an economic, social and environmental resource.
- Policies should focus on both demand and supply.
- Government regulatory frameworks are critical for developing a sustainable water resource.

Figure 6.26: Travelling with a reusable drinking container prevents plastic pollution and minimises waste.

Sustainable water extraction

Sustainable water extraction should consider the conservation of resources, and the changing behaviour of water users. Sustainable water management focuses on the effective use of water and aims to reduce demand on existing water supplies. It can include technology such as low flow shower heads and low flush toilets. In the agricultural industry, sustainable water management covers the move from pivot irrigation to drip irrigation, with night time irrigation when evaporation rates are lower.

Water can also be saved by educating people about reducing water consumption. If people consume smaller amounts of water, less water will need to be extracted.

To gauge the amount of water they are using, water users can measure their use with a water meter. A water meter (Figure 6.27) measures the flow of water through a pipe and shows how much water is being used.

When a water user abstracts water from a borehole, water use should be both licensed and metered. By knowing how much each water user is taking out of the ground water, it is possible to start managing the groundwater resource. Data from water metering is then combined with measurements of the depth of the water table to determine if overuse is occurring. If the water table drops, more stringent water regulations can be put in place while the aquifer recharges.

Figure 6.27: A water utility worker checks the water meter to make sure that it is working correctly. The water meter measures the amount of water flowing through the pipe, helping to manage water usage.

Surface water is easier to manage, as it is possible to see water levels. Putting a full management system in place to monitor surface water levels and quality ensures that a water source is protected. Where water is well-managed, it is possible to respond rapidly to drought conditions. During drought, water users can be asked to reduce their water consumption. This ensures a stable water supply for a longer period of time.

The water supply can also be protected by monitoring it for water pollution. If pollution is identified early, it can be managed quickly and efficiently. If it is detected too late, a long-term, large-scale clean-up project may be needed.

Too often, the overall complexity of managing the water resource is misunderstood, putting even further pressure on supply.

Improved supply

Both hard and soft engineering options are available for improving the quality and quantity of fresh water supplies. Hard engineering uses structures built by humans to supply water, for example, pipelines, dams and bore holes. Soft engineering protects water resources from becoming degraded. It may be used to prevent pollution or salt water intrusion into water resources. Hard and soft engineering options can be used together to optimise effectiveness. However, both require significant investment and long-term planning.

Hard engineering solutions include the storage (Figure 6.28, 6.29), transport and purification of water supplies. Water supply strategies can include stream diversion, new reservoirs, new pipelines, desalination, water recycling plants and the purification of urban water runoff.

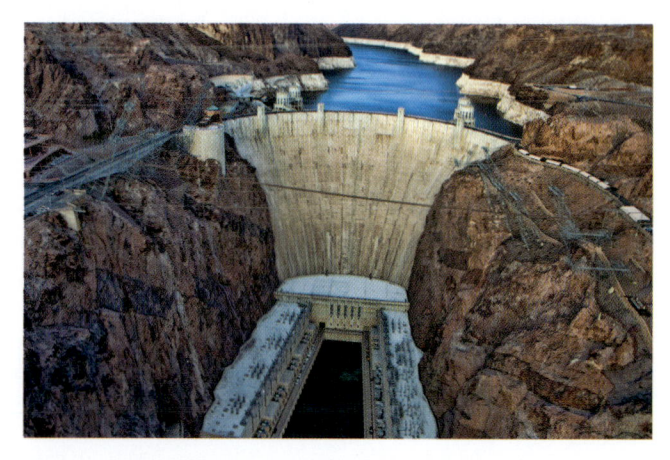

Figure 6.28: The Hoover dam on the Nevada–Arizona border in the USA is an example of hard engineering. It is designed to hold water for long term supply and the generation of electricity.

Figure 6.29: A dam wall in the Swiss Alps with a reservoir behind it. The dam is designed to maximise the amount of water stored in the valley behind it. Water can be pumped via pipelines from the reservoir to towns or farms.

Developing new reservoirs or increasing the holding capacity of existing dams and reservoirs helps to protect water supply during periods of increased water stress. Cities and towns need to know what their demand is to ensure that long term water storage is developed to meet increasing demand.

All reservoirs need water-tight pipelines to take water to towns and cities. Water loss from leaks in a pipeline adds to the water stress of an area.

Where the water source is a great distance from the receiving area, a water transfer system can be developed. Water transfer schemes involve the diversion of water from one drainage basin to another. This can be done by either diverting a river, or by constructing storage dams and piping water from one catchment area to another. Such schemes can negatively impact both the source and the receiving area. The source area will experience a decline in water levels as water is diverted, which may result in greater levels of pollution as less water will be available to dilute pollutants. Receiving areas may continue to overconsume, for example, developing and expanding cities and their associated activities due to the increased availability of water. This type of system can therefore be unsustainable.

Where necessary, streams can be diverted into reservoirs to increase the water supply. However, the environment's water needs must be met during this process if the water supply is to be sustainable. Aquatic ecosystems require water, so streams must be left with sufficient water to meet their **ecological demands**.

Gravity-fed schemes can also be employed to supply water. In these schemes, small upland rivers, streams or springs can be dammed within a protected catchment area to supply water downstream. This is an example of a sustainable water supply that, when well-managed, does not require any water treatment or advanced technology to run.

Any planning for the use of water as a resource requires planning to determine the carrying capacity of the system. If it is exceeded, then the water supply will become unsustainable. With large hard engineering solutions, there is the additional challenge of the cost of the project. Large infrastructure requires significant amounts of financing. Many LICs are unable to afford these types of schemes, limiting their ability to improve their water supply.

> **KEY TERM**
>
> *****ecological demand:** the amount of water an ecosystem requires to continue functioning

Reduced water usage and education

Reduced water usage is achieved through a combination of factors. These include the well-managed extraction of water, technology which limits the use of water and education of the end user.

Management of water extraction limits the amount of water being used. Educating end users (whether individuals or industries) gives them the knowledge about and, therefore, the responsibility for, their own water consumption. Technology can also be used to reduce the amount of water being used. By combining these factors, it is possible to reduce water usage in an area.

Domestic use

It is possible to install a number of water saving systems in the home. These include, but are not limited to, the following:

- Home owners can reduce the water they use by being aware of their water consumption. For example, shortening shower times, not leaving taps running and monitoring water consumption all help to limit water wastage.

- Low flow shower heads (Figure 6.30) can be installed, along with low flush toilets. Grey water systems recycle grey water for use in flushing toilets or watering the garden.

- Rain water tanks can be installed to harvest rainwater and supplement the municipal supply, reducing consumption from the main water source.

- Home owners can ensure that there are no leaks in their water system where water is being lost.

- Swimming pools can be filled from rain water collected off the roof rather than from a municipal source. In extreme cases, swimming pools can be removed.

- Gardens can be made 'water wise' (Figure 6.31). Indigenous plants adapted to the local environment often need less water than plants imported from wetter climates. Choosing garden plants carefully can therefore reduce water consumption.

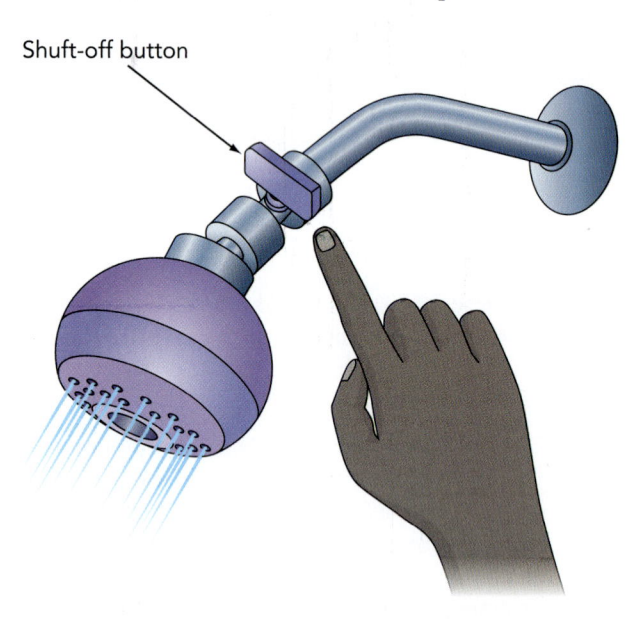

Figure 6.30: A water-saving shower head. Shower heads can be designed to limit the amount of water being used. This lowers individual water consumption.

- During drought periods or in water stressed areas, domestic usage can be further reduced by limiting the flow of water to different parts of a city at different times of the day. Water usage can also be limited by charging more for water (both domestic and industrial), as this focuses the end user's attention on their consumption.

Figure 6.31: A water wise garden design, planted to minimise the need for irrigation.

Agricultural water use

Agriculture accounts for approximately 70% of all fresh water consumption. This means it is an area of key concern in water conservation. In agriculture, education and the improved use of technology can also achieve reduced consumption (Figure 6.32).

Figure 6.32: Farming tunnel with plastic covering the soil around the plants. Plastic reduces evaporation from the soil. It also prevents the growth of weeds that would compete with the crop for water and nutrients.

6 Managing Water Supplies

Water conservation can be achieved through the use of drip irrigation, night time irrigation and computerised systems. Drip irrigation (Figure 6.33) ensures the supply of water directly to the plant that requires it. In doing so, it limits the amount of water lost to the surrounding area. Irrigating at night time reduces the rate of evaporation, allowing the soil and plant more time for the absorption of water. Computerised systems are designed to irrigate only when water is needed. They measure all the factors that influence water in the soil (wind, temperature, precipitation) and irrigate when water is required.

Agriculture can also reduce water usage by growing crops in closed tunnels. This reduces the rate of evaporation along with the need for irrigation.

Figure 6.33: Drip irrigation used to supply water directly to the plants. This minimises the loss of water.

ACTIVITY 6.5

Calculating runoff from a roof area

A farmer relies on rainwater for their supply of domestic water. There are four 10 000 litre tanks installed on the property to capture rain water. The roof area of the property is 600 m^2.

1. Using the following criteria, calculate how much rainfall the farmer requires to fill all four rainwater tanks (assuming they are empty at the start).

   1 mm of rainfall on 1 m^2 of roof gives 1 litre of water.

   At the end of the wet season the tanks are full. During the four-month dry season the farmer recorded the following weekly rainfall.

Dry season rainfall records (mm).

Week 1	Week 2	Week 3	Week 4	Week 5	Week 6	Week 7	Week 8	Week 9	Week 10	Week 11	Week 12
6	3	0	0	1	0	1	0	0	5	4	6

2. The farmer's household is using 5000 litres a week. Using 40 000 litres as the starting volume, and the rainfall recorded in the dry season, calculate the point at which the farmer has insufficient water stored to supply water to the house.

3. Calculate the amount of water the household should be using for the water (40 000 litres) to last through the dry season if:

   a. there was zero rainfall

   b. there was the 26 mm rainfall recorded over the 12 weeks.

4. What action could the farmer take to ensure that the water supply lasts through the dry season?

5. As a group, discuss how much water you think you use in your household each month. Find out the amount. Then, work out how much water is used person per month.

   Average global water consumption per person per day is estimated to be 185 litres. In HICs like the USA, it is estimated to be around 574 litres per person a day. In LICs, which experience drought, it can be as low as 11 litres a day.

6. How does the water consumption in your household compare?

> CAMBRIDGE INTERNATIONAL AS LEVEL ENVIRONMENTAL MANAGEMENT: COURSEBOOK

> **SELF ASSESSMENT**
>
> - Compare your answers to those of others in your group. Were your answers the same?
> - If you had different answers, did you take the time to work out why?
> - Do you need to make any corrections to your answers? Do you understand what you need to do and why?

These technologies all require financial investment, so it is not always possible for smaller farms to implement them. In HICs, these techniques are becoming more and more common, while in some areas of LICs irrigation is still manual.

Farmers can also choose to grow crops more suited to the climate, soils and local environment. By doing so their crops will be more resilient to periods of low rainfall and require less irrigation. Indigenous crops are more likely to be naturally adapted to local conditions. Genetically modified crops or crops produced from selective breeding could be used to grow crops in areas were other crops fail (due to soil being too saline, too wet or too dry).

Industrial water use

In industry, water is used for many different tasks. These range from cooling equipment to being used in products, providing water to livestock for meat farming, to fracturing rock for the release of oil and gas. Industry requires large amounts of water. However, it does not always require water of the same quality as drinking water. What specific technological steps are taken to save water depends on the industry. However, there are standard water saving steps that can be applied across industries:

- Performing routine maintenance and checking for leaking equipment and pipes.
- Replacing small components, for example, using high pressure, low volume hoses and nozzles for cleaning. Automatic shut-off nozzles minimise waste.
- Reusing/recycling water that is clean enough to reuse without treatment.
- Protecting water sources in the environment by ensuring that any water released is clean enough not to harm receiving waters. This can be achieved through filtration, biological treatment, distillation or absorption.
- Upgrading existing equipment to more efficient technologies.
- Installing monitoring devices to alert operators to periods of excess use. This can also identify where use of water can be reduced.
- Collecting steam when boilers are used in the industry, known as boiler condensate.
- Investment in closed-loop systems, where used water is captured for reuse in the same industrial cycle.

Poverty reduction

Millions of people live without access to safe drinking water. Billions still lack access to basic sanitation services, such as toilets, in their homes. The divide between HICS and LICs is evident in the supply of clean water and sanitation systems. Even within countries, divides exist. Urban dwellers are more likely to have access to clean, safe drinking water than rural populations.

A safe, clean supply of water and a sound sanitation system helps to reduce poverty for a number of reasons:

- The population will be healthier. There will be less disease (such as cholera), making it easier for people to work and earn a living.
- A secure water source allows the development of agriculture. Developments in agriculture results in a more secure food supply that further strengthens the health of the worker. A better water supply makes it possible for farming to move from subsistence to commercial methods. This means more jobs for the local population, and a stronger local economy.
- A secure water source allows industry to develop and produce manufactured goods. These goods have higher value than raw materials and help money flow of money into an area, creating jobs and improving the quality of life.

However, putting policies in place and ensuring the effective delivery of services to those in need is still a significant challenge for many LICs. Designing a solution to the water problem is the first step. In order for that design to be effective, service delivery problems need to be solved. Constraints are not necessarily technological: sometimes they are administrative or political.

6 Managing Water Supplies

Rationing

Rationing, or limiting the amount of water any individual, household or industry may use daily, is a method of water supply management. It can be employed permanently in areas of extreme water scarcity, or temporarily during periods of drought.

Water rationing is usually combined with specified water use restrictions, such as banning the following activities: the irrigation of lawns, washing cars, filling swimming pools or washing down paved areas. Rationing can also limit the amount of water available, either in terms of the time it is available, the volume available, or both. Severe water rationing usually includes a temporary suspension of the water supply.

Rationing can extend the time before water supplies run out. It can also ensure that the water resources available are fairly distributed.

Rationing does not lead to a long-term change in the behaviour of the population. Once the rationing or restrictions are lifted, the use of water will rise again. For areas where persistent water scarcity exists, other more permanent water saving strategies/water supply strategies need to be implemented.

CASE STUDY

Water insecurity and sustainability solutions

The East African water crisis, 2011

Between July 2011 and July 2012, the entire East African region (Figure 6.34) was affected by the worst drought in 60 years. The countries impacted included Somalia, Djibouti, Ethiopia and Kenya. Tens of thousands of people died and over 13 million people were affected.

Figure 6.34: A map of the location of the countries affected by the 2011–12 drought.

Causes

Although a decline in precipitation and the resulting drought sparked the east African crisis, human factors added to the disaster.

- Weather conditions in the Pacific, which affect global climate, included a strong La Nina event which interrupted rainfall for two consecutive rainy seasons.

- Climate change was thought to have a long-term impact on an already dry area. A long-term decline in rainfall resulted in a lower supply of ground water and less water resilience during periods of low rainfall.

- Old migratory routes for nomadic farmers had been closed by international borders. This limited the farmers' ability to move their animals from one traditional feeding area to another during periods of low rainfall.

- Ongoing conflict within the region worsened the impacts of the drought conditions. Many people were displaced from their homes, abandoning farms to move to areas of safety. Large numbers of people settled in refugee camps, and the spread of disease within these crowded conditions added to the death toll. Providing these camps with resources (water, food and sanitation) was challenging. War made supplying aid risky as there was the threat of attack on water and food convoys.

CONTINUED

Figure 6.35: Nomadic farmers moving their livestock across a drought stricken landscape. The lack of rainfall resulted in the vegetation dying off. Finding food for livestock became very difficult.

Figure 6.36: Cattle that died in Kenya due to drought. A lack of water and food resulted in cattle, goats and other livestock dying.

- The area was experiencing ongoing rapid population growth. This had caused greater impoverishment, and threats to water availability even in normal rainfall years. During the drought, the population size outweighed the carrying capacity of the available water.

- Western governments were slow to send aid to the southern parts of Somalia due to conflict in the area. A six-month delay in responding to the situation meant that the drought had a greater impact than if earlier intervention had taken place. Early warning systems had identified the developing problem as early as August 2010, but a full scale response was not launched until July 2011.

- Fighting disrupted aid delivery in some areas.

Impacts

The impacts on the region included a high death toll. Over 29 000 children under five in a 90-day period from May to July 2011 were estimated to have died. The overall death toll was estimated at 100 000 people. Livestock starved to death (Figure 6.36), and crops failed. This resulted in the destruction of livelihoods and local markets. Crop production was down by as much as 25% and food prices rose by 40%, making food too expensive for the impoverished population to buy. Even education was affected, as schools shut down because there was no food for the children.

Solutions

The main short-term solution was the supply of aid, both financial and supplies. Once the food and water crisis was no longer at emergency level, aid agencies shifted their focus to implementing strategies that would ensure sustainable water supplies. These included both short-term and long-term solutions:

- Digging irrigation canals to move surface water. This is a short-term solution. When no surface water is available, these canals will serve no purpose.

- Supplying pumps to move water from rivers to crop lands, especially rivers that continue to flow during periods of drought.

- Increased investment in public infrastructure to help supply water from storage dams or areas of high rainfall during periods of drought.

- Implementation of water-wise irrigation solutions such as drip irrigation or clay pot irrigation.

- Planting of drought tolerant trees to stabilise soils.

> **CONTINUED**
>
> - Educating the population on how to manage their water supply while producing crops or livestock.
>
> - Switching from drought-sensitive crops like maize to more resilient crops such as cassava. This reduces the risk of starvation during periods of low rainfall.
>
> - Reducing livestock numbers to a sustainable level, and retaining emergency wells for periods of drought. Preserving areas of feed for livestock that can be used during periods of drought.
>
> - Accessing deep aquifer stores. This has to be done in a sustainable manner to ensure the aquifer is not depleted.
>
> - Utilise early warning systems (mapping areas with water and food stress) to ensure that aid is implemented in the right areas.
>
> - Conflict management is key to any of the plans above working. As long as conflict is present, providing a clean, safe water supply will be difficult to achieve.
>
> **Case study question**
>
> 1. With reference to a case study that you have researched yourself, evaluate how effective solutions to an existing water shortage are.
>
> The essay should:
>
> - Explain what water security means and, therefore, what is meant by a water shortage.
>
> - Identify potential causes of the water shortages in your case study.
>
> - Provide detailed solutions to the existing water shortage.
>
> - Conclude by assessing how effectively these solutions address the water security crisis. This conclusion should be supported by your discussion on solutions, and should address counter arguments. For example, solutions to the problem may incur costs too high for the local population to meet, making them unachievable. Or perhaps local culture may dictate that the number of cattle you own reflects a person's wealth, making it difficult to persuade people to keep fewer cattle.
>
> Your essay should have a clear introduction, body and conclusion.

International agreements and water related aid

International borders are often drawn along major geographical boundaries, such as deserts, mountain ranges and rivers. These natural 'fence lines' make it easy to identify a border. However, one of the many problems associated with drawing a boundary through the middle of a river valley is negotiating a common management plan for the shared usage of the resource. When this resource comes under pressure, either due to natural changes in water supply or due to increased population pressures, conflicts can arise.

According to the United Nations, water resources that cross international borders account for an estimated two thirds of global fresh water. It is therefore vital that countries cooperate over shared water resources. This will ensure the protection and sustainable supply of these trans-national water supplies.

International agreements

Water cooperation within and between countries helps to drive transboundary water policies, reducing the risk of conflict, and increasing the potential for water security.

The Water Convention of 1996 (The Convention on the Protection and Use of Transboundary Watercourses and International Lakes), aims to facilitate and promote cooperation between nations to ensure fair and sustainable transboundary water usage.

The convention is designed to be implemented in water-rich as well as water-scarce areas, and can be applied in different settings, conditions and levels of development. It defends the rights and obligations of both upstream and downstream countries. In addition, the convention provides a legal and intergovernmental platform for transboundary water co-operation, with more than 110 countries participating. Its key goals are outlined in Table 6.1

An example of a successful international water use agreement, despite conflict over other issues, is that of Cambodia, Laos, Thailand and Vietnam. These countries have had water cooperation since 1957. However, the Mekong River that supplies water to that region flows out of China. For this cooperation to be fully effective, China, which manages the source of the Mekong, also needs to be involved in the management of this transnational water source.

The Three Pillars of the Convention	
1	Prevent, control and reduce transboundary impacts. Impacts on the environment, human health and safety and socioeconomic conditions have to be managed by participating parties.
2	Ensure reasonable and equitable use. This depends on many factors, and they all need to be considered so that water can be used sustainably.
3	Cooperate through agreements and joint bodies. Pillar 1 and 2 are put into practice through agreements between participating parties. The agreement ensures co-operation, regular consultation, joint monitoring, assessment and action plans.

Table 6.1: The Three Pillars of the Water Convention.

Transborder water agreements can also take place within a country and between states. An example of this is the Murray-Darling River Basin in south-eastern Australia, where the Murray River forms the political boundary between the states of New South Wales and Victoria. The same model that was developed here was adopted for the use in the Mekong River Commission in Southeast Asia. Mounting pressure on the water resource in the Murray River Basin posed the potential for conflict between irrigators and environmentalists. To avoid conflict, a common regulatory body was established to manage water rights. In the current agreement, the neighbouring state of South Australia is guaranteed specific quantities of water from the Murray River each month. These volumes have to take both human and environmental needs into consideration. Each state then needs to manage their allocated resources according to what is available.

Water-related aid

Access to water and sanitation are crucial mechanisms for reducing infant mortality and poverty. This knowledge has resulted in the development of a worldwide donor community with the specific target of supplying a safe water supply. By helping to ensure a safe and sufficient water supply, aid agencies are trying to bridge the gap between the wealthiest and poorest countries in the world. Globally, it is estimated that one in three people do not have access to safe drinking water, while two in five do not have access to handwashing facilities with soap and water.

The underlying goal in water aid is to develop economies that focus on equity, sustainability and strengthened systems. This in turn will help to:

- further reduce infant mortality
- increase opportunities for education (young women who are usually busy fetching water will be able to attend school and improve their opportunities)
- improve food production, reduce food insecurity and develop the economy.

Water aid can also be supplied during emergency periods, such as after a natural disaster like an earthquake or flood event (Figure 6.37). This is short-term aid that is put in place under unusual conditions. Such aid may be required by people from all economic backgrounds.

Figure 6.37: Local residents from the informal settlement of Mhlaseni Village, north of Durban in South Africa, receive water from a water tanker in May 2022 during the supply of water aid by volunteers from the South African Red Cross Society and the Qatar Red Crescent Society.

The Millennium Development Goal and Sustainable Development Goals

The United Nations (UN) designed eight Millennium Development Goals (MDGs) in order to tackle a wide range of social issues. These included the goals listed in Table 6.2.

1	Eradicate extreme poverty and hunger
2	Achieve universal primary education
3	Promote gender equality and empower women
4	Reduce child mortality
5	Improve maternal health
6	Combat HIV/AIDS, malaria and other diseases
7	Ensure environmental sustainability
8	Develop a global partnership for development

Table 6.2: Goals of the MDGs starting in 1990 with the target year of 2015.

The concept underlying the Millennium Development Goals was the access to a safe and sufficient water supply. Many of the targets set in the MDG were not met by the target year of 2015. However, a new set of Sustainable Development Goals (SDGs) were developed to continue the work started by the MDG. Unlike the MDG, the SDGs identify clean water and sanitation as a key goal. This goal is, of course, essential if the other goals are to be met. These goals are worked towards through funding from international aid programmes. They involve inter-governmental, NGOs (non-governmental organisations) and charity co-operation.

Figure 6.38: Sanitation workers protect aquatic plants on a floating island in the Yingtai ecological safety buffer zone designed to promote pollution control an environmental restoration. Zones like this one support the SDG of protecting water resources.

NGOs such as UNICEF (United Nations International Children's Emergency Fund) and charities such as One Drop, work to implement the supply of clean water, sanitation and education. Education is focused on as it is key in changing behaviour and reducing the wastage of water.

One such programme is the HSBC Water Programme. This programme is a collaboration between WaterAid, WWF and Earthwatch Institute. It is specifically designed to tackle water provision, protection, education and scientific research. Between 2012 and 2019, WaterAid supplied 1.72 million people with clean water access and 2.7 million with sanitation. It helped the most vulnerable people in countries achieve a healthier life.

CAMBRIDGE INTERNATIONAL AS LEVEL ENVIRONMENTAL MANAGEMENT: COURSEBOOK

6.3 Questions

1. Explain why water resources should be managed at the lowest appropriate level.
2. Describe the potential cause of a dropping water table. Explain what action can be taken to manage the problem.
3. Explain why climate resilience is so important with regard to water security.
4. Explain the importance of education in water management (consider individual, agricultural and industrial consumption).
5. **Assess** the roles international agreements play in increasing water security and decreasing conflict and poverty.

> **COMMAND WORD**
>
> **assess:** make an informed judgement

EXTENDED CASE STUDY

Arsenic pollution: natural water contamination in Bangladesh

Bangladesh (Figure 6.40) is located east of India on the Bay of Bengal. It is a low-lying country that has the Himalayan mountains to the north and the Indian Ocean to the south. It has a humid, warm climate with a high annual precipitation that falls mainly in the monsoon season. Large inland rivers make it vulnerable to flooding and sea level rise.

Figure 6.40: Map of Bangladesh showing the large river system. These rivers flow in from India and the Himalayan mountains. They carry large volumes of water that flow through soils with naturally occurring arsenic. This results in pollution of the water resources.

Due to high rates of precipitation, 98% of the population has access to water. However, only 35% of this water is safe for consumption. Surface water quality is compromised due to pollution and the associated bacterial contamination that causes illnesses such as diarrhoea. This has led the population to adopt groundwater as a safe, clean source of drinking water. Unfortunately, the groundwater of the Ganges Delta contains naturally high levels of arsenic. Chronic (long-term) exposure to arsenic is toxic to humans.

In the 1990s, an estimated 50 million people in Bangladesh were at risk of exposure to arsenic through the consumption of water from contaminated wells. Eating food grown or cooked using water from the wells also poses a health risk, as food sources are then also contaminated.

The causes of arsenic contamination

Arsenic in groundwater is common in many countries. Excess quantities of arsenic in drinking water has been reported from countries worldwide, and on every continent. The World Health Organisation estimates that at least 140 million people in 50 countries are exposed to arsenic through groundwater sources.

Bangladesh lies in the Bengal Basin which has the Ganga-Brahmaputra-Meghna river system running through it and into the delta. Both the basin and delta have been formed by the deposition of large volumes of naturally occurring sediments containing arsenic.

Arsenic leaches from the sediments into the groundwater. Exactly how the leaching occurs is not fully understood. Some theories suggest that microbial activity in the sediments releases the arsenic, while others suggest that chemical reactions release the arsenic.

6 Managing Water Supplies

CONTINUED

The impacts of arsenic contamination

Chronic exposure to arsenic causes skin problems (Figure 6.41) and cancers. It compromises brain development and results in miscarriages or malformation of foetuses. Ultimately, it causes multiple organ failure and eventually death. Every year an estimated 43 000 people die from arsenic poisoning in Bangladesh.

Figure 6.41: One of the symptoms of arsenic poisoning is dark marks on the hands. The hands of this person show evidence of arsenic poisoning.

As with many long term illness, patients suffering from arsenic poisoning are unable to work. Unable to afford medical treatment or a safe supply of drinking water, they find themselves trapped in the poverty cycle.

Solutions to the contamination of groundwater with arsenic

The solution to this problem is to ensure a sustainable, arsenic-safe water supply while educating the population about the health dangers of water which contains arsenic. The problem is that without proper diagnosis, people do not know if they are suffering from a common disease or one caused by long term exposure to arsenic. As illnesses only develop with long-term exposure to arsenic, it is difficult to convince people that prolonged contact with contaminated water may cause illness or death.

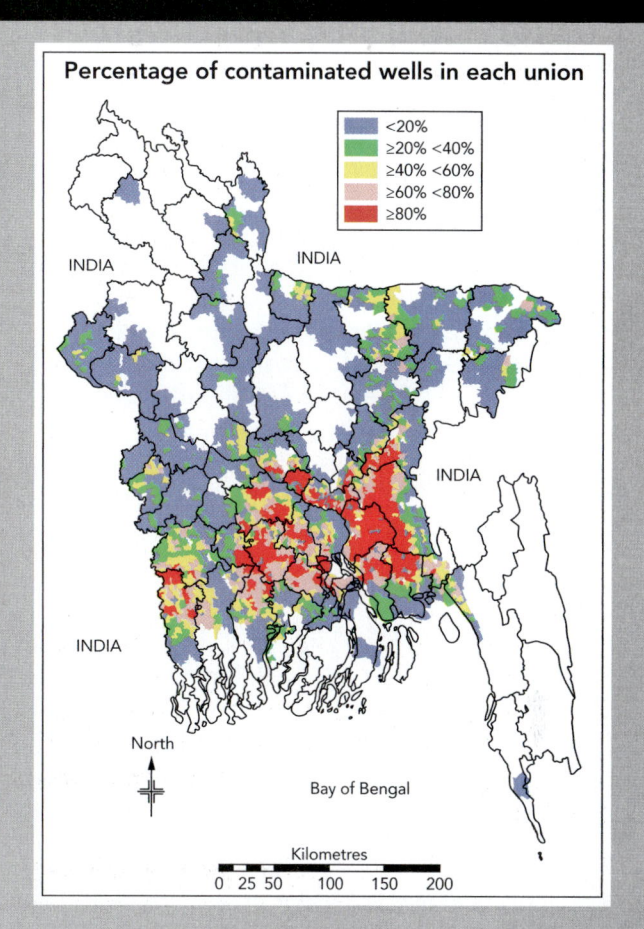

Figure 6.42: Arsenic contamination map of Bangladesh. This map shows where the highest levels of arsenic contamination in the groundwater of Bangladesh are.

Arsenic is normally found in boreholes 15–50 metres deep where water is stored in the arsenic containing sediments. However, arsenic contamination is not normally found in boreholes with depth greater than 150 metres. One solution is to install deeper wells, as this will avoid the water contaminated with arsenic. Another solution is to pipe water in from arsenic safe zones.

Several other programs were undertaken to provide a range of arsenic-safe options to affected areas of Bangladesh. By 2005, a total of 106 939 safe drinking water options were provided to assist those at risk. These options included rainwater harvesting, slow sand filters, arsenic iron removal filters,

263

CONTINUED

pond sand filters and deep bore holes. By 2009, almost 700 000 safe water supply options had been installed in arsenic contaminated areas. This supplied approximately 55% of the population.

To reduce the risk of unsafe water supplies, the Bangladeshi government continues to implement a wide range of initiatives. In response to the scale of the problem, the government formed national committees to undertake research and take action. This research was not just successful in determining where the highest risk areas were, and what water solutions were required. It also identified cases of arsenic poisoning (arsenicosis) and put public awareness programs in place to educate those at risk.

Extended case study questions

1. Discuss the long-term social and economic impacts of groundwater's contamination with arsenic on Bangladesh and its population.

2. The table gives statistics on the arsenic pollution in Bangladesh.

| In 1993 concentrations of arsenic were found to be up to 0.388 mg/litre. The Bangladesh safety standard for arsenic in drinking water is 0.05 mg/litre. |
| Arsenic pollution of the ground water was found in 62 of the 64 districts of Bangladesh. |
| Of the 4.95 million wells tested, 1.44 million contained arsenic at dangerous levels. |
| In 15% of the villages, more than 80% of the wells were contaminated. |

a. Calculate the percentage of districts that were found to have arsenic pollution in their groundwater.

b. Assuming that 200 villages were tested, and each had only one well, calculate how many wells were contaminated.

c. How far above the safe drinking standard was the arsenic concentration in 1993?

Extended case study project

In small groups, carry out research on the impacts of arsenic poisoning. Create an informational poster or presentation that could be used to inform the population of Bangladesh about the dangers of contaminated water. The poster should include:

a. a description of what arsenic is

b. where it is found in their environment

c. the impacts of long term consumption of arsenic

d. how to avoid drinking contaminated water

e. symptoms that may suggest arsenic poisoning.

SUMMARY

Water is distributed unevenly across the surface of Earth. It is found in the atmosphere, on the surface, under the surface as ground water and frozen in ice sheets and glaciers.
Only 2.5% of water is fresh.
The amount, location and quality of surface water changes over time.
Aquifers are underground water stores made of porous rock.
Water insecurity is the inability to access sufficient clean, safe water.
Climate change, natural disasters, pollution, population growth, competing demands, conflict and mismanagement are causes of water scarcity.
There are differences between urban and rural areas' access to water.
Impacts of water insecurity include food shortages, water related illnesses, poverty, and gender inequality.
Water management strategies include a wide variety of management options that protect water quality and quantity.
Water is an economic, social and environmental resource. Both demand and supply must be considered.
Governments must have regulatory frameworks in place to protect the quality and supply of water.
Water must be managed at the lowest appropriate level to help with poverty reduction.
The Water Convention is designed to facilitate cooperation between nations that have access to water that crosses international boundaries.

REFLECTION

Did the practical water filtration activity make it easier for you to visualise how an aquifer helps to filter and clean water?

Did you find the mathematical activity in 6.5 challenging? If so, consider what you need to do to get better at applying mathematical equations to real life problems. If you found it easy, did you assist others to understand the task? How did that help their learning and your understanding?

PRACTICE QUESTIONS

1 The article below discusses a community in Kenya that suffered a period of prolonged drought.

> ### The struggles of farmers in the Tana River region
>
> The farmers in a camp made up of a few round huts, in the drought stricken, remote region of the Tana River, watch as a truck full of bananas comes bouncing down the dirt track. How, in this dry, dusty landscape is it possible that some farmers flourish while others continue to struggle?
>
> Some 77% of the farmers in the camp are from the Tana North District. They live below the poverty line and have a long-term reliance on food aid. The truck belongs to farmers who have been supported by the Kenya Red Cross Society sustainable livelihoods project. Despite the deepening drought, they are able to grow a good crop of bananas.
>
> The driver of the truck has 60 bunches of bananas to sell. He makes the trip twice a week and is able to earn money to support his family. He is doing good business and is able to support his family.

   a Explain why the farmers developed a reliance on food aid. [4]

   b Is the production of bananas in a water scarce area sustainable in the long term? State your answer, giving reasons. [3]

   c Outline **three** ways that population growth increases the pressure on water resources. [6]

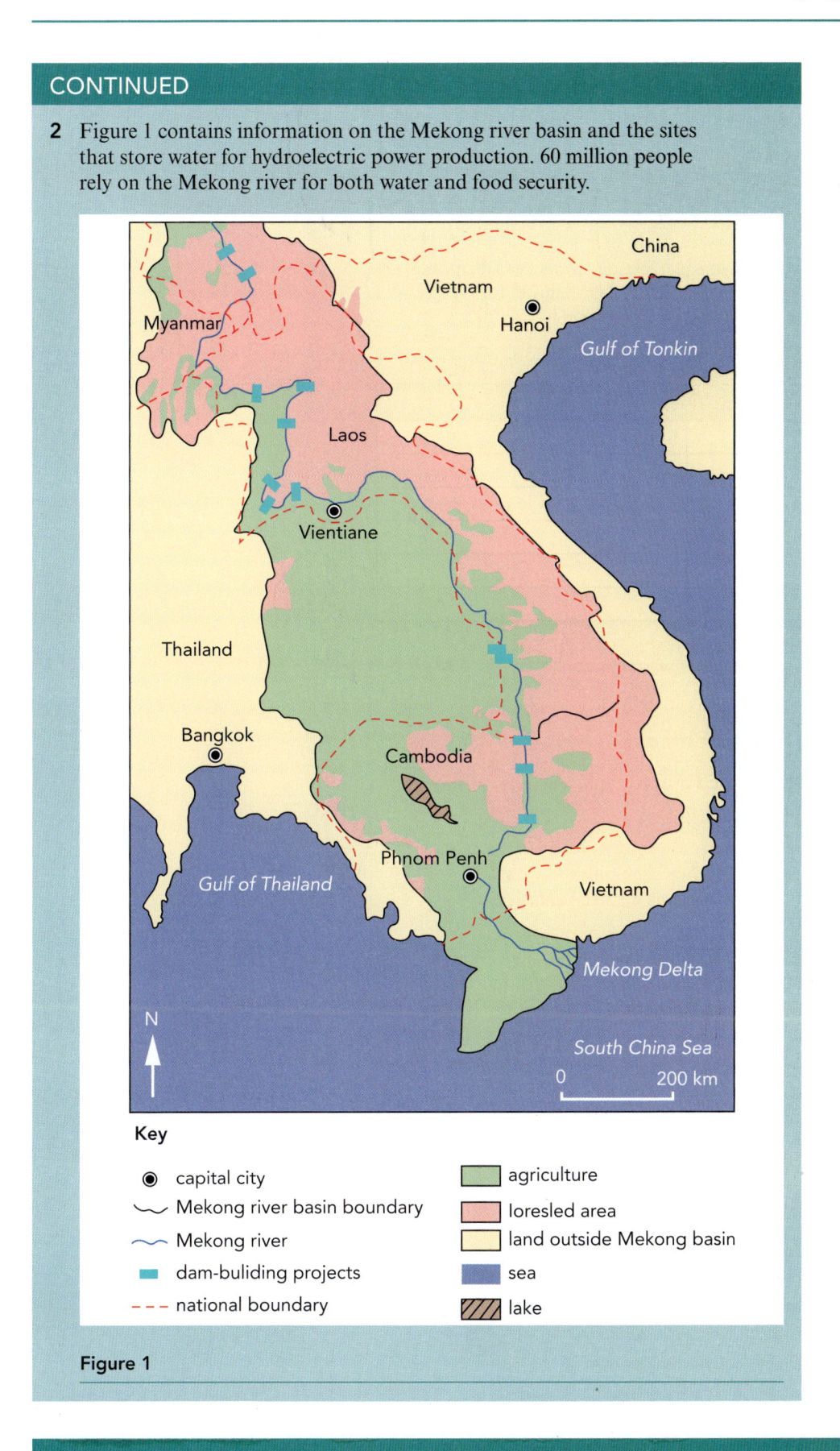

CONTINUED

2. Figure 1 contains information on the Mekong river basin and the sites that store water for hydroelectric power production. 60 million people rely on the Mekong river for both water and food security.

Key

- ⊙ capital city
- ∽ Mekong river basin boundary
- ∼ Mekong river
- ▬ dam-buliding projects
- --- national boundary
- ▨ agriculture
- ▨ loresled area
- ▨ land outside Mekong basin
- ▨ sea
- ▨ lake

Figure 1

CAMBRIDGE INTERNATIONAL AS LEVEL ENVIRONMENTAL MANAGEMENT: COURSEBOOK

CONTINUED

a Define the term *water security*? [2]

b With reference to Figure 1 explain why there is a risk of water-related conflict arising. [5]

c The Mekong river basin suffered severe drought in 2019/20 when the monsoons failed. This lack of rainfall, in combination with the damming of the river, resulted in the Mekong river's flow dropping to its lowest level in more than 100 years. Table 1 shows the difference in level of in the Mekong over six years.

Mekong river water levels	
Year	Level (metres)
2016	5.0
2017	4.5
2018	3.0
2019	2.5
2020	2.8
2021	2.5

Table 1: Water level in the Mekong river.

d Plot the data in table as a bar chart. [4]

e Describe the trend shown by the chart. [2]

f With reference to Figure 1, the data in Table 1 and your own knowledge, describe the impact of the extremely low flow on farmers in the Mekong river delta. [3]

3 Under normal conditions, the seaward movement of fresh water prevents saltwater intrusion into aquifers. The point at which fresh and salt water meet normally remains near the coast or deep underground.

Figure 2 shows a map of a Bangladesh that suffers from salt water intrusion, while the data and graph show the increase in land area affected by salt water intrusion. Bangladesh has a population density of 1265 people/km^2 and suffers from flood events, cyclones and storm surges. The country is also experiencing loss of mangroves due to deforestation.

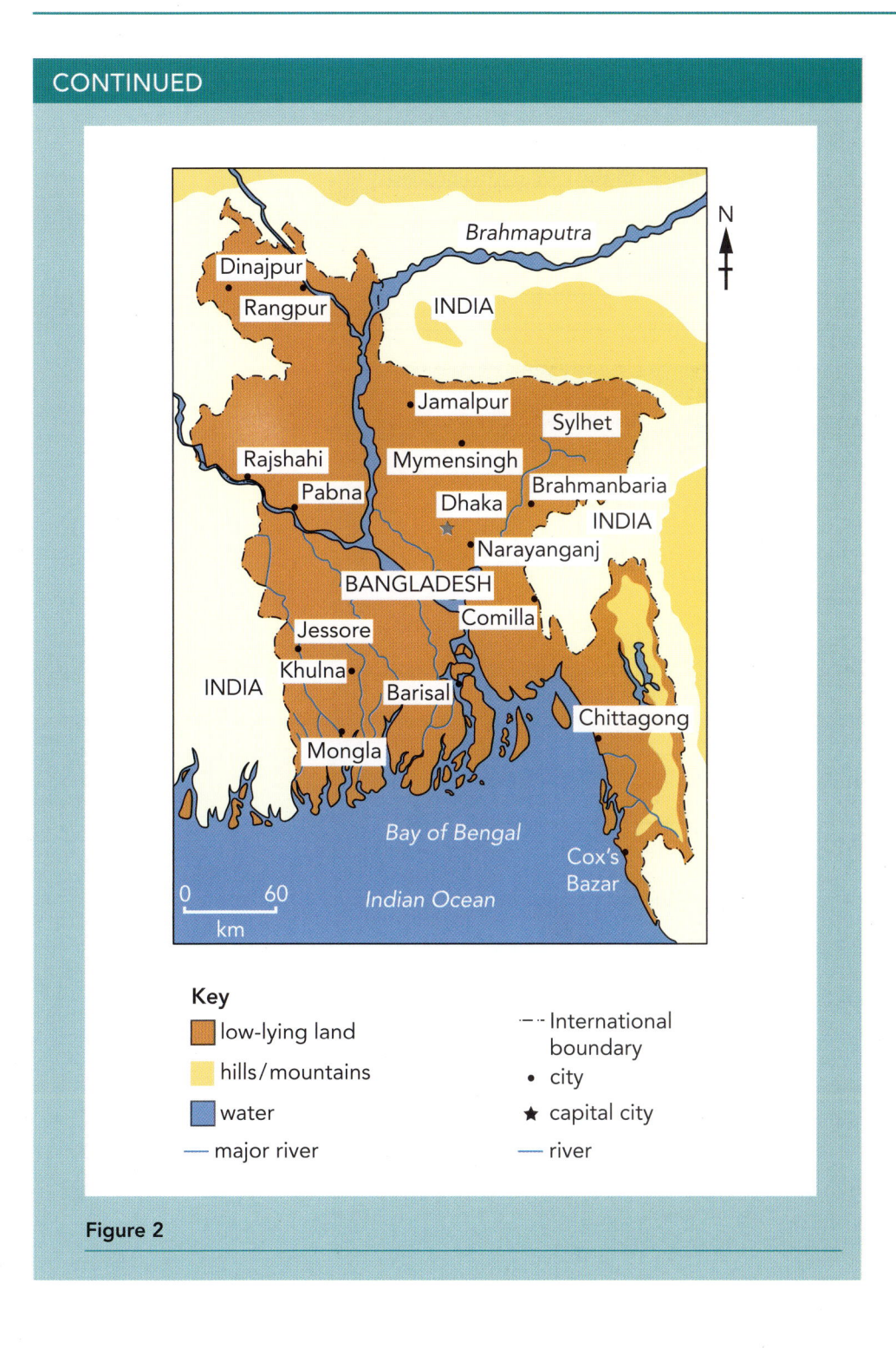

CONTINUED

Figure 2

CONTINUED

Year	Area affected by salt water intrusion (million hectares)
1973	83.3
2000	102.0
2009	105.6
2018	111.2
2030 (predicted)	119

Table 2: The area affected by salt water intrusion in Bangladesh between the years 1973 and 2018, with predicted areas of intrusion for 2030 (area in million hectares).

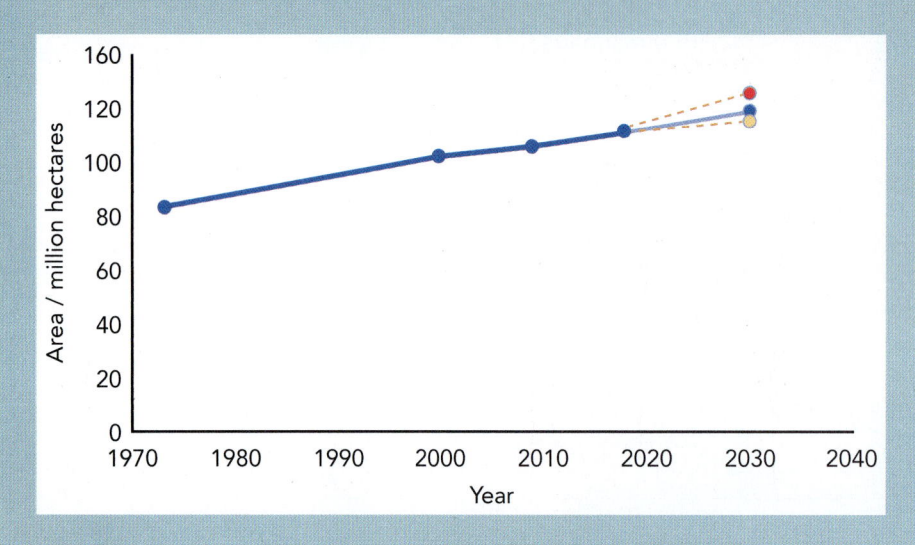

Figure 3

a Describe *salt water intrusion*. [2]
b Using Figure 3, calculate the percentage increase in area affected by salt water intrusion between 1973 and 2018. [2]
c Figure 3 shows an additional two possibilities for the area affected by salt water intrusion for 2030. Explain why there is a variation in future estimates. [2]
d With reference to Figure 3 and your own knowledge explain why Bangladesh is at risk from salt water intrusion. [5]

CONTINUED

Figure 4 shows the differences in access to improved water sources and sanitation in rural and urban areas (Global vs LICs) between 1990 and 2004.

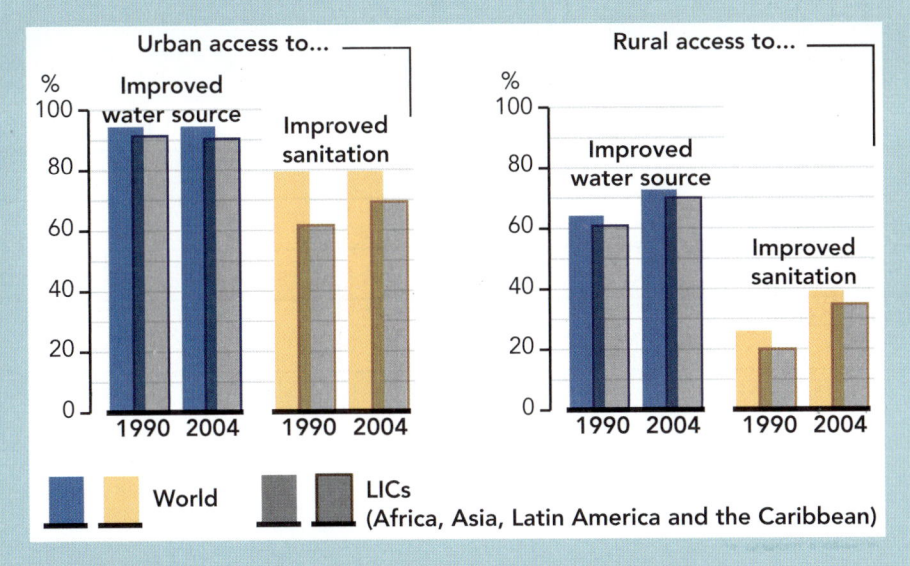

Figure 4

a State which sector showed the greatest increase in access to sanitation between the years 1990 and 2004. [1]

b With reference to Figure 4, explain why greater death rates and shorter life expectancy may be expected in rural areas compared to those in urban areas. [3]

c 'Urban areas always have better access to a safe water source than rural areas.' Evaluate this statement. [6]

4 'The most significant cause of water insecurity is climate change.' How far do you agree with this statement? Give reasons and include information from relevant examples to support your answer. [20]

SELF-EVALUATION CHECKLIST

After studying this chapter, think about how confident you are with the different topics.
This will help you to see any gaps in your knowledge and help you to learn more effectively.

I can	Needs more work	Getting there	Confident to move on	See section
Describe the distribution of Earth's water.				6.1
Define the term water security.				6.1
Explain the causes of water insecurity.				6.2
Explain the impacts of water insecurity.				6.2
Describe and evaluate strategies for managing water security.				6.3

Chapter 7
Managing the Atmosphere

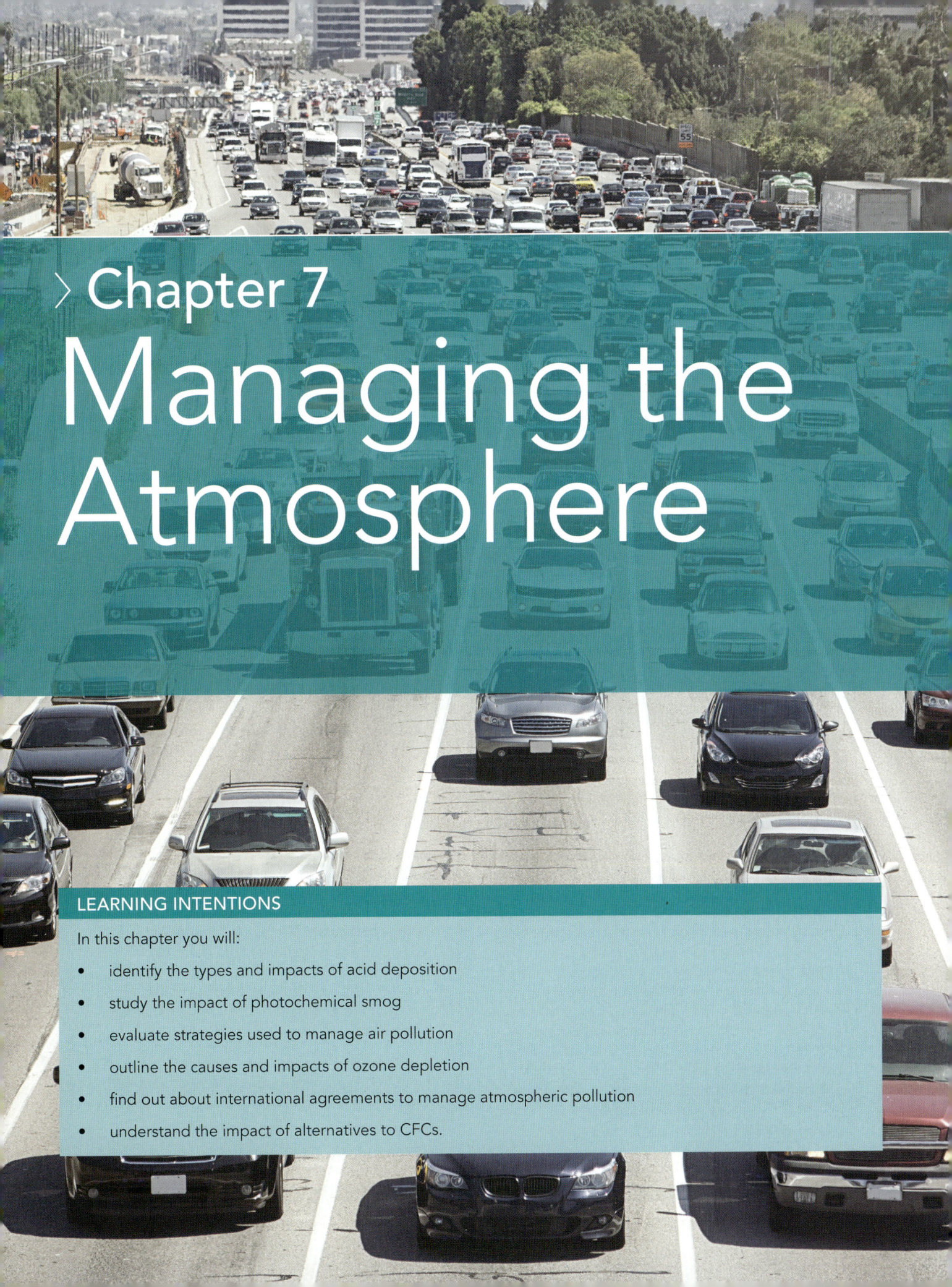

LEARNING INTENTIONS

In this chapter you will:

- identify the types and impacts of acid deposition
- study the impact of photochemical smog
- evaluate strategies used to manage air pollution
- outline the causes and impacts of ozone depletion
- find out about international agreements to manage atmospheric pollution
- understand the impact of alternatives to CFCs.

GETTING STARTED

In small groups discuss the following questions.

1. Why is air pollution usually worse in cities than in rural areas?
2. What sources of air pollution have you noticed where you live?
3. Does weather affect air pollution in the area where you live?
4. Describe the effects that air pollution has on people.
5. Consider what could be done to manage air pollution.

As a class discuss some actions that can be taken to reduce local air pollution.

ENVIRONMENTAL MANAGEMENT IN CONTEXT

Air Quality Index and real time air pollution monitoring

An Air Quality Index (AQI) is a measure of air quality which governments use to assess polluted the air is or how polluted the air is expected to become. Different countries have their own AQIs. AQI data is collected using air quality monitoring equipment. Levels of air pollution increase due to traffic, fires, or even changes in the weather. As air pollution increases, the risk to public health increases. Real-time monitoring of air quality helps governments give advice to people in affected areas.

Real-time air quality monitoring is the continuous testing and assessment of air quality. This process identifies when air pollution exceeds the AQI, posing a risk to the surrounding environment and people. Real time air quality monitoring requires the establishment of a complex, dense network of monitoring equipment in order to collect as much information as possible. Feedback from these monitors is useful when planning the management of air pollution in a specific area.

There are many real-time air quality monitoring systems. They are used to monitor the critical environmental parameters (measurable factors) related to air quality, including noise, smell, weather, radiation, pollutants and pollen levels. Key pollution parameters such as **particulates (PM$_1$, PM$_{2.5}$ and PM$_{10}$)** carbon monoxide (CO), CO_2, SO_2, NO_2 and **ground-level ozone** are monitored using these systems in cities around the world. Governments and urban management teams use information from these systems to determine if the AQI has been exceeded and if the quality of the air poses a threat.

Figure 7.1: Smog hanging over the city of Shanghai, China in 2016.

KEY TERMS

particulates: solid particles and liquid droplets in the air. Generally, these come from any type of burning or dust-generating activities

*PM$_{10}$: particulate matter of diameter of 10 micrometres or smaller. The particles can be inhaled. The number next to the PM indicates the size of the particulate matter being measured, e.g. PM$_1$ is 1 micrometre in diameter or smaller

ground-level ozone: forms when NO$_x$ and volatile organic compounds (VOCs) react in the presence of sunlight

7 Managing the Atmosphere

CONTINUED

Air quality varies from one place to another. Different communities are exposed to different types and levels of pollutants. In order to improve air quality and protect the population when air quality is poor, it is important to know what the pollutants are and where they are coming from. Most monitoring systems around the world are not densely spaced enough to identify how the air quality differs spatially, and therefore how different communities are impacted. This limits their effectiveness.

Figure 7.2: Air pollution in a rural area in India. The prevailing wind is blowing pollutants into surrounding valleys.

Discussion questions

1. How could authorities use collected data to manage air pollution in an urban area?
2. Explain why a network of air monitors has to be relatively dense to be useful for air pollution management.
3. Describe the limitations that an LIC may face when developing an air monitoring system.
4. Explain why monitoring air pollution is important.

7.1 Acid deposition and photochemical smog

Acid deposition and photochemical smog are both forms of air pollution that are a by-product of industrial activities. Some gases in atmospheric pollution come from natural sources, such as volcanoes. However, the major sources of air pollution are from the combustion of fossil fuels, which releases gases such as SO_2 and NO_x into the atmosphere.

Wind can blow these pollutants long distances, causing pollution in areas that are not responsible for the release of the pollutants. This makes air pollution a global problem.

Figure 7.3: Air pollution from industrial activity.

Primary air pollutants are the direct result of an industrial process or volcanic activity. Primary pollutants are released from the source and have not undergone any change. An example of primary pollution is sulfur dioxide (SO_2) emitted from coal power stations.

KEY TERMS

acid deposition: a mix of air pollutants that deposit from the atmosphere as acidic wet deposition (with a pH < 5.6) or acidic dry deposition

*primary air pollutant: an air pollutant emitted directly from the source

Secondary air pollutants come from chemicals in primary air pollution reacting either with each other or sunlight. This creates a secondary compound. **Smog** is an example of a secondary pollutant.

Acid deposition is the deposition of either wet pollutants, **acidic wet deposition**, or dry pollutants, **acidic dry deposition**, from the atmosphere onto the surface of Earth. These pollutants are acidic in nature, and have a pH of 5.6 or less.

> KEY TERMS
>
> *secondary air pollutant: an air pollutant that forms when pollutants react in the atmosphere
>
> *smog: a mixture of smoke and fog; smog intensifies atmospheric pollution
>
> wet acid deposition: atmospheric pollution deposited on Earth's surface as precipitation, such as snow, rain, hail and fog
>
> dry acid deposition: atmospheric pollution deposited on Earth's surface in the absence of moisture, like dust or gases

Types of acid deposition

Acidity is the measure of the pH of a chemical compound. The pH scale runs from zero (the most acidic) to 14 (the most alkaline or basic). A substance that is neither acidic or alkaline, such as pure water, is referred to as neutral (pH 7 only).

Wet deposition is acidic atmospheric pollution deposited as rain, snow, sleet, fog or hail. Wet acid deposition is also broadly referred to as acid rain. Wet acid deposition has been recorded with a pH as low as three, which is similar to the pH of vinegar.

Dry deposition is when gases and dust particles become acidic in the atmosphere and are deposited onto surfaces without the addition of moisture. When dry deposition is washed off by the next rain event, the resulting acidic water flows over and through the ground, harming plants and wildlife.

Both wet and dry deposition contain acidic components such as sulfuric or nitric acid. They can be carried over long distances and across international borders by the wind. This makes acid deposition an international challenge, not just a problem for those living near the source. Both wet and dry deposition can land on trees, soil, water bodies, cars and buildings. Both are corrosive. In lakes and rivers, deposition makes the water more acidic.

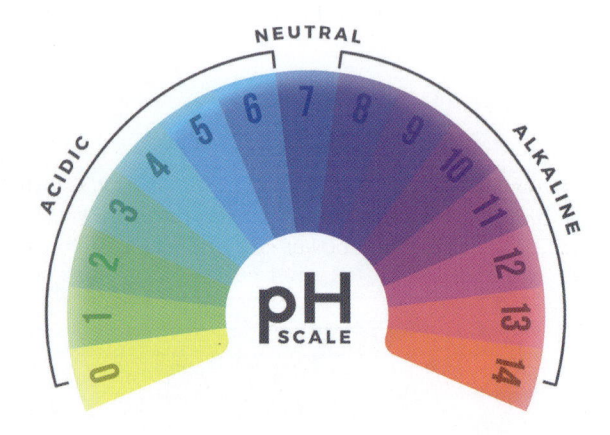

Figure 7.4: The pH scale ranges from extremely acidic at 0 to very alkaline at 14. Only 7 is neutral.

Figure 7.5: Pollution from a factory can be blown by the wind and rain back down as wet acid deposition in another area.

Formation of acid deposition

Acid deposition is formed when sulfur dioxide (SO_2) and nitrogen oxide (NO_x) are emitted into the atmosphere and transported from the point of pollution by winds and air currents. SO_2 and NO_x react with water, oxygen and other chemicals to form sulfuric and nitric acids.

A small portion of SO_2 and NO_x originates from natural sources such as volcanic eruptions. However, most of it comes from the burning of fossil fuels (coal, oil and gas).

Fossil fuels contain sulfur compounds. When fossil fuels are burned, SO_2 gases are released. SO_2 reacts with water and oxygen in the atmosphere to form sulfuric acid (H_2SO_4). Nitrogen from the atmosphere reacts with oxygen in the high temperatures in vehicle engines to form nitrogen monoxide gas. Therefore, nitrogen monoxide (NO) gas is released into the atmosphere from vehicle emissions. Nitrogen monoxide reacts with oxygen and water in the atmosphere to form nitric acid (HNO_3). These processes are summarised in Figure 7.6. In each of these cases, wind blows the pollution from its source to distant locations.

It is estimated that two thirds of the SO_2 and one quarter of the NO_x in the atmosphere comes from diesel-powered generators. Coal burning power stations create the most acid deposition.

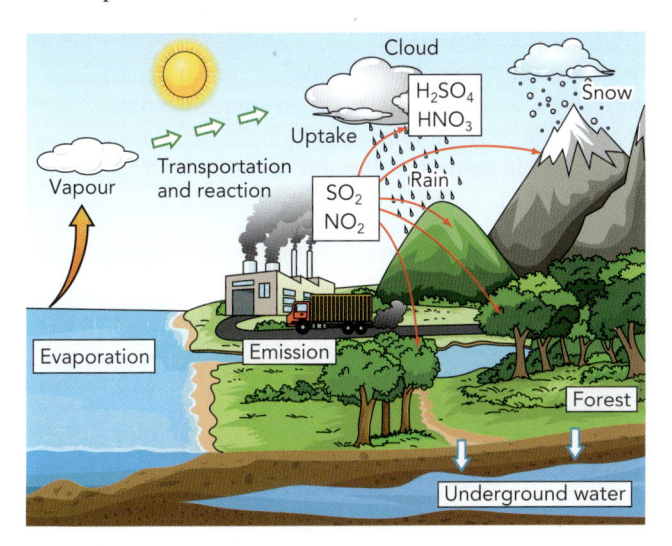

Figure 7.6: The formation and movement of atmospheric acid deposition and acidic deposition.

Impacts of acid deposition

Aquatic environments

The harmful effects of acid deposition on ecosystems can be clearly seen in freshwater aquatic environments such as lakes, streams and wetlands. Many fish have a 'critical pH range' in which they are able to thrive. A pH that drops below this critical range is harmful.

When acidic water flows through soil, it leaches aluminium from clay particles and transports them to bodies of water. Aluminium is particularly toxic for gill-breathing organisms like fish. In acidic waters, aluminium can reduce gill function and make it hard for fish to breathe. High levels can even kill gill cells completely, causing the fish to die.

In general, most young organisms are more sensitive to environmental conditions than adults. At a pH of 5 or lower, many fish eggs do not hatch, and at lower pH levels most adult fish die. In many instances, species that can tolerate the lower pH (like ducks, Figure 7.7) and elevated aluminium levels die because the food chain has been disrupted and their food source has disappeared.

Figure 7.7: A scientist collecting water samples to test them for pollution and levels of acidity.

In shallow coastal waters, acid deposition can have very harmful effects. The acidification of seawater prevents marine organisms from creating their calcified exoskeletons. Corals are particularly sensitive to the pH of the water as their skeletons are made of calcium carbonate ($CaCO_3$), which dissolves at lower pH levels. The loss of corals affects the entire food web and biodiversity of coastal zones.

The nitrogen in acid deposition also causes problems in aquatic ecosystems as it adds to the nutrient load in the water. This can cause eutrophication which puts further stress on the ecosystem.

Vegetation and crops

A common sight in areas affected by acid deposition is dead or dying vegetation. Increased acidity causes aluminium to seep out of the soil which can be harmful to plants and animals. This increased acidity also removes beneficial nutrients and minerals from the soil, reducing the growth of plant life.

At higher elevations, low-lying acidic fog coats vegetation with an acidic film. As a result, nutrients are stripped from plants' foliage which causes them to turn brown. When this is severe, it results in the loss of leaves (Figure 7.8), without which plants are unable to photosynthesise. This results in plants being weakened and unable to withstand cold temperatures. For farmers, this may lead to reduced crop yields.

Figure 7.8: Vegetation stripped of foliage by acid deposition.

An example of extreme acid deposition pollution occurred in the 1970s and 1980s. The so-called Black Triangle in Europe (Figure 7.9), which included areas of the Czech Republic, Germany and Poland, received heavy acid deposition. Entire forests died, while railway tracks became corroded. Emissions from coal-burning factories in Eastern Europe were the cause of this problem. In 1979, the Geneva Convention on Long-range Transboundary Air Pollution put strict pollution regulations in place to control the acid precipitation. This significantly reduced the acid deposition in the region.

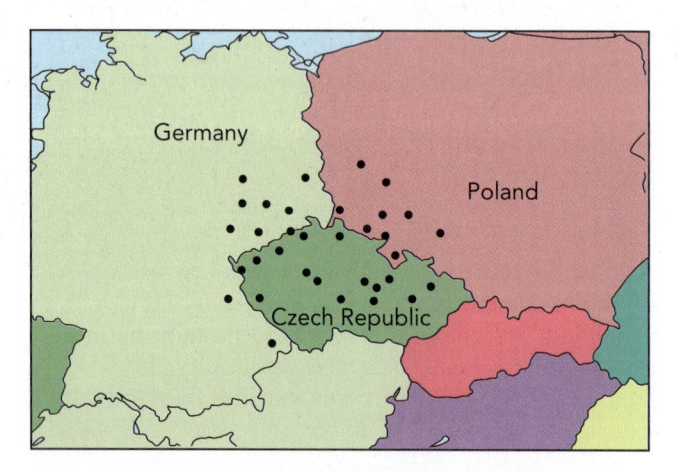

Figure 7.9: A map of Northern Europe showing the area most affected by the air pollution during the 1970s and 1980s, known as 'The Black Triangle'.

Stone and brick buildings

Some of the most visible damage from acid deposition occurs on human structures. Limestone and marble buildings, monuments, tombstones and statues all deteriorate more rapidly when exposed to acid deposition. The calcium carbonate in limestone and marble reacts with sulfur dioxide (SO_2) to form a softer material known as gypsum. This causes the rock to flake off or dissolve.

Damage caused by acid deposition can be seen on roofs, stone carvings and the limestone or marble blocks used for building (Figure 7.10). This damage has occurred in many countries around Europe, including the Black Triangle.

Acid deposition can be expensive, as ruined materials need to be repaired or replaced. In some instances, the damage is irreparable. For example, many stone carvings hold cultural value and cannot be replaced or easily fixed.

Figure 7.10: Visible damage caused by acid deposition to sandstone sculptures on a building in Leipzig, Germany, in 1990. Acid deposition dissolves the rock and sculptures lose their definition and shape.

7 Managing the Atmosphere

Photochemical smog

The main sources of pollution in cities are motor vehicles (Figure 7.11), wood burning heaters and industry.

> **KEY TERMS**
>
> **photochemical smog:** a mixture of air pollutants and particulates, including ground level ozone, that is formed when oxides of nitrogen and volatile organic compounds (VOCs) react in the presence of sunlight
>
> **volatile organic compounds (VOCs):** compounds that have a high vapour pressure and a low water solubility; VOCs are emitted as gases from certain solids or liquids including paints, paint strippers, cleaning supplies, pesticides and building materials
>
> **ozone:** O_3, a colourless, odourless gas found naturally in the stratosphere and formed from oxygen by UV light

Figure 7.11: Ayrton Senna Highway at rush hour, São Paulo, Brazil, South America.

These factors all release pollutants that form the base of photochemical smog which can leave a brown haze above a city (Figure 7.12). **Photochemical smog** is the mixture of air pollutants, particulate matter and **volatile organic compounds (VOCs)** that are changed from primary pollutants to secondary pollutants in the presence of sunlight. An example of a secondary pollutant is ground-level **ozone** which is formed when chemical reactions between NO_x and VOCs occurs. Photochemical smog tends to occur more often in the summer when there is a greater number of sunlight hours each day.

Figure 7.12: Photochemical smog over Naples, Italy, in 2016.

INVESTIGATIVE SKILLS 7.1

Air pollution investigation

This investigation uses water to represent the atmosphere, and food colouring and other materials to represent various air pollutants. It gives a visual representation of how our own activities contribute to air pollution in our region.

You will need:
- a glass beaker
- liquid food colouring for each table (blue, red and green)
- ground charcoal
- cocoa powder
- powdered drink with a strong colour
- clean water
- a large container to dispose of waste.

Getting started

First decide whether you think your local air quality is good or bad, and whether there is any evidence of air pollution.

Do you ever experience a burning sensation in your eyes, an itchy throat or difficulty breathing?

CONTINUED

Method

Work in groups of between 3 and 6.

Check your station has the following items before you start:

- Food colouring, ground charcoal, cocoa powder and powdered drink.
- A glass beaker three-quarters full for each learner in the group.
- The teacher will read out the list of activities from Table 1. For each activity that you have participated in within the last week, add one drop/pinch of the corresponding 'pollutant' into your beaker of water.
- If you did not do the activity mentioned, do not add anything to your beaker.

All of the activities involve the burning of fossil fuels or a release of VOCs.

Air pollutant	Corresponding colour/mix
Sulfur dioxide (SO_2)	Pinch of drink mix
Nitrogen oxide (NO_x)	Pinch of cocoa powder
Carbon monoxide (CO)	One drop of red food colouring
Lead (Pb)	One drop of green food colouring
VOCs	One drop of blue food colouring
Particulate matter (PM_{10})	Pinch of ground charcoal

Table 1

Activities 1–8:

1. You came to school in a private car
   - a   1 drop red colouring (CO)
   - b   1 drop blue colour (VOC)
   - c   1 pinch drink mix (SO_2)
   - d   1 pinch cocoa powder (NO_x)

2. You enjoyed a warm shower.
   - a   1 drop green colouring (Pb)
   - b   1 pinch drink mix (SO_2)
   - c   1 pinch cocoa powder (NO_x)

3. You got ready for school and used nail polish or hair spray.
   - a   1 drop blue colour (VOC).

4. You use a computer, laptop, cell phone
   - a   1 pinch of ground charcoal (PM_{10})
   - b   1 drop green colour (Pb)
   - c   1 pinch drink mix (SO_2)

5. Your family burned wood or garden waste
   - a   1 drop red colouring (CO)
   - b   1 pinch of ground charcoal (PM_{10})

6. You travelled down a dirt road
   - a   1 pinch of ground charcoal (PM_{10})

7. Your family used some paint or solvent
   - a   1 drop blue colour (VOC)

8. You or your family used a petrol-powered lawn mower or leaf blower in the garden.
   - a   1 drop blue colour (VOC)
   - b   1 pinch drink mix (SO_2)
   - c   1 pinch cocoa powder (NO_x)

Discussion

- How 'polluted' is the 'air' in your cup? Would you be happy to breathe it in?
- What other sources of air pollution occur in day-to-day life?
- What could you do to reduce the number of pollutants released each day?

At the end of the activity, dispose of your water sample safely in the container provided.

7 Managing the Atmosphere

> **SELF ASSESSMENT**
>
> - Could you have been more precise in your addition of 'pollutants' to the 'air'?
> - Did you follow the instructions clearly to get results that were comparable to others?
> - Did other students have worse or better 'air quality' results than you did? Have you considered why their results were different from yours?

Impacts of photochemical smog

Photochemical smog has a number of harmful effects on both humans and the environment. In humans and animals it can cause painful irritation to the respiratory system, difficulty breathing and reduced lung function. High levels of smog can also trigger asthma attacks.

Figure 7.13: Asthma attacks can be aggravated by photochemical smog which irritates the lungs.

Smog is also an irritant to eyes. In areas where ground level air pollution is prevalent, eye conditions like conjunctivitis and eye inflammation are more common. Ground-level ozone found in smog can cause a condition known as dry eyes. This occurs when tear ducts do not produce enough liquid to keep eyes moist.

Crop yields are also affected by smog and ground-level ozone. Ozone damages crop production by entering the leaves during normal gaseous exchange. Ozone is a strong oxidant that causes harm to plants, and results in the yellowing of leaves, creating light brown spots, reddening and bronzing. As a result, plant growth is damaged and crop yields decline.

In 2014, a study in India found that air pollution was severely reducing yields of wheat and rice crops. Compared to 1980, yields were 36% lower than they should have been in 2010. This equates to a reduction in yield of approximately 24 million tonnes of wheat in 2010. In 2019 it was estimated that between 5% and 12% of the global staple crop yield (rice, maize, wheat and soya) was lost due to the effects of air pollution.

Plastics and rubbers are subject to deterioration when exposed to pollutants and UV light. Particles and dust stick to plastic and rubber surfaces, causing chemical damage and disfiguration. Abrasive air pollutant particles can scratch plastics, causing significant weakening. Plastics and rubbers are also susceptible to developing sticky surfaces.

Figure 7.14: Cracking in plastic sheeting, which will deteriorate and break into smaller pieces.

Acidic atmospheric pollutants such as SO_2 and NO_x accelerate the decay of rubbers and plastics, especially in humid environments where the pollutants mix with moisture to form acidic water. In addition, ground-level ozone can cause rapid oxidation of the plastics and rubbers, weakening them. Different types of plastics respond differently to pollutants. However, evidence such as stress fractures and cracks (Figure 7.14) indicate plastics that have been damaged by pollutants, ozone or UV exposure.

7.1 Questions

1. Explain the difference between wet and dry acid deposition.
2. Describe the impacts of acid deposition on the environment.
3. Explain the impact of acid deposition on humans.
4. Describe photochemical smog and how it is formed.
5. Give some of the impacts of photochemical smog.

CASE STUDY

Air pollution in the Greater Cairo region

Cairo, Egypt lies on the Nile Delta. It is one of the most densely populated regions of the world, with approximately 52 200 people per km^2 and a population of over 20 million. Air pollution is a significant environmental, social and economic issue in this megacity.

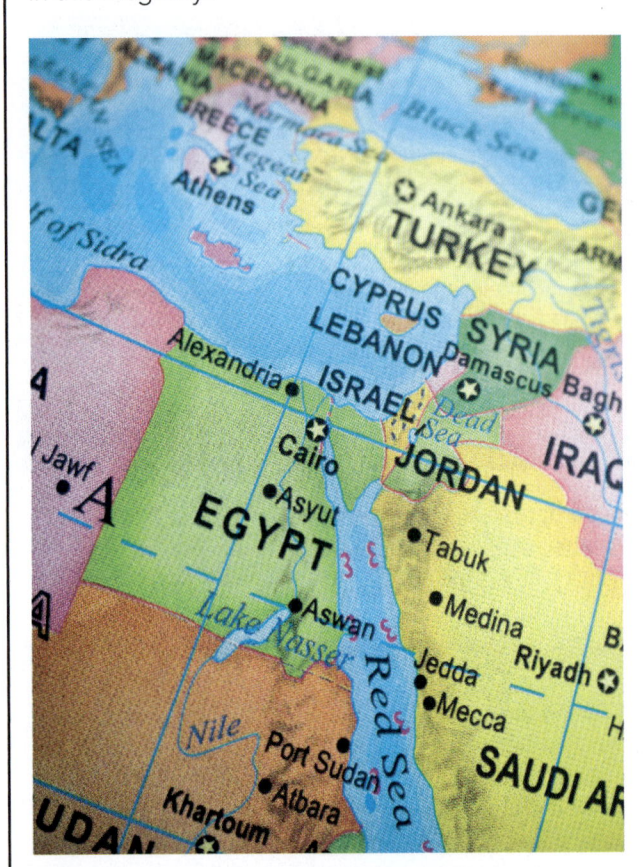

Figure 7.15: Map of Egypt in North Africa, showing the location of Cairo at the apex of the Nile delta.

Causes of air pollution in Cairo

Atmospheric pollution in Cairo has many causes. These are linked to the city's large population size, geographical relief and climate. They include:

- Very little vegetation remains.
- Dust from the surrounding desert and low levels of precipitation.
- The Mokattam hills to the southeast of the city prevent prevailing northerly winds from blowing the pollution away.
- Narrow streets with tall buildings which trap pollutants.
- A large number of private cars, with many old vehicles that emit high levels of pollutants.
- The narrow streets easily become congested.

Figure 7.16 Traffic congestion in Cairo increases air pollution in the city due to vehicle emissions.

> **CONTINUED**

- The annual burning of rice straw waste by rice farmers in the Nile delta, between September and November every year. This results in toxic smoke drifting over Cairo.

- Emissions from the industry surrounding the city.

Impacts of atmospheric pollution

Thousands of deaths due to atmospheric pollution are recorded each year, and the number of people suffering from respiratory (lung) disease is increasing annually.

The economy is also being hit hard. Poor air quality affects the ability of the population to work, and this has an impact on the productivity of industry. Air pollution even affects the population structure of Cairo, with young families choosing to leave the city and move to outlying cities that have lower levels of pollution. As a result, the city's population is an aging one.

Figure 7.17: Pollution over the pyramids in the Greater Cairo area.

Solutions to Cairo's air pollution challenges

- Monitoring urban air pollution allows policy makers to evaluate the level of pollution and try to minimise its occurrence.

- Developing public transport systems to reduce the number of vehicles on the road. A third metro line is being developed, and new bus companies mean accessible public transport. Cycling has also become more popular in recent years.

- Controlling the illegal burning of agricultural waste (rice straw), which causes a black cloud over the Nile Delta every year. This can be done by seeing rice straw as an asset rather than a waste product. Entrepreneurs are using the straw to create and sell artistic products. Others are using it to build houses. The government is also paying farmers to deliver their straw to collection depots, so the straw can be put to other uses.

- Greater regulation and management of vehicle emissions would help to reduce atmospheric pollution in Cairo.

- Tight regulation of emissions from the large industries that surround the city would help to control pollution. Where industry is found to be breaking the law, administering fines and shutting factories down would increase compliance.

Figure 7.18: Farmers burning their agricultural waste to prepare the land for the next crops. Smoke from the burning crops adds to air pollution over Cairo.

Cairo is not unique; only 12% of the urban areas around the world meet the WHO air quality standards. LICs tend to suffer the most as there is rapid urbanisation and industrialisation with limited controls or environmental regulation.

> CAMBRIDGE INTERNATIONAL AS LEVEL ENVIRONMENTAL MANAGEMENT: COURSEBOOK

CONTINUED

Case study questions

1 Using the information in the table below:

Old car	New car
For every 24 000 km 1890 litres of fuel will be burnt	For every 24 000 km, 1419 litres of fuel will be burnt
Over the course of a year the old car will emit 4.35 tonnes of CO_2	Over the course of a year the newer car will emit 3.25 tonnes of CO_2
Older cars need more servicing, replacement parts, and more engine oil	
Vehicles from the 1990s or earlier emit far more gases than new cars	

a Explain why old cars create greater urban air pollution problems than newer ones.

b Calculate the percentage difference between the emissions of older cars and the newer cars over 24 000 km.

2 How do public transport and cycling reduce urban air pollution?

3 Explain why rice farmers would want to burn their rice straw each year.

4 What do you think the greatest challenge is for the authorities in Cairo when considering the management and control of atmospheric pollution?

7.2 Managing air pollution

Strategies for managing air pollution

To control air pollution, the source of pollution needs to be identified. The pollution then needs to be reduced, eliminated or prevented. This can be done by using fewer toxic raw materials or fuels, by using pollution reducing technology in industry, or by improving the efficiency of production in factories.

Air pollution does not recognise geographical boundaries. It can travel great distances, so local, regional, national and international strategies need to be employed to combat air pollution. This means that strategies to improve air quality need to be managed at government level to ensure controls are established countrywide and ultimately globally.

Reducing fossil fuel use and renewable energy

One of the most significant strategies employed to control atmospheric pollution and acid deposition is using a reduced amount of fossil fuels. This has been achieved through the development of renewable energy sources and more efficient technologies. This long-term strategy requires time to implement fully as many daily operations still rely on burning fossil fuels. The transition to clean energy cannot stand alone. It has to be combined with other technologies to reduce the ongoing impact of burning fossil fuels.

Figure 7.19: Solar panels supply renewable energy on roof areas in Singapore, allowing energy generation without habitat destruction.

Renewable energy (detailed in Chapter 5) is often referred to as clean energy. It is generated through harnessing natural sources that are constantly being replenished (wind, tides, sunlight) so it can be used sustainably. Renewable energy production is expanding as it has become more cost effective thanks to the development of new technology. The lack of emissions, such as SO_2, means that the production of clean energy does not add to the acid deposition problem. This helps to combat air pollution.

Figure 7.20: An electric car charging at a public charging station in Berlin, Germany.

Electric vehicles are another example of technology that has been introduced to reduce the use of fossil fuels. This technology is most effective where vehicles are charged from renewable energy sources. The availability of public charging stations encourages people to purchase electric cars. These vehicles do not burn fossil fuels and so they do not produce emissions. This has a positive impact on city air quality. However, creating the cars and retiring old cars does have a significant environmental impact.

Reducing emissions

Regulating emissions released by industry is also key in reducing atmospheric pollution. By cutting emissions of sulfur, nitrous oxides and particulate matter from heavily polluting industries, it is possible to stop or reduce the problem at the source.

Reducing pollution at its source can be achieved through a number of techniques. For example, sulfur emissions can be cut by using low sulfur coals, washing the coal or by installing technology called 'scrubbers' in flues. Power plants can also be converted from coal burning to ones which use low-sulfur fuels such as natural gas or alternative energy sources.

Flue-gas desulfurisation

Flue-gas desulfurisation (FGD) eliminates SO_2 from gases leaving industrial chimneys. This method can remove as much as 95% of the SO_2 from gases emitted by industry. This helps to reduce pollutants entering the atmosphere.

Electrostatic precipitators

Electrostatic precipitators are devices that use an electric charge to remove particulates (solid particles and liquid droplets) from the gases emitted in industrial smoke. Precipitators can be used for:

- removing oil mist in machine shops
- removing acid mist in chemical plants
- removing bacteria and fungi in medical facilities
- purifying air in air-conditioned systems.

The precipitator works by applying energy to the particulate matter being collected. Charged particles are attracted to collection plates in the precipitator. They are removed by cleaning the plates when enough particles have collected. The particles are then recycled or disposed of.

Catalytic converters

Catalytic converters are devices fitted to vehicle engines to reduce air pollution. They were introduced to lower NO_x emissions from exhaust systems. Catalytic converters, in combination with low-sulfur fuels and lead-free fuels, have reduced atmospheric pollution from vehicles.

Fuel desulfurisation

Fuel desulfurisation is designed to remove sulfur content from a fuel source before it is burnt.

- Coal washing is the use of water and chemicals to remove sulfur and impurities from coal to reduce the level of air pollution when it is burnt.
- Bio-desulfurisation uses microorganisms to eliminate sulfur compounds in fuel before it is burnt.
- Low-sulfur fuels are fossil fuels used in the transport industry. Sulfur is reduced before the fuel is used to power a vehicle engine. This reduces emissions of CO, NO_x and SO_2 into the air.

Often, the costs associated with new technology make it too expensive for LICs. It is necessary to develop more cost-effective mechanisms that use less water.

Volatile organic compounds

Volatile organic compounds (VOCs) have a high vapour pressure and low water solubility. VOCs are emitted from certain solids and liquids as gases. They are often easy to identify as they give off a strong smell (for example, the smell of petrol).

High-VOC products used regularly in homes include paint, paint strippers, solvents, wood preservatives, aerosol sprays, cleansers, disinfectants and fuels. They are even used in the dry cleaning of clothing.

VOCs can have both short and long-term health effects, which include headaches, nausea, loss of co-ordination, cancer and damage to the liver, kidneys and central nervous system.

Steps to reduce exposure to VOCs include:

- increased ventilation when using products with VOCs
- reading and meeting all requirements on product labels
- not storing open containers of products with VOCs
- disposing of empty or little-used containers safely following the disposal guidelines on the container (many products need to be disposed of as hazardous waste)
- keeping out of reach of children and animals
- purchasing limited quantities and not buying more than is needed (to avoid having to store the product)
- storing VOC products safely, and in a way that prevents them mixing with other products to limit hazards.

Figure 7.21: Paints, like petrol or diesel, are high in VOCs. Many modern paints are now low in VOCs.

> **ACTIVITY 7.1**
>
> **Managing air pollution**
>
> In pairs or small groups, investigate an HIC city of your choice. Produce a newspaper article which discusses air quality in the city.
>
> Consider the following factors in your investigation:
>
> - the sources of pollution
> - actions being taken to minimise the pollution
> - health risks faced by the population
> - solutions that could help control the air pollution problems.
>
> The article should be written with the purpose of educating the public.

Legislation

Local, regional and international legislation is necessary in order to manage the production and impact of atmospheric pollution. The underlying principle of any atmospheric pollution legislation is that 'the polluter must be responsible'. This is also known as the 'polluter pays principle' (Figure 7.22). This means that the individual or entity that generates the pollution must be responsible for reducing or preventing it. They must also incur any costs associated with repairing the harm done.

Figure 7.22: Many regions of the world have put a charge on plastic bags. This reflects the 'polluter pays principle' and aims to reduce the number of plastic bags being manufactured and ending up in landfills.

However, unless the source of pollution is easy to identify, it is difficult to determine a single source. The widespread use of coal power stations and vehicles means that, in many instances, governments rather than individuals are responsible for putting controls in place to manage pollution.

International cooperation to address air pollution and acid deposition began in 1972 with the United Nations Conference on Human Environment in Stockholm, Sweden. In 1979, the Geneva Convention on Long-range Transboundary Air Pollution created the framework for controlling and reducing acid deposition and air pollution in Europe. This was the first international agreement to reduce and manage air pollution on a regional basis. Several protocols have refined this agreement since it was first introduced.

In the USA, the Clean Air Act of 1970 and the amendments to this act in 1990 worked towards developing an agreement between the USA and Canada to reduce air pollution and acid deposition. However, the agreement was only formalised in 1991. This agreement placed permanent caps (limits) on SO_2 emissions and put guidelines in place to reduce NO_x emissions. This agreement led to a 88% reduction in SO_2 emissions between 1990 and 2017. NO_x emissions declined by 50% during this period.

Both the European and USA agreements resulted in significant reductions in acid deposition in Europe and North America.

At a more local level, there is legislation which puts a limit on the number of days that vehicles can travel into a town or city. By limiting the days on which cars may travel, there are fewer vehicles on the road which means reduced air pollution. Other laws charge motorists for driving into a specific area (this is called 'congestion charging', Figure 7.23). Congestion charging results in people using public transport rather than their own cars as it costs them less money. The resulting reduction in cars on the road causes a decrease in levels of air pollution in the town. It is important to note that congestion charges are more likely to be effective in a city with a well-developed and affordable public transport system. People then have the choice to pay the fee or use the less expensive public transport to get in and out of town.

Figure 7.23: Road markings in London indicate to drivers that they are entering a congestion charge zone and will be charged for accessing this area.

London uses congestion charges to manage the number of vehicles on its roads and the air pollution in the city. This congestion charge system was set up in 2003. Drivers are charged if they access a congestion charge zone between 7 a.m. and 6 p.m. from Monday to Friday, and between 12 noon and 6 p.m. on Saturdays, Sundays and bank (public) holidays.

7.2 Questions

1. Give four uses for electrostatic precipitators.
2. How does reducing the use of fossil fuels manage atmospheric pollution?
3. Describe three steps that can be taken to reduce the impacts of VOCs.
4. Briefly explain the challenges of managing atmospheric pollution on a global scale.

7.3 Ozone depletion

Approximately 90% of the ozone (O_3) in Earth's atmosphere is found in the stratosphere. This forms a more concentrated layer than anywhere else in the atmosphere. The total concentration of ozone in the atmosphere is measured in Dobson units. One Dobson unit is equivalent to a layer of pure ozone 0.01 mm thick at a standard temperature and pressure. A column of ozone measuring less than 220 Dobson Units is considered to be a region of ozone depletion, while a measurement of 100 Dobson units indicates a hole in the ozone layer.

Causes of ozone depletion

The main cause of ozone depletion and the hole (thinning) in the ozone layer is the human activity of manufacturing chemicals known as chlorofluorocarbons (CFCs). These were extensively used in refrigerants, solvents, propellants and foam-forming agents up until the late 1980s.

CFCs are stable synthetic chemical compounds which speed up the breakdown of ozone. They contain chlorine molecules which are insoluble in water and unreactive and stable in the lower atmosphere. However, CFCs break down in the presence of UV radiation at higher altitudes, releasing chlorine atoms. Ozone (O_3) is broken down by chlorine atoms to form chlorine monoxide (ClO) and oxygen (O_2).

$$Cl + O_3 \rightarrow ClO + O_2$$

Ozone naturally regenerates in the stratosphere. However, its rate of depletion from the action of CFCs is greater than the rate it can regenerate. The presence of molecules such as methane (CH_4) in the atmosphere, which naturally break down ozone, along with the fact that the ozone layer is able to recover from this damage, shows that the ozone layer can recover if CFCs are no longer produced.

Figure 7.24: Aerosol cans used to have CFCs as a propellant. Modern spray cans do not use CFCs.

Ozone depletion over Antarctica

In the early 1980s, scientists identified damage to the ozone layer caused by CFCs. A hole in the ozone layer had opened up over Antarctica and further ozone depletion was identified at the mid latitudes. This discovery caused great concern. As a result, scientists urged politicians to put global regulations in place to control the use of CFCs.

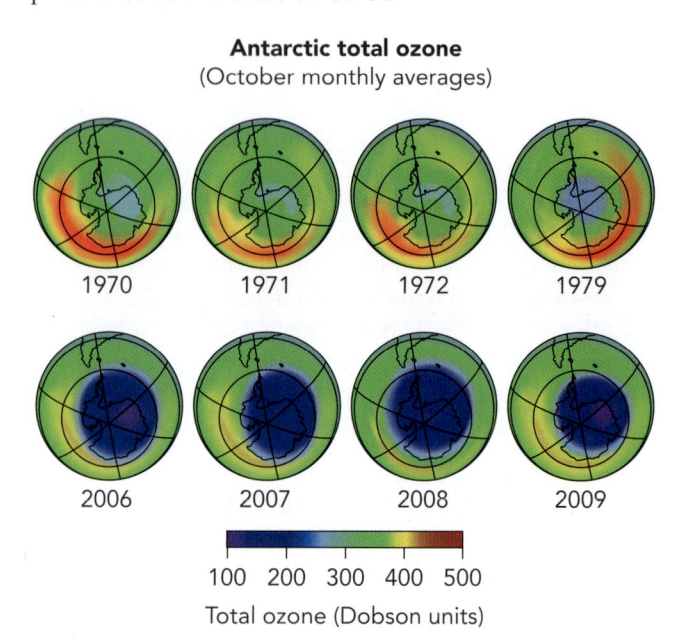

Antarctic total ozone
(October monthly averages)

1970　1971　1972　1979

2006　2007　2008　2009

100　200　300　400　500
Total ozone (Dobson units)

Figure 7.25: Ozone depletion over Antarctica. Blue indicates a low level of ozone and red a higher concentration.

Ozone-depleting substances are found throughout the stratosphere because they are transported great distances by the wind in the atmosphere. The hole in the ozone over Antarctica occurred because of atmospheric conditions specific to that part of the planet.

Very low temperatures at different levels in the stratosphere over a large area of the Antarctic, for a long period of time in winter, cause **polar stratospheric clouds (PSCs)** to form. Low temperatures and moisture at high altitudes allow liquid and solid PSCs to develop. Chemical reactions on the surfaces of liquids and solids substantially increase the abundance of reactive chlorine (chlorine gas). Chlorine gas, which reacts with the ozone,

> **KEY TERM**
>
> **polar stratospheric clouds (PSCs):** stratospheric clouds that form over the poles in winter at altitudes of between 15 000 metres and 25 000 metres. One of the main types of PSC is mostly made up of supercooled droplets of water and nitric acid

resulted in the formation of the **ozone hole** over the Antarctic in late winter and early spring.

Figure 7.26: High clouds in the skies over Antarctica in the summer, adding to the rate of ozone depletion in the atmosphere over Antarctica.

ACTIVITY 7.2

Ozone depletion

1. CFCs are only one of a number of ozone-depleting substances. Carry out research to find out what other natural gases or human made chemicals are linked to the depletion of the ozone layer. List their sources.
2. Create a table like the one below to record your findings. Share the information with your class.

Ozone depleting substances	Sources
Chlorofluorocarbons (CFCs)	Refrigerants, air conditioners, solvents, dry cleaning agents
Human chemicals	
Natural gases	

The problem is made worse by the fact that stratospheric air in the Antarctic region is relatively isolated from other stratospheric regions over a long period of time. This is due to the winds that encircle the poles during the winter months, creating a **polar vortex**. This polar vortex prevents the air over the Antarctic mixing with air outside of this zone. Therefore, once chlorine gases are present in this region during winter, they persist for long periods of time, which causes ongoing damage to the stratospheric ozone layer.

Impacts of ozone depletion

The ozone absorbs UV radiation from the sun, preventing it from entering the lower atmosphere. Ozone layer depletion increases the amount of UV radiation reaching Earth's surface. Studies in humans have shown that this type of radiation can cause damage like skin cancer, immune system suppression and the formation of cataracts.

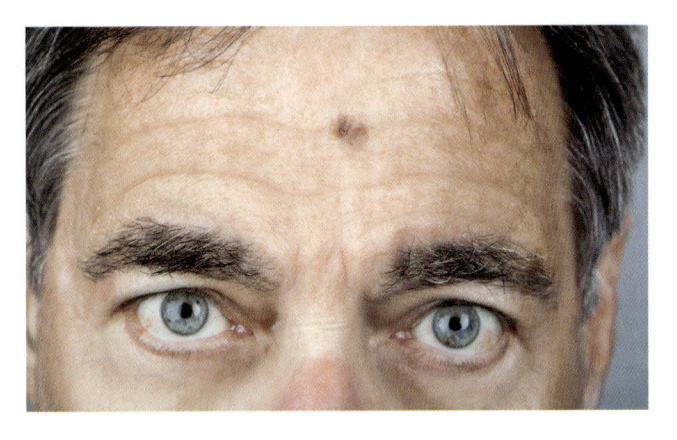

Figure 7.27: Skin damage from exposure to UV radiation, which increases the incidence of sunburn and skin cancer in humans.

UV directly affects plant growth, how plants form and how they utilise their nutrients. These changes have important implications for the competitive ability of plants, their resistance to plant diseases, and their resilience in a changing global climate. Damage to plant growth directly impacts people as it results in declining harvests.

KEY TERMS

ozone hole: an area where the average concentration of ozone is below 100 Dobson Units

polar vortex: a large, long-lasting rotating low-pressure system located over the north and south poles. It weakens in summer and strengthens in winter

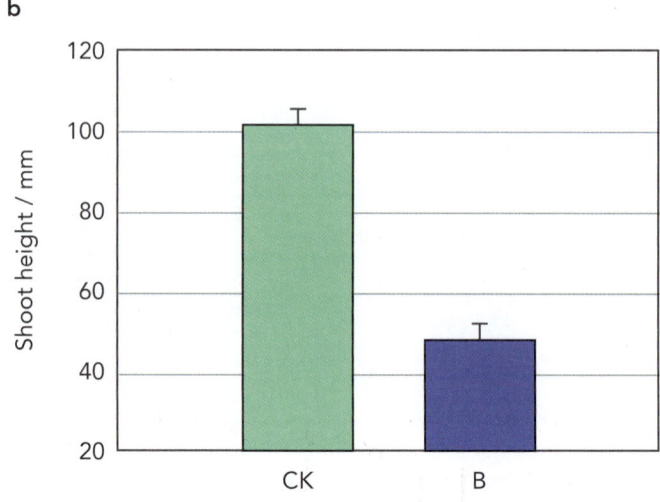

CAMBRIDGE INTERNATIONAL AS LEVEL ENVIRONMENTAL MANAGEMENT: COURSEBOOK

Figure 7.28: Stunted plant growth due to higher levels of UV radiation. Plants in the petri dish labelled B were exposed to higher rates of UV radiation than plants in the petri dish labelled CK.

Plant damage has a direct impact on habitat health and biodiversity. With the loss of healthy vegetation, the habitat of organisms who rely on that vegetation declines or disappears.

Marine organisms are also affected by ozone depletion. Phytoplankton form the foundation of marine food webs and play an important role in marine biodiversity. There is evidence of a direct link between increased rates of UV radiation and decreased phytoplankton numbers. UV radiation also has a negative impact on the early developmental stages of many marine organisms.

Many materials that are commercially used (both synthetic and natural) are negatively affected by UV radiation. They can lose strength, become less flexible, crack and disintegrate. Clothing materials, building materials like silicon and some types of glass break down when exposed to UV radiation. This limits their usefulness outdoors.

The impact of CFC alternatives

CFCs played a significant role in a wide range of industrial applications. These included refrigeration, air conditioning, electronics cleaning and even firefighting (foams). In order to protect these industries, it was necessary to find substitute chemicals that performed the same function without the same environmental impacts.

Figure 7.29: Ozone-friendly products with no ozone-harming chemicals.

290

Replacement chemicals are hydrochlorofluorocarbons (HCFCs) and Hydrofluorocarbons (HFCs) also known fluorinated gases (F-gases).

HCFCs were used as refrigerants while industry phased out the use of CFCs. Because HCFCs contain hydrogen, they are not as stable as CFCs and break down more quickly in the atmosphere. Although they still impact the ozone layer, they do so less than CFCs. In addition, they have less global warming potential as greenhouse gases than CFCs do.

F-gases/HFCs are manufactured gases often used in place of ozone-depleting substances because they do not harm the ozone layer. However, F-gases are a powerful greenhouse gas, and have a significant global warming potential. F-gases have a global warming effect 23 000 times greater than that of CO_2. The rapid increase in the use of fluorinated gases is a cause for concern as they may compound the impact of an already warming atmosphere.

The emissions of F-gases almost doubled between 1990 and 2014 in the EU. All other greenhouse gas emissions were reduced.

Ozone destruction hypothesis

The ozone destruction hypothesis started as an educated guess based on observation and knowledge. However, every hypothesis must be testable if it is to have any value. A good hypothesis can be tested and potentially proven incorrect. Any scientific investigation of the hypothesis must produce useful and reliable information.

Failure to support a hypothesis is common in science. However, hypotheses that are not supported by scientific data are problematic when trying to convince an audience that they are important, for example, ozone depletion or climate change. This can cause long-term doubt in real scientific findings.

Rowland–Molina ozone depletion hypothesis

In 1974, Sherwood Rowland and Mario Molina were the first to discover that Earth's atmospheric ozone layer could be depleted by CFCs. They suggested that long-lived CFCs could reach the stratosphere where they would be broken down by UV radiation, releasing chlorine atoms. Their hypothesis was that 'Earth's protective ozone layer was being damaged by synthetic chemicals called CFCs'.

The breakthrough that led to this hypothesis was unusual. Normally, new hypotheses are based on critical new evidence. However, in this instance Molina did not carry out experiments to gather new data. He looked at existing information about chemical reactions and processes in the atmosphere. He then used this information to show that all the factors together posed an environmental threat.

Figure 7.30 shows the evidence that the Rowland-Molina hypothesis was based on. One of their supporting (or auxiliary) hypotheses was how fast chlorine reacts with ozone molecules. If one of these auxiliary hypotheses proved to be incorrect, then the Roland-Molina hypothesis would also be incorrect.

Hypothesis: CFCs will cause significant ozone depletion

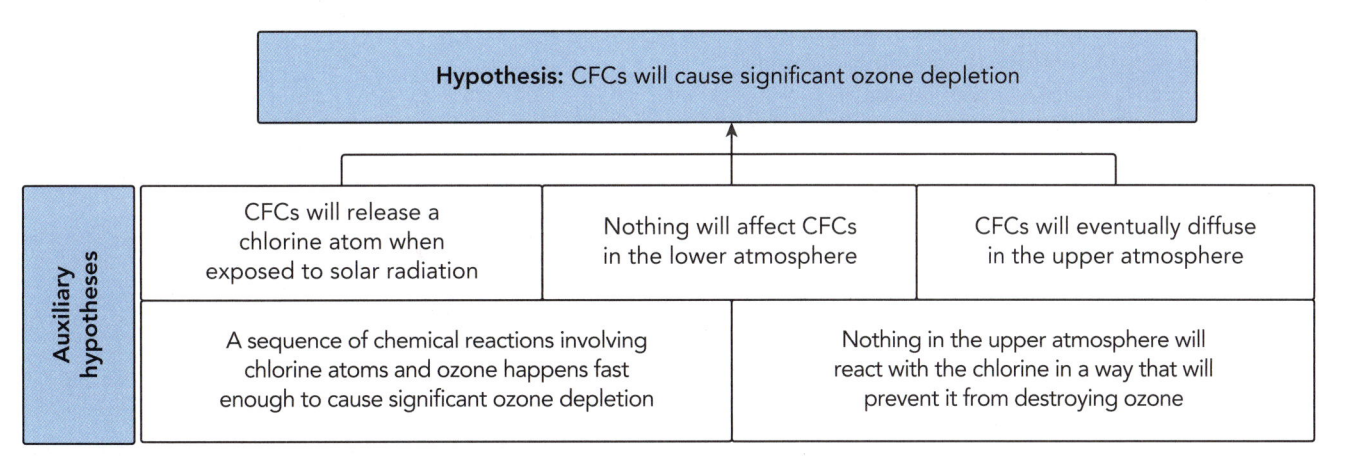

Auxiliary hypotheses:
- CFCs will release a chlorine atom when exposed to solar radiation
- Nothing will affect CFCs in the lower atmosphere
- CFCs will eventually diffuse in the upper atmosphere
- A sequence of chemical reactions involving chlorine atoms and ozone happens fast enough to cause significant ozone depletion
- Nothing in the upper atmosphere will react with the chlorine in a way that will prevent it from destroying ozone

Figure 7.30: The Rowland-Molina hypothesis was based on many auxiliary hypotheses.

Originally, their hypothesis was strongly disputed by industry, because it was not supported by scientific data.

However, scientists studied the phenomenon and made further investigations. They provided reliable data on how the chemical reaction between CFCs and ozone molecules was destroying ozone. Within three years, laboratory measurements and observations in the stratosphere

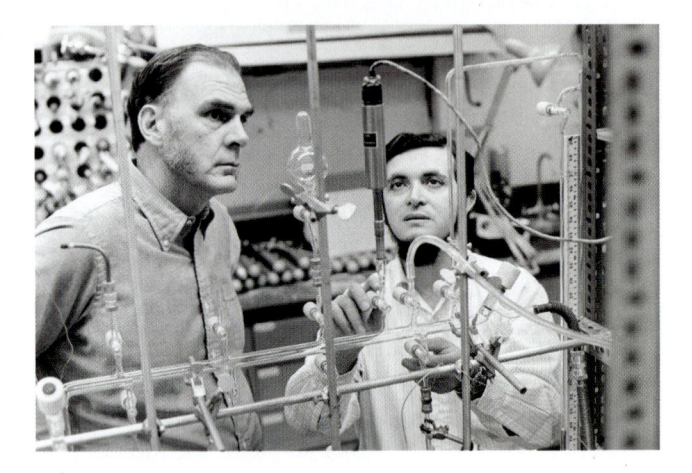

Figure 7.31: Rowland and Molina were awarded a one-third prize share each in the 1995 Nobel Prize in Chemistry for their research on the depletion of the ozone layer, along with a third scientist, Crutzen.

showed that CFCs were indeed a major source of stratospheric ozone depletion. These observations also showed that most CFCs produced would eventually reach the ozone, damaging the ozone layer.

Ultimately, the research showed that CFCs should be banned and that the ozone layer must be protected. The challenge was to persuade those who had doubted the initial hypothesis due to lack of evidence.

International agreements

The Montreal Protocol was created to reduce CFC use. This was the first time that an environmentally driven global agreement was reached. The world was starting to realise that pollution from one country could damage the whole planet. A global agreement was needed to control the problem.

Despite Rowland and Molina's work in the mid-1970s, it was not until 1985 that international research into the ozone layer and the effects of ozone damaging chemicals was undertaken. It was also in this year that the hole over the Antarctic was discovered. As a result of this discovery, the Montreal Protocol (1987) was negotiated by 24 countries and the European Economic Community.

The aim of the Protocol was a 50% reduction in CFC emissions by 2000. This was the first time that entire countries were legally bound to reducing and ultimately phasing out the use of CFCs. Failure to do so carried heavy financial penalties. The protocol was supplemented in 1990 (London) and 1992 (Copenhagen), where 191 countries signed an agreement to stop using CFCs by 1995.

LICs were given more time to achieve these targets. This is because CFCs are less expensive than replacement chemicals. However, they were not allowed to show a significant increase in the use of CFCs. To help LICs make the transition to more environmentally friendly technologies and chemicals, financial assistance was provided.

This protocol has been further expanded in recent years to include newer replacement chemicals to CFCs which have since been proven to harm the stability of the atmosphere. For example, the Kigali Amendment (2019) added F-gases to its list of controlled substances. All members of this Protocol agreed to reduce their production and use of F-gases.

7.3 Questions

1. Describe the main human-related cause of ozone depletion.
2. Describe the impacts of ozone depletion.
3. Are the alternatives to CFCs a long-term solution to the ozone layer issue? Give reasons for your answer.
4. Explain why the ozone destruction hypothesis was not believed at first.
5. Why was an international agreement the only way to effectively manage the damage being done by CFCs?

7 Managing the Atmosphere

EXTENDED CASE STUDY

Atmospheric pollution disaster

In India, on 3 December 1984, more than 40 tonnes of methyl isocyanate (MIC) gas leaked from a pesticide plant in Bhopal. At least 3800 people died in the immediate aftermath and thousands more died prematurely. Over 600 000 workers were affected. This event came to be known as the Bhopal disaster.

In 1969, a Union Carbide chemical factory was built in Bhopal to produce a pesticide known as Sevin. By 1976, trade unions were complaining about the pollution associated with the plant. A few years later, a worker died after accidently inhaling toxic gas. There were warnings that the plant posed a high risk.

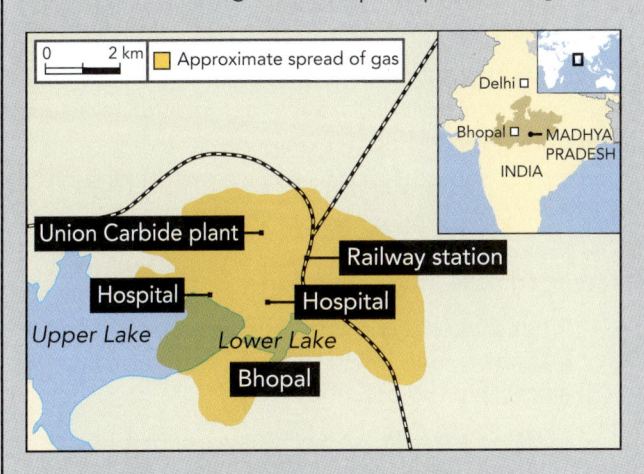

Figure 7.32: A map of the Bhopal Region in India, showing the area affected by the gas leak.

Again in 1982, approximately 45 workers were hospitalised after being exposed to phosgene gas. Between 1983 and 1984, a further three toxic leaks were recorded.

The cause of the Bhopal disaster

On the day of the accident, one of the tanks holding the MIC failed. Safety regulations stated that storage tanks must not be filled to more than 50% capacity and that the tanks were to be kept pressurised with nitrogen gas. This meant that MIC could be pumped out of the tanks. However, one of the tanks lost the ability to contain the nitrogen gas pressure. As a result, it was overfilled to approximately 75% capacity. This failure resulted in a shut down in the production of MIC while the tank was repaired. However, the attempt to repair the tank failed and other safety systems malfunctioned. On 2 December, water entered the malfunctioning tank. A chemical reaction occurred and pressure rose rapidly in the tank. This rapid increase in pressure resulted in MIC escaping from the tank and into the atmosphere.

This factory was associated with a **transnational corporation** (TNC) from a HIC. HICs often have more stringent health and safety regulations than LICs. This means that a TNC may apply lower regulatory requirements when managing a factory in an LIC, and less robust safety precautions may be in place. Clearly, this increases the risk of a disaster occurring. In the aftermath of the Bhopal disaster, accusations were made regarding the role of the TNC. However, the TNC was never legally found to be responsible for the incident.

KEY TERM

*****transnational corporation (TNC):** a company that has its head office in one country, and has branches or factories in other countries. It works across international borders (*trans*-national)

The impacts of the Bhopal disaster

In the area where people were exposed to the toxins, more than 15 000 people died. Over 600 000 workers were affected by the disaster. In the population exposed to the toxins, the stillbirth rate and neonatal mortality rate increased by up to 300% and 200% respectively. The pollution event resulted in an illness that became known as Bhopal Gas disease which caused tingling, numbness, and muscle aches.

The natural environment was also impacted. Within two days, nearby trees lost all their leaves. Over 2000 livestock animals died and their bodies had to be disposed of. Heavy gas was absorbed into the local water bodies. This made the water unsafe to drink, and poisoned aquatic life.

CAMBRIDGE INTERNATIONAL AS LEVEL ENVIRONMENTAL MANAGEMENT: COURSEBOOK

CONTINUED

Figure 7.33 Rusting tanks inside the derelict Union Carbide factory compound. The years since the disaster have witnessed the decay of the remaining infrastructure.

Solutions to the Bhopal disaster

Not much was known about the toxin. It is known to react with water at high temperatures and can release as many as 300 other highly toxic chemicals. Not knowing exactly what the gas contained meant that medical staff were not sure about what treatment to give the injured people. In addition, medical staff could not work safely in the area, as they did not know how to protect themselves from the toxins.

Even in 2014, the Indian Council of Medical Research was still unable to identify the toxin that was causing Bhopal gas disease. Independent studies have indicated that serious health issues such as cancer and birth defects continue to occur in the population. Since there is no medical study of the illness, it is easy to dismiss the illnesses found in the population as being linked to poverty and lack of sanitation.

The parent company did eventually pay US$470 million in 1989 as compensation to the families who suffered due to the disaster. In return, all civil and criminal cases against the company were dropped.

The pollution of surface and groundwater stores with MIC means that second and third generations of people living in the area continue to suffer from the effects of the disaster. More recently, the issue of decontamination has been addressed. The aim is to remove the toxic waste that remains at the plant and dispose of it safely.

The disaster had impacts far beyond the area of the spill and its people. Growing awareness of industrial accidents resulted in a worldwide change to the way chemical and hazardous waste management was enforced. Stronger health and safety legislation was implemented to prevent another such disaster occurring. Environmental protection acts were put in place to give authority to central decision-making bodies. Authorities now have the power create rules, regulation and pollution norms for hazardous waste. Even though the Bhopal disaster itself has yet to be cleaned up, it has helped to reduce the risk of such a disaster occurring again.

Extended case study questions

1. A transnational corporation (TNC) is a company that has a head office in one country, and many offices in other countries.

   Explain how enforcing globally acceptable environmental and health and safety regulations would limit the risk of a disaster like this one.

2. Suggest what could have been done differently to prevent an event like this occurring.

3. Consider the problems caused by withholding information about the chemicals in the spill.

Extended case study project Creating a calendar

In pairs or small groups, work together to research the timeline of the Bhopal disaster. Note the key dates and events that occurred on that date.

1. Consider events before the day that the Bophal disaster occurred.

2. A timeline can give more specific details of times on the day that the event occurred.

CONTINUED

3   Consider the events after the day that the Bophal disaster occurred.

4   Make short notes on each date and/or time that you include on the Bophal Disaster timeline.

Create an annotated timeline of the Bophal disaster. Your timeline should be informative so that another learner reading it can clearly understand the events of the days, weeks and years relating to the event.

SUMMARY

Atmospheric pollutants are deposited onto the Earth surface as wet or dry acid deposition.
SO_2 and NO_x come mainly from burning fossil fuels. They can also come from natural sources such as volcanic eruptions.
Acid deposition is harmful to organisms in aquatic and terrestrial ecosystems.
SO_2 can cause health problems in humans linked to skin, eyes and lungs.
Photochemical smog forms when ground-level pollutants react with sunlight. It is harmful to living organisms and building materials.
Managing air pollution is done through identifying its source and then using a variety of methods to reduce or eliminate it.
Legislation, at both a local and international level, needs to be implemented to control the release of atmospheric pollutants.
Stratospheric ozone which blocks incoming UV radiation is depleted by the use of CFCs.
Global regulations were put in place to reduce and stop the use of CFCs.
Alternatives to CFCs, such as HCFCs, are less stable and break down more quickly. This means they have a lower impact on the ozone.
The ozone destruction hypothesis by Roland Molina (1974) was, at first, disputed due to insufficient data. However, further research supported the hypothesis.
The Montreal Protocol (1987) was established to control or limit the use of CFCs globally.

REFLECTION

- Did you find that writing a newspaper article for activity 7.1 helped you consider the material from a different view point? Did that style of writing clarify your understanding of the material?

- Discuss one of the topics you have learned about in this chapter with students who do not take this subject. How much were you able to teach them about the material?

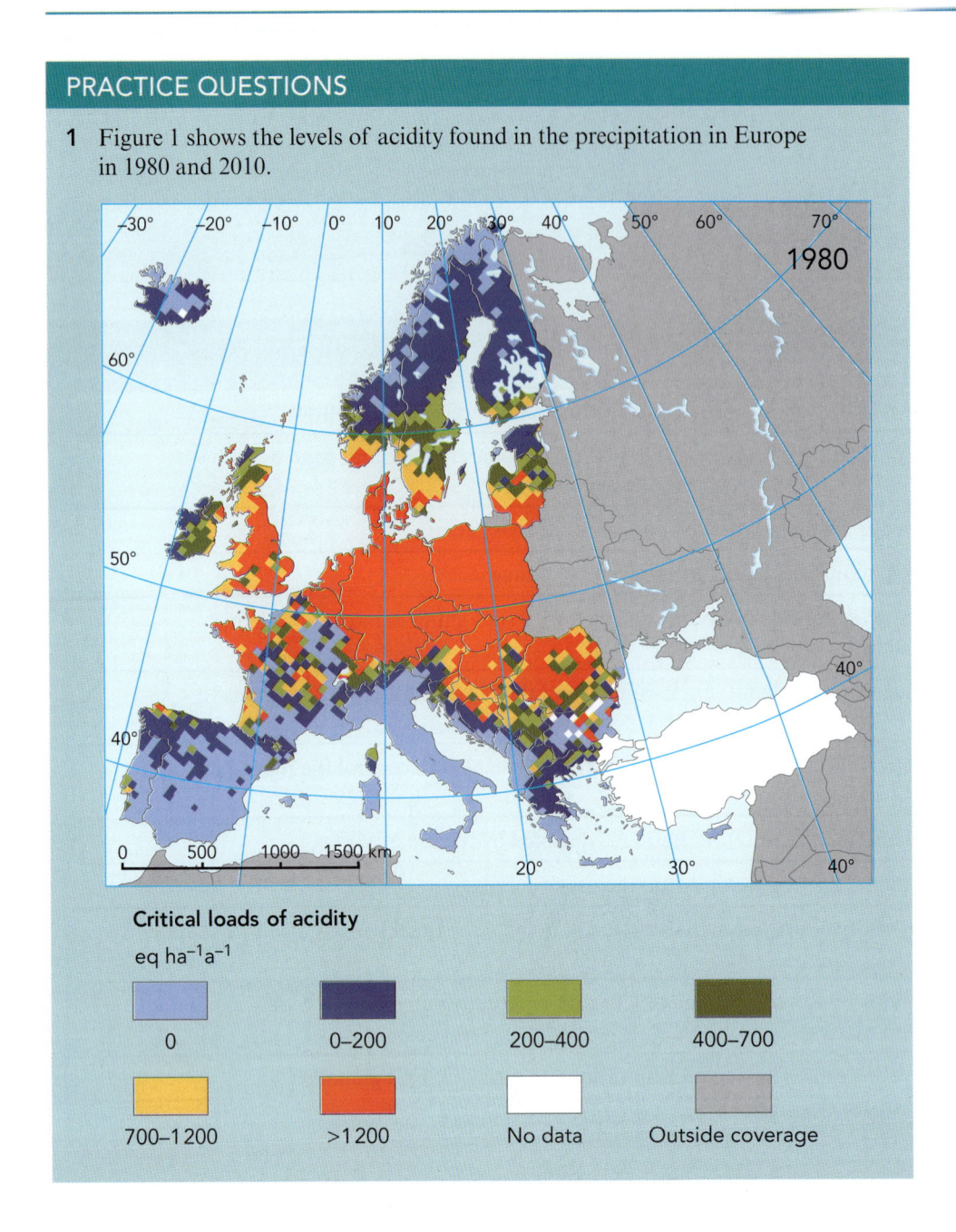

CAMBRIDGE INTERNATIONAL AS LEVEL ENVIRONMENTAL MANAGEMENT: COURSEBOOK

PRACTICE QUESTIONS

1 Figure 1 shows the levels of acidity found in the precipitation in Europe in 1980 and 2010.

Critical loads of acidity

eq ha^{-1}a^{-1}

- 0
- 0–200
- 200–400
- 400–700
- 700–1200
- >1200
- No data
- Outside coverage

7 Managing the Atmosphere

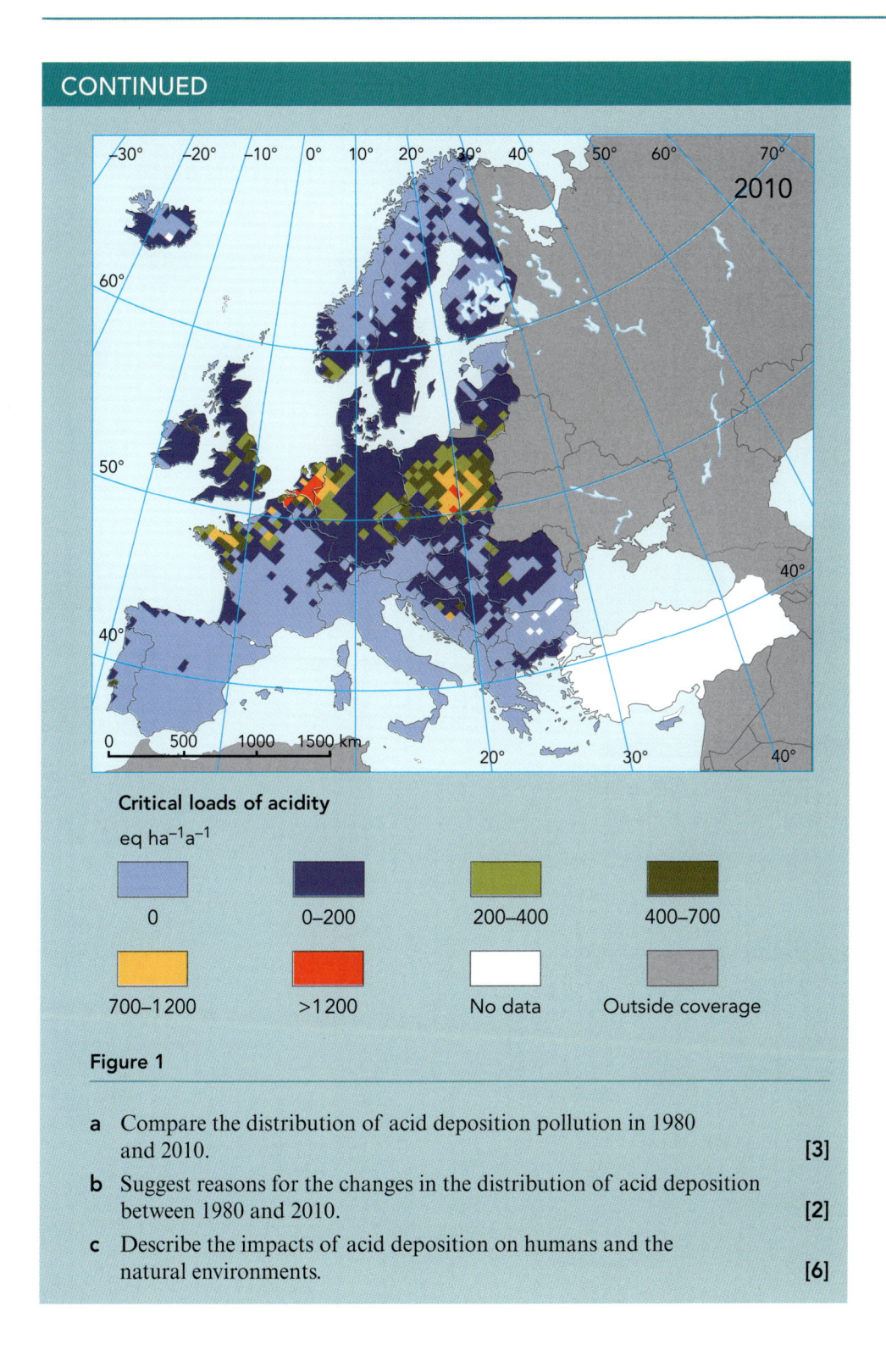

CONTINUED

[Map of Europe showing critical loads of acidity in 2010]

Critical loads of acidity

eq ha⁻¹a⁻¹

- 0
- 0–200
- 200–400
- 400–700
- 700–1 200
- >1 200
- No data
- Outside coverage

Figure 1

a Compare the distribution of acid deposition pollution in 1980 and 2010. [3]

b Suggest reasons for the changes in the distribution of acid deposition between 1980 and 2010. [2]

c Describe the impacts of acid deposition on humans and the natural environments. [6]

CONTINUED

2 Figure 2 shows the expansion of the hole in the stratospheric ozone over Antarctica between 1970 and 2009.

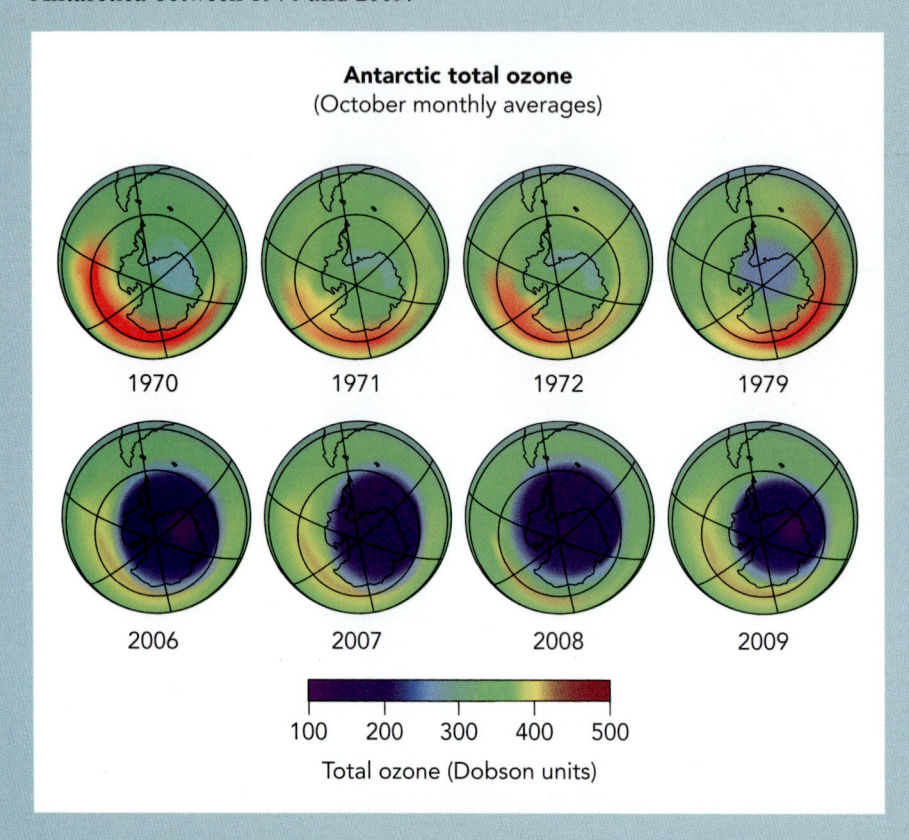

Antarctic total ozone
(October monthly averages)

1970 1971 1972 1979

2006 2007 2008 2009

100 200 300 400 500
Total ozone (Dobson units)

Figure 2

a Describe the changes in the density of ozone over Antarctica between 1970 and 2009. [3]

b In the 1970s the ozone depletion hypothesis was proposed by Rowland and Molina.
  i State what the Rowland–Molina hypothesis proposed. [1]
  ii Explain why the hypothesis was not initially accepted by society. [1]

c The Montreal protocol (1987) was established to address the problems with the ozone layer. State the main goal of this protocol. [1]

d Give **two** impacts of ozone depletion. [2]

e Explain how the ozone gas (O_3) in the stratosphere becomes depleted. [3]

CONTINUED

3 Figure 3 and Table 1 show the number of days that the metropolitan area of Los Angeles suffers from unhealthy air quality (smog).

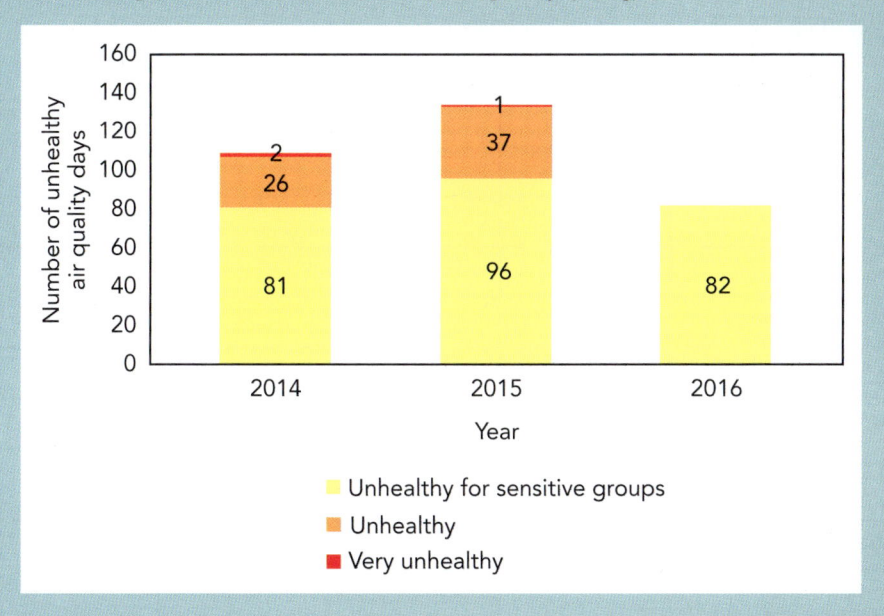

Figure 3

	Unhealthy for sensitive groups	Unhealthy	Very Unhealthy	% of very unhealthy out of total unhealthy days
2014	81	26	2	1.83
2015	96	37	1	0.75
2016	82	21	4	

Table 1: The number of unhealthy days in the Los Angeles Metropolitan area for 2014, 2015 and 2016.

- **a** Calculate the percentage of very unhealthy days out of the total unhealthy days for 2016. [1]
- **b** Suggest methods that can be used by a city to manage air pollution and to reduce the risk of smog formation. [5]
- **c** Suggest methods a city could use to manage air pollution and reduce the risk of smog formation. [5]

4 'It is not possible to manage atmospheric pollution'. To what extent do you agree with this statement? Give reasons and include information from relevant examples to support your answer. [20]

CAMBRIDGE INTERNATIONAL AS LEVEL ENVIRONMENTAL MANAGEMENT: COURSEBOOK

SELF-EVALUATION CHECKLIST

After studying this chapter, think about how confident you are with the different topics.
This will help you to see any gaps in your knowledge and help you to learn more effectively.

I can	Needs more work	Getting there	Confident to move on	See Section
Define acid deposition.				7.1
Describe the two types of acid deposition.				7.1
Outline the impacts of acid deposition.				7.1
Define photochemical smog.				7.1
Describe the impacts of photochemical smog.				7.1
Describe strategies for managing air pollution.				7.2
Outline how ozone depletion occurs.				7.3
Define the term ozone hole and why depletion is the greatest over Antarctica.				7.3
Describe the impacts of ozone depletion.				7.3
Evaluate the international agreements used to reduce ozone depletion.				7.3
Outline the impacts associated with the use of some alternatives to ozone depleting substances.				7.3
Outline the importance of experimental evidence supporting the hypothesis that ozone was being depleted.				7.3
Describe the strategies for managing air pollution.				7.2

Chapter 8
Managing Climate Change

LEARNING INTENTIONS

In this chapter you will:

- explore the meaning of climate change
- outline the causes and impacts of climate change
- identify the impacts of climate change on the environment and humans
- describe and evaluate strategies for managing climate change

CAMBRIDGE INTERNATIONAL AS LEVEL ENVIRONMENTAL MANAGEMENT: COURSEBOOK

GETTING STARTED

1 Discuss your views on the following questions in small groups. Write down your answers so that you can revisit them later in the chapter and see what you have learned.

   a   What is your understanding of the terms climate change and global warming?

   b   What are the causes of climate change?

   c   What impact does climate change have on the natural environment?

   d   Can we do anything to help manage climate change?

ENVIRONMENTAL MANAGEMENT IN CONTEXT

Sustainable architecture

When the top priorities in designing, building and using a building are the environment and sustainability, this is known as sustainable architecture. As consumers become more aware of the impacts humans have on the environment, sustainable architecture is becoming more common.

The purpose of sustainable architecture is to:

1 Reduce the CO_2 emissions created during the construction and operation of the building

2 Reduce the impact of urbanisation on the environment

3 Be both functional and aesthetically pleasing.

Sustainable buildings should last a long time without requiring additional resources. They should be able to be repurposed when they are no longer required for their current use.

Figure 8.1: A green building incorporates vertical gardens. This increases the green space in a city.

Sustainable buildings help to minimise the release of greenhouse gases in a number of ways:

- They are built with materials that minimise their **carbon footprint**.
- They are designed in a way that minimises carbon footprint (for example, window size is considered to optimise cooling and heating).
- They minimise waste going to landfills.
- They use renewable energy sources.
- They include areas of vegetation around or on the buildings.
- They minimise water usage.

All of these factors help to reduce the release of greenhouse gases into the atmosphere and combat their potential impact on global climate.

Discussion questions

1 Why is it important that architects have a strong environmental understanding?

2 Is the architect the only member of a team constructing a building that needs to have environmental knowledge? Give reasons for your answer.

3 How does minimising water usage and waste going to landfill reduce the carbon emissions of a building?

KEY TERM

carbon footprint: the amount of CO_2 released into the atmosphere as a result of the activities of an individual, organisation, community or activity

8.1 Climate change

The climate on Earth is constantly changing in a natural cycle of warming and cooling. This process is known as climate change. However, in more recent years, a significant shift in temperatures has been noted. As of 2022, the hottest years on record are now 2015, 2016, 2017, 2018, 2019, 2020 and 2021. According to the UN weather agency, 2021 was the seventh consecutive year where global temperatures were more than 1°C warmer than pre-industrial average temperatures. Records also show that since the 1980s each decade has been warmer than the one before. This trend is expected to continue.

Greenhouse gases are gases within the atmosphere that absorb the **infrared radiation** emitted by Earth's surface. Greenhouse gases include water vapour, carbon dioxide (CO_2), methane (CH_4), nitrous oxides (NO_x), ozone (O_3) and some artificial chemicals such as chlorofluorocarbons (CFCs). Changes in the concentration of greenhouse gases within the atmosphere result in changes to the atmospheric temperatures. **Global warming** is the result of an increase in the concentration of greenhouse gases in the atmosphere.

Climate change predictions look at how climate may change in the future due to global warming. These predictions are estimated using climate change computer models. Currently, predictions vary greatly between models, but overall they suggest further increases in global temperatures.

Human sources of greenhouse gases

We emit greenhouse gases into the atmosphere through their daily activities. Three of the main sources of these greenhouse gases are the burning of fossil fuels such as coal, gas and oil, cattle and rice farming, and the disposal of waste in landfill sites.

Combustion of fossil fuels

Fossil fuels are formed from the decomposition of buried carbon-rich life forms. They are then extracted to burn for energy. The **combustion** of fossil fuels results in the release of large amounts of CO_2 from underground into the atmosphere. In addition to this, when fossil fuels undergo combustion, water is created and released as vapour. Both CO_2 and water vapour are greenhouse gases. Changes in their concentrations are responsible for the increase in global temperatures, as they trap infrared radiation in the atmosphere.

Coal is the most polluting fossil fuel. It is responsible for an estimated 0.3°C of the 1°C increase in global average temperatures so far. The combustion of coal is estimated to make up over 40% of the global CO_2 emissions, while oil makes up about 30% and natural gas makes up approximately 20%.

Figure 8.2: Emissions from a car exhaust contain greenhouse gases.

Rice farming and livestock farming

The greenhouse gas methane has an atmospheric lifetime (the amount of time it remains in the atmosphere) of approximately 12 years, while the atmospheric life time of CO_2 is more than 100 years. However, methane is 28 times more effective at retaining heat than CO_2. Methane

> **KEY TERMS**
>
> **infrared radiation:** a type of radiation not visible to the human eye, but felt as heat
>
> **global warming:** the increase in the average surface temperatures on Earth's surface due to rising levels of greenhouse gases
>
> **climate change predictions:** estimates of future climate conditions, taking into account both time (months, years and decades) and place (global, regional or local)
>
> **combustion:** the process of burning fossil fuels

is primarily released through activities such as rice and livestock farming. Organic material **biodegrades** when it is broken down by other living organisms, such as bacteria and during digestion by animals. This process releases methane.

Figure 8.3: A dense herd of cattle at a commercial cattle farm.

Traditional rice farming is carried out in flooded fields. The amount of water needed for rice farming blocks oxygen from getting to the soil where biodegradation is taking place. This creates the conditions for the bacteria decomposing vegetation to release methane. Rice farming accounts for up to 19% of total greenhouse gas emissions. Its significant impact needs to be considered when managing climate change. Changing farming methods could achieve this, for example, changing from flooded fields to techniques such as alternate wetting and drying. Not only would this reduce greenhouse gas emissions, it would also save water.

Figure 8.4: A farmer harvests rice from flooded rice fields known a rice paddies.

Figure 8.5: Cattle in a pasture that has been cleared in the Amazon jungle. The loss of trees adds to the problem of global warming.

Livestock farming also has a significant impact on greenhouse emissions. There are two main reasons for this. Firstly, livestock eat large amounts of vegetation. During the digestion process, enzymes breakdown the vegetation, releasing methane. In addition to this, the deforestation of large areas of forest in order to farm beef cattle for global markets adds to CO_2 levels within the atmosphere. Trees act as a carbon sink. Fewer trees means that forests' ability to absorb CO_2 from the atmosphere is decreased. Cattle farming also has a tangible impact on biodiversity, particularly when forests are removed to make space for the livestock.

Methane is also released through melting permafrost. In regions such as Alaska and Russia, the warming climate is causing the methane that was trapped in permafrost to be released. This is a knock-on effect, where one event causes another to occur, and the second event makes the first one worse. Atmospheric warming is accelerated in a positive **feedback mechanism**. This means that the more methane that is released, the more the permafrost melts, releasing more methane.

KEY TERMS

*biodegrade: when organic material is decomposed by bacteria or other living organisms

climate feedback mechanism: a process that speeds up or slows down the trend of climate warming.

This is currently not fully understood. It is also referred to as a climate change loop.

8 Managing Climate Change

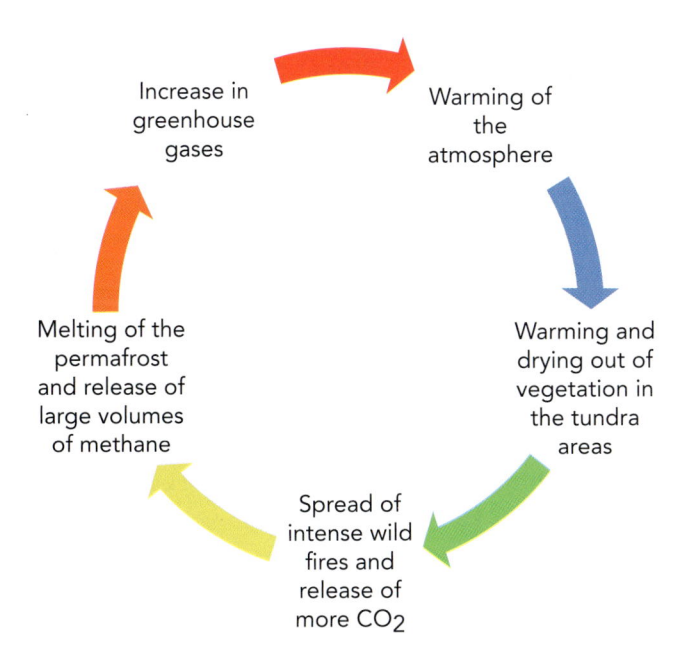

Figure 8.6: The global warming feedback mechanism, where one change triggers another change in the balances of greenhouse gases within the atmosphere.

Landfill sites

Landfill gas (LFG) is a natural by-product of the decomposition of organic material found in buried waste. LFG is predominantly composed of methane and CO_2 in equal proportions. In 2019, LFGs were found to be the third-largest source of human-related methane emissions in the USA, contributing as much as 15% of the country's annual greenhouse gases.

However, many countries now recover methane from landfill sites as it can be used as a renewable energy source. This has helped to reduce the emissions from some landfill sites.

Figure 8.7: Methane being captured from a landfill site in Alicante, Costa Blanca, Murcia, Spain.

INVESTIGATIVE SKILLS 8.1

The Greenhouse effect

This investigation looks at how a 'greenhouse' traps heat. It represents how the atmosphere acts to trap heat around the planet.

Getting started:

Consider how placing a plastic bottle over a jar may change the temperature inside the jar. Discuss what you think causes this to occur.

You will need:

1. a large glass dome (bowl or beaker)
2. two glass jars small enough to fit inside the glass bowl or beaker.
3. scissors
4. two thermometers

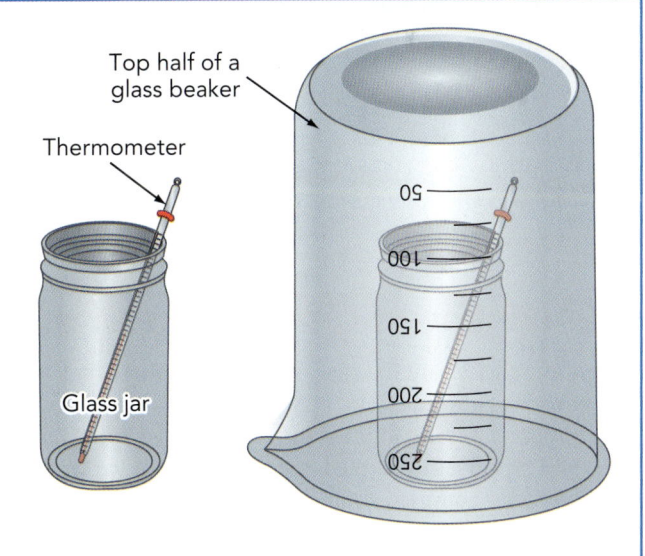

Figure 8.8: Experiment setup: one glass jar needs to be placed under a dome while the second one is left open.

305

> CAMBRIDGE INTERNATIONAL AS LEVEL ENVIRONMENTAL MANAGEMENT: COURSEBOOK

CONTINUED

Steps

1. Set up your experiment as shown in Figure 8.8.
2. Stand the thermometers inside the two glass jars. The thermometers show the difference in temperature between the two jars.
3. Record the temperature in each jar at the start of the experiment.
4. Stand both of the glass jars in the sun, placing the plastic soda bottle over one of the glass jars.
5. Leave the jars in the sun for 45 minutes to an hour. Then check and record the temperatures.

Questions

Write out answers to the following questions:

1. Calculate the change in temperatures in the jars from the start (when the thermometers were first put in the jars) to the end (when the jars had been standing the sun for an hour).
2. Compare the change in temperature in the two jars.
3. Explain why the temperatures differed.
4. How do you think the temperature would have changed inside the bottle covered with the dome if CO_2 had been added? Explain your reasoning.
5. How could this experiment be improved? Explain your answer.
6. A learner decided to carry out the experiment in the following way for steps 4 and 5:
   a. Stand the thermometer inside a glass jar, place it in a sunny spot and check the temperature an hour later.
   b. Put the dome over the jar and leave for another hour. Check the temperature an hour later and compare it with the earlier temperature recorded.

   Explain the limitation of this approach in comparison with the experiment you carried out.

SELF ASSESSMENT

- How successful was your experiment? Were there obvious differences in the results?
- How could you change what you did to get another set of results?
- Did your results differ greatly from those of another team? If so, consider why the results were so different. Which factors changed?

The enhanced greenhouse effect

Greenhouse gases are the gases in the atmosphere that influence the balance of energy within Earth's atmosphere. Greenhouse gases are found naturally. However, they have increased significantly in the last century due to various anthropogenic (human) sources. This is referred to as the enhanced greenhouse effect. The increase in the amount of infrared radiation being retained in the atmosphere in more recent times is linked to human activities, which increase the concentration of greenhouse gases in the atmosphere. (see Figure 8.9).

The enhanced greenhouse effect is a disruption in the balance of atmospheric greenhouse gases. In recent years, this has led to an increase in global average surface temperature, known as global warming.

Difficulties monitoring and predicting climate change

Monitoring and predicting climate change presents a wide range of difficulties and this has resulted in scientific debate. However, it has also encouraged an improvement in both research and data sharing.

8 Managing Climate Change

Greenhouse effect

Step 1: Solar radiation reaches Earth's atmosphere – some of this is reflected back into space.

Step 2: The rest of the sun's energy is absorbed by the land and the oceans, heating Earth.

Step 3: Heat radiates from Earth towards space.

Step 4: Some of this heat is trapped by greenhouse gases in the atmosphere, keeping Earth warm enough to sustain life.

Enhanced greenhouse effect

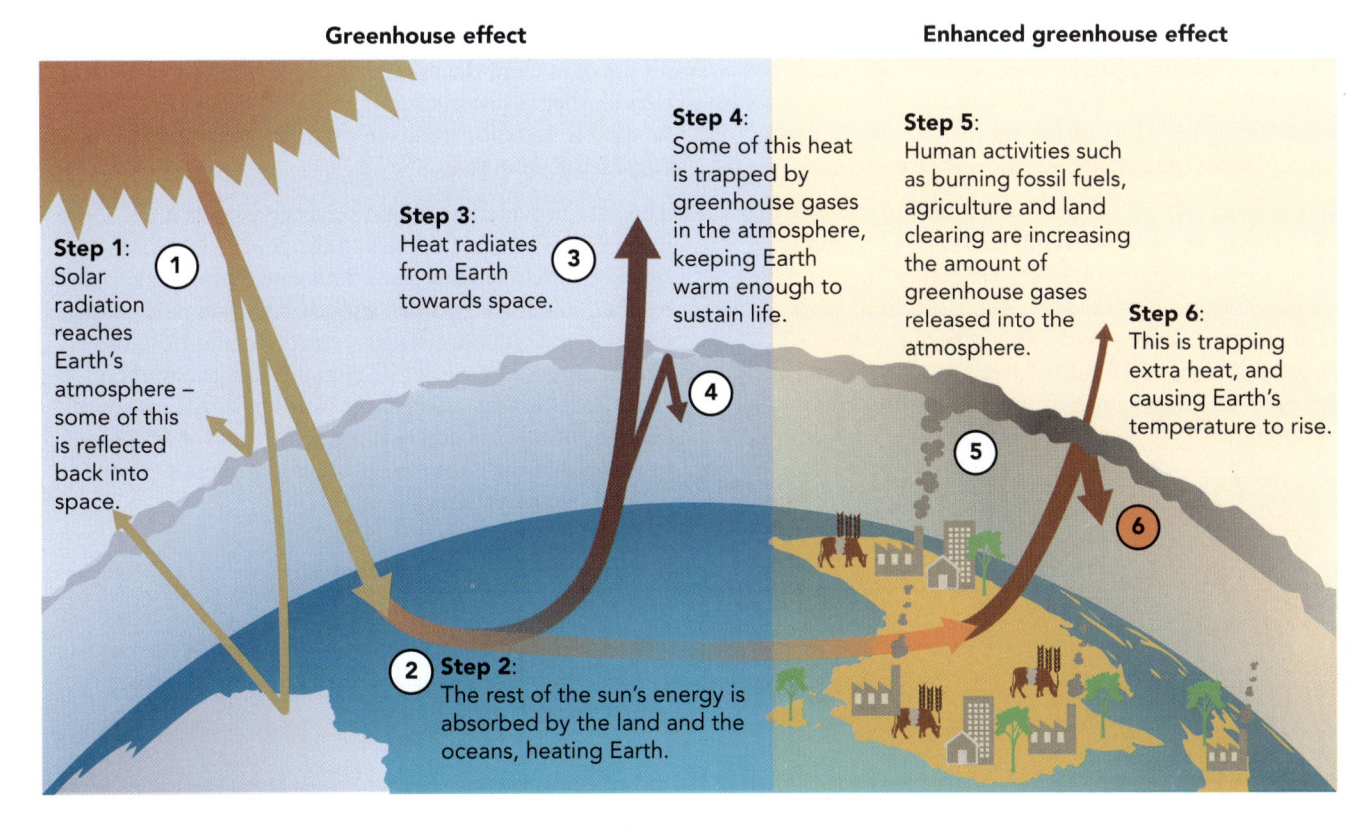

Step 5: Human activities such as burning fossil fuels, agriculture and land clearing are increasing the amount of greenhouse gases released into the atmosphere.

Step 6: This is trapping extra heat, and causing Earth's temperature to rise.

Figure 8.9: The greenhouse and enhanced greenhouse effect.

> ### KEY TERM
>
> ***proxy data:** preserved physical characteristics of the environment that stand in for measurements of the actual climate information

Historical data

Historical data is the study of past climates. Scientists use sources such as ice cores, tree rings, fossilised pollen, ocean sediments and corals to obtain data on past climates. This type of data is referred to as **proxy data**, as it relies on natural records from the environment. In addition to this data, scientists also look at historical human records from ships and farmers' logs, travellers' diaries and newspapers. These documents provide both qualitative and quantitative data that can be used to determine climate conditions in the past.

Ice cores (Figure 8.10) are taken from near the poles show snowfall that has accumulated over very long periods of time. Scientists drill cores from deep in the ice and then analyse the ice for information on past temperatures, precipitation, atmospheric composition and volcanic activity.

Figure 8.10: Scientists drilling for ice cores in Antarctica in order to collect samples that can be analysed.

Corals build their skeletons out of calcium carbonate. This material contains elements that can determine the water temperature in which the coral grew. If scientists know the temperature of the sea historically, they can identify what the climate was like at the time.

Tree rings (Figure 8.11) are formed during the growth of a tree. Each year the tree adds a layer of growth to the outside of its trunk. The rings that form as a result can be used to determine climatic conditions at the time of their growth. This is because climate directly influences the rate of tree growth. The width, density and composition of tree rings reflect changes in the climate.

Figure 8.11: A slice through a tree trunk showing the annual growth rings that form when the tree grows.

Reliability of historical data

Reliable modern human climate records only began in the late 1800s. Therefore, historical data is important when trying to determine past climate. However, any individual proxy data is not a reliable indicator of historical climate. It is important that more than one set of proxy data is assessed when estimating historical climates.

If we only look at one set of proxy data, such as tree-rings, the data alone is not a reliable source. This is because data shows that in the 20th century, tree rings grew more slowly than they would have been expected to, given the recorded temperatures. However, if we combine various sources of historical climate data, we can increase the reliability and our understanding of what the climate was like in the past. For example, if all the data from the different sources suggests a colder period, then it becomes more reliable. Proxy data can be used to look for trends in data, but it cannot yet be considered definitive.

The data provided by climate change models is uncertain. The same applies to the conclusions drawn from historical climate data. This uncertainty has resulted in debate between supporters of climate change theories and climate change sceptics. In addition, the delay between the causes of climate change and their effect have further cast doubts over the actual causes of climate shifts. This makes it more difficult for scientists to prove that the current shift in global climates is due to anthropogenic activities.

Computer climate models

Computer climate models (see Figure 2.19 in Chapter 2) are used to predict future climate. These models use data on a wide variety of variables which include, but are not limited to:

- temperature.
- precipitation.
- wind direction and speed.
- ocean temperatures and currents.
- changes in greenhouse gases.
- rates of melting of land and sea ice.
- rate of melting of permafrost.

The complexity of combining these variables along with the wide range of possibilities as to how these variables are going to change over time, makes creating accurate computer models very difficult. Models use data that is already known about the climate and mathematical equations predict climate change. However, several factors result in computer climate models remaining unreliable.

Climate feedback mechanisms

Climate feedback mechanisms or loops (Figure 8.6) are processes that can either speed up or slow down the factors that drive climate change. This results in a chain reaction that repeats over and over again. A positive climate feedback mechanism results in an acceleration of the rate of climate change. The melting of permafrost can be used as an example.

308

1. As more heat is trapped in the atmosphere due to the increasing concentrations of greenhouse gases, including methane gas, the atmosphere warms up.
2. This triggers melting of the permafrost in tundra regions.
3. Melting permafrost releases more methane into the atmosphere.
4. This additional methane causes the atmosphere to heat up more.
5. The heating up of the atmosphere results in more melting of the permafrost.

Predicting these feedback mechanism and the impact they are going to have on the climate is difficult. Feedback mechanisms may be missed by computer models, making their predictions less reliable.

Time delay on cause and effect

There is a time delay between events (cause) and the result (effect) in the global climate. Studies estimate that there may be decades between an increase or surge in greenhouse gases (the cause) and most of the warming it causes (the effect). This means the changes in climate we are now experiencing are due to greenhouse gas emissions that occurred decades ago.

This delay is due to the time it takes for oceans to show a change in temperature. Oceans heat up far more slowly than the atmosphere. This is because the ocean's mass is around 500 times that of the atmosphere. Therefore, the immediate effects of changes to greenhouse gases in the atmosphere take a long time to be seen in the oceans. It is difficult to calculate the rate of change in the oceans as the mixing up of their upper warm layers and lower cold layers is not fully understood.

This makes computer climate modelling both complicated and uncertain.

Data uncertainty

The doubt over predictions made by computer climate models is caused by the uncertainty of a range of factors such as:

- historical data
- feedback mechanisms
- delays between the cause and effect of climate change
- Missing data linked to processes in the climate system that occur on a small scale
- Climate mechanisms so complex that the data is not available to reproduce them in the models.

These factors combine to create the uncertainty that has resulted in differences in opinion on future potential changes in the climate, within both scientific and political communities around the world.

8.1 Questions

1. Define the terms global warming and greenhouse gas.
2. Describe two factors that create the different climate zones.
3. Explain how rice and livestock farming contribute to the enhanced greenhouse effect.
4. Explain the limitations of historical climate data.
5. Explain it why is so difficult to slow down the rate of global warming.

8.2 The impacts of climate change

Impacts of climate change on the environment

Climate change impacts many aspects of the natural environment. These include temperature, precipitation, sea levels, ocean and wind circulation and global ice stores. These changes have a direct impact on habitats around the world, affecting both species distribution and biodiversity.

Temperature

Figure 8.12: People cool off in a fountain during a heat wave in Brooklyn, New York, June 2021.

One of the most visible impacts of climate change is an increase in temperatures, with the occurrence of hotter days and heatwaves. 2020 was one of the hottest years on record, 2015–2019 were the warmest five years on record and 2010–2019 was the warmest decade on record. The planet is now 1.1 °C warmer than it was in the 1800s, prior to the Industrial Revolution.

Europe and the UK suffered from extreme heatwaves during the summer of 2022, with Pinhão, Portugal recording 47 °C in July.

Precipitation

Figure 8.13: A cloud burst, showing the large amount of rainfall that can be released by a well-developed storm cloud.

Changes in temperature result in changes to the intensity and frequency of precipitation such as:

- There is an increase or decrease in the amount of precipitation.
- Storms become more severe, with heavy precipitation and potential flooding (Figure 8.13).
- Runoff from flood water can wash away aquatic organisms, decrease water quality and reduce aquatic biodiversity as turbidity increases.
- Areas that normally experience relatively high rates of precipitation may see an increase in precipitation.
- Areas located in rain shadows or which have low precipitation risk more severe and prolonged drought conditions.
- A potential decrease in winter snowfall may result in a decline in the formation of annual snow packs. These are vital in the provision of water to surface springs and rivers through drier periods, as they melt and release water.

Sea level

Figure 8.14: A woman boards a bus through flood waters caused by unusually high tides, which some have linked to global warming.

Melting land ice and ice sheets is resulting in more water entering the oceans. This, in combination with the expansion of the oceans as they warm, is causing sea levels to rise around the globe:

- Records from 1900 to 1990 show an estimated 10–12.5 cm rise in sea levels.
- From 1990–2015 there was an increase of approximately 7 cm.
- Sea levels are currently rising at about 0.3 cm per year.

This sea level rise threatens coastal regions.

Ocean circulation

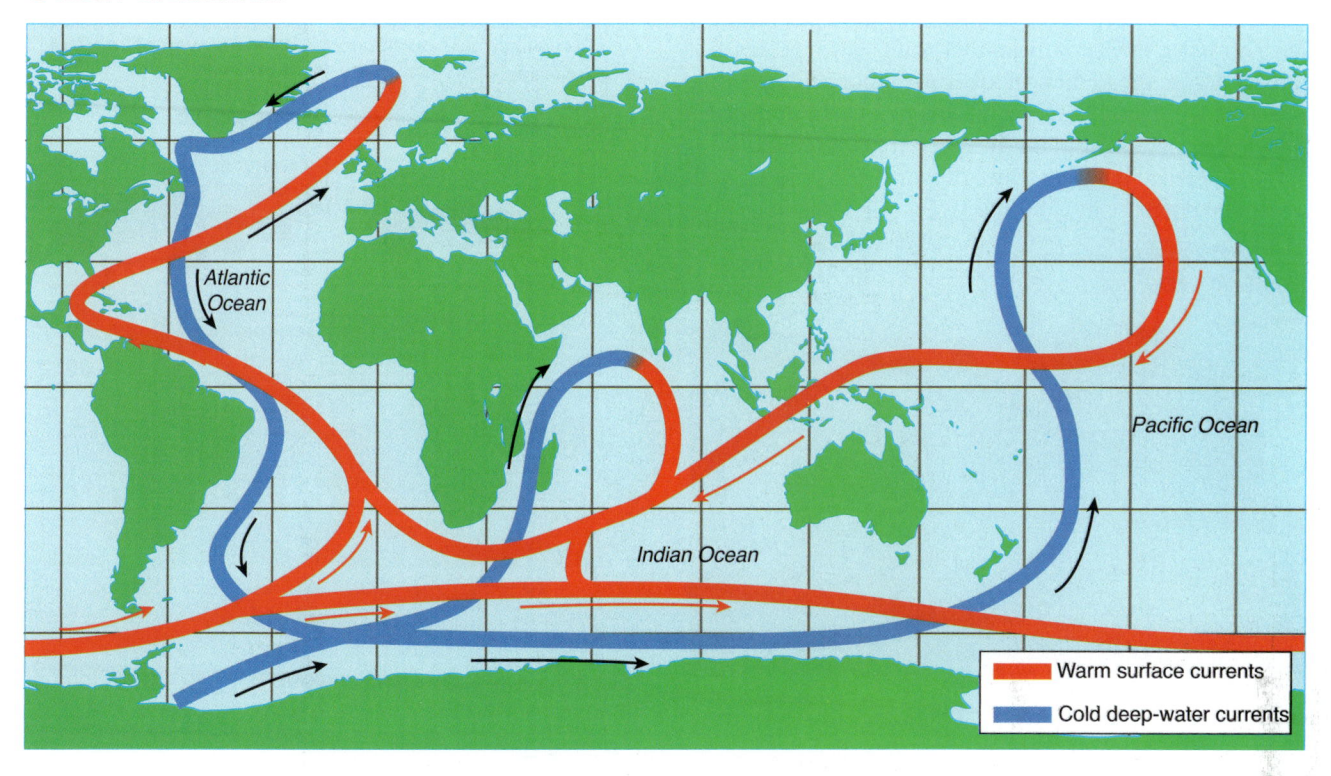

Figure 8.15: The Ocean Conveyor Belt, showing how warm surface water moves from the equator towards the poles, while cold, dense polar water moves from the poles towards the equator at depth.

Oceans regulate the global climate, and ocean currents move energy around the planet. Warm surface currents take the heat from the equator towards the poles, while cold deep ocean currents carry cooler water from the poles towards the equator. This is carried out through an oceanic mechanism called the ocean conveyor belt (see Figure 8.15).

Research suggests that ocean currents can be disrupted. If this occurs, the climate is likely to change significantly. If ocean currents change, expected climate changes include

- an increase in precipitation in the North Atlantic
- melting of polar ice sheets and glaciers
- an influx of warm fresh water into the sea
- the inability of polar ice to form.

Wind circulation

Figure 8.16: Strong winds in Krakow, Poland (2022) uprooted trees in the city.

Wind is the movement of air. It is caused by the uneven heating of Earth's surface by the sun, and the rotation of Earth. Therefore, the changing temperature of the oceans and atmosphere impacts the winds that circulate the planet.

A study in 2019 found that global wind speeds had increased by nearly 6% over a period of nine years. This, however, does not mean that winds have increased equally across the planet. In areas with more energy available, wind speeds are increasing due to the contrast between warm and cold air. In other areas, a reduction in the difference between zones of warm and cold air will result in a decrease in wind speeds.

Melting ice

Figure 8.17: An Iceberg melting in Greenland.

Since the mid-1980s, the surface temperatures of the Arctic have warmed at a rate of at least double the global average, causing ice sheets to shrink and become thinner (Figure 8.17). Sea ice, glaciers and ice sheets on Greenland have also seen a significant decline.

The heating and melting of frozen water stores, has a direct impact on the planet:

- The melting of surface ice reduces the **albedo** effect. This is the ability of ice to reflect light. The reduction of the albedo effect increases the amount of solar energy being absorbed.
- The melting of zones of permafrost results in increased rates of methane release.
- Melting land ice is causing sea levels to rise.

> **KEY TERM**
>
> **albedo:** the ability of a surface to reflect light. White surfaces have a high albedo

Species distribution and biodiversity

Arctic-like traits dominate:
Small size
Slow growth
Feeding at sea floor
Specialist

Warmer water-like traits dominate:
Large size
Fast growth
Feeding at sea floor + water column
Generalist

Higher temperature
Retreating sea ice

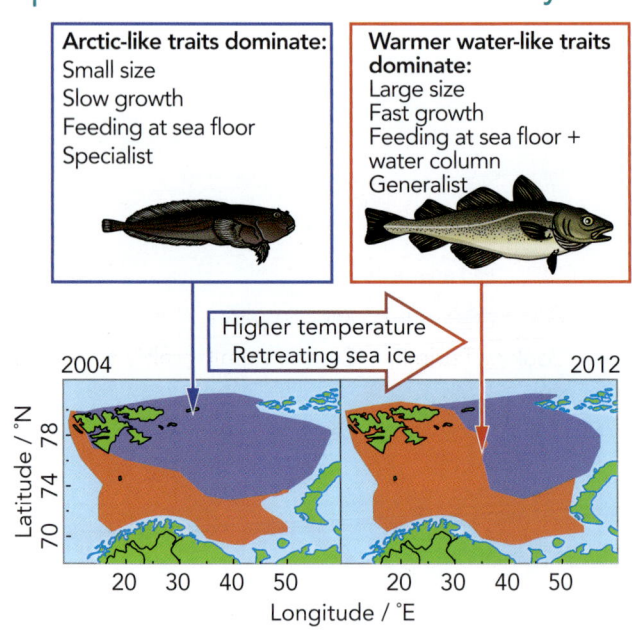

Figure 8.18: Changes to the ranges of fish species in the Barents Sea (Northwest Russia) between 2004 and 2014.

Climate change is altering habitats around the world and affecting where species can live. Where temperatures are increasing, mobile species may seek cooler regions.

- Analyses of species of arthropods, birds and butterflies found a poleward migration of approximately 16.9 km per decade.
- The rate of climate change can limit species' survival. If trees are unable to colonise new locations fast enough, then climate change may result in those vegetation types disappearing.
- Loss of vegetation is also loss of habitat, which will impact biodiversity, as other species disappear along with the shrinking habitat.
- Decreased biodiversity reduces the resilience of ecosystems, making them more vulnerable to collapse as food webs fail.
- Food webs fail when one or more species disappear and remaining organisms need to find an alternate source of food. This is not always available.
- In marine ecosystems, species move to cooler waters in search of temperatures they can tolerate. Species that prefer warmer waters are expanding their territories and outcompeting the colder water species (see Figure 8.18).

8 Managing Climate Change

> **ACTIVITY 8.1**
>
> **The impact of sea level rise**
>
> 1. Working in pairs or small groups, investigate the impact of sea level rise on a low lying coastal region of your choice. Consider the impacts on
>   - species and biodiversity
>   - towns or cities in that region
>   - fresh water resources
>
>   Are any solutions being implemented to reduce the impacts of sea level rise in that area?
> 2. Look at Table 8.1. Create a similar table to explain the impact of sea level rise on ecosystems and people within the area you have studied.
> 3. Discuss your findings with the rest of the class and compare your information.

Impacts of climate change on humans

Climate change and the shift in climate patterns affected approximately 68 million people in 2020, and caused an estimated US$131 billion in damage. Of that damage, approximately 95% was caused by storms, floods, wildfires and drought. Food, water, health and human habitat and infrastructure have been identified as being most at risk from climate change.

> **ACTIVITY 8.2**
>
> **Impacts of climate change**
>
> 1. Research the impacts (both positive and negative) of climate change within your region.
> 2. Collect images from magazines or online that represent these changes.
> 3. As a class create a story board showing the impacts that you have discovered. Discuss your findings as a group.
>   a. Why is it important to understand these impacts?
>   b. Does your story board show a specific trend?
>   c. Are your findings supported by scientific data? Why is this important?
>
>   You will develop your storyboard in the next section.

Increased frequency and severity of extreme weather

Figure 8.19: San Juan during a hurricane. Note the strong winds created by the storm.

Areas normally exposed to strong storm systems face an increase in frequency and severity of extreme weather events. This leads to increased:

- flooding
- strong winds
- loss of land
- damage to property and infrastructure
- loss of lives

Areas exposed to less precipitation and increased temperatures face:

- more frequent droughts
- wildfires starting more easily, burning hotter and faster.

Forced migration

Figure 8.20: Signs at COP26 seek protection rights for climate refugees.

Natural disasters associated with climate change, like flooding, wildfires, droughts, sea level rise, and increased storm size/frequency, put pressure on communities. As some parts of the world become more difficult to live in, people are forced to migrate, and are referred to as 'climate refugees'. The extended case study on the Marshall Islands discusses the concept of **forced migration** in more detail.

> **KEY TERM**
>
> **Forced migration:** the involuntary movement of a person or people away from their home or home region. Causes include natural disasters, wars, and food or water insecurity.

Food security

Figure 8.21: The drought in Guangxi, China, led to the failure of crops.

Changes in rainfall patterns, frequency and duration affect the health and productivity of crops. Higher temperatures impact the length of the growing seasons and the speed at which crops mature.

- Where temperatures continue to increase around the equatorial zones, maize yields are expected to decline.
- In areas of both increased storm intensity and drought, crop yields will be significantly impacted.
- Crops are damaged during storms, and some areas lose their entire annual yields. Where crops do survive, they often rot in standing water, or are contaminated by septic waste.
- In areas of drought, persistent crop failure occurs due to insufficient water (Figure 8.21). Global production of staple food sources such as wheat and rice is shrinking, especially in food-insecure countries.

Food security is severely impacted in many regions. Weakened crops are more susceptible to the spread of disease. The changing climate is triggering outbreaks of insects such as locust swarms, which have been linked to unusual weather events. Increases in cyclones and wetter, warmer spells have resulted in the outbreaks of locust swarms across Africa, Pakistan and India. The worst swarming events in 30 years were witnessed in 2019–2020.

Water security

Figure 8.22: Glossop, England, 2021. High temperatures, drought and high population pressure puts water security in the UK at risk.

Water scarcity is expected to affect 1–3 billion people by 2100. This has the potential to cause conflict in the form of 'water wars'. Many areas that are predicted to face water shortages are in the least economically developed regions. Water scarcity is due to:

- increasing rates of evaporation as the atmosphere warms
- a decline in precipitation by up to 20% in many areas, as climates change.

In areas of increased precipitation, floods can contaminate drinking water, causing diseases such as cholera. Contaminated water cannot be consumed as it is a health risk. This puts pressure on water security.

Energy security

Figure 8.23: Cities like Tromsø in Northern Norway need a secure energy source to light and heat the city.

Energy production needs to be addressed in relation to climate change. However, it is difficult to balance a moving away from fossil fuels with the need for energy.

Energy production accounts for approximately 60% of greenhouse-gas emissions globally. Reducing emissions from energy production is essential in combatting climate change. However, the gap between energy produced by renewable sources and the burning of fossil fuels is still problematic. Finding ways to manage energy shortages is challenging, as global pressure on energy supply is already significant.

8.2 Questions

1. Explain how ocean currents and atmospheric circulation help to balance global energy budgets.
2. Discuss how biodiversity and species distribution will be impacted by a change in the global climate.
3. Explain why crop yields are expected to decrease if climates continue to change.
4. With reference to Figure 8.24:
   a. Describe the trends for the number of extreme weather events recorded.
   b. Explain the potential limitations with the data from the years before 1960.

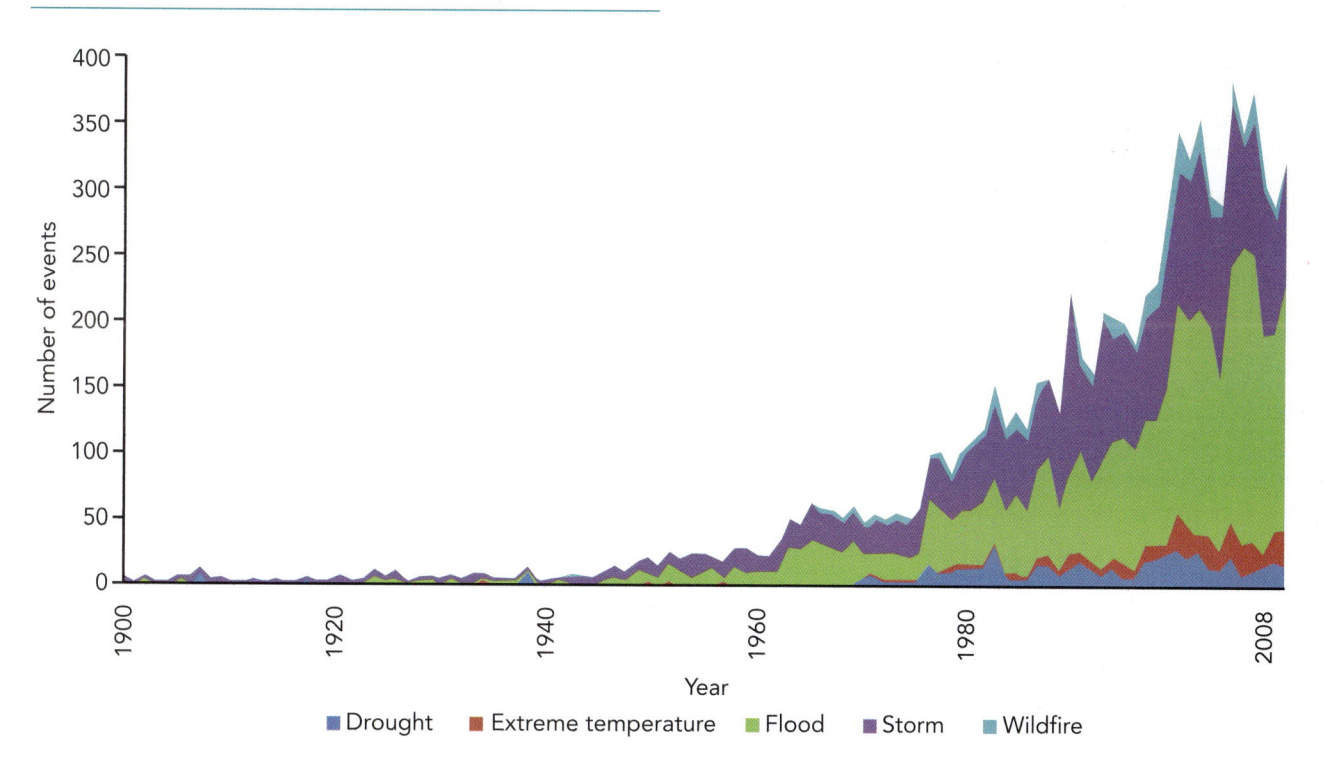

Figure 8.24: Graph showing the number of extreme weather related events recorded between 1900 and 2008.

CASE STUDY

The impact of climate change on the Great Barrier Reef, east coast of Australia

The Great Barrier Reef is the longest and largest reef complex in the world. It extends more than 2000 km in a northwest–southeast direction along the east coast of Australia. Ranging between 16 km and 160 km offshore, it is between 60 km and 250 km wide.

The reef is high in biodiversity and made up of approximately 2200 reefs and 800 fringing reefs. The corals are made up of living and previously living organisms. They have skeletons of calcium carbonate which is susceptible to changes in ocean acidity.

Figure 8.25: A section of healthy coral reef on the Great Barrier Reef, Australia.

The ideal living conditions for most corals is clear saline water of a temperature between 23 and 29 °C. Most corals are restricted to the **euphotic zone**, the region that descends to approximately 70 m, where sunlight penetrates and allows for photosynthesis.

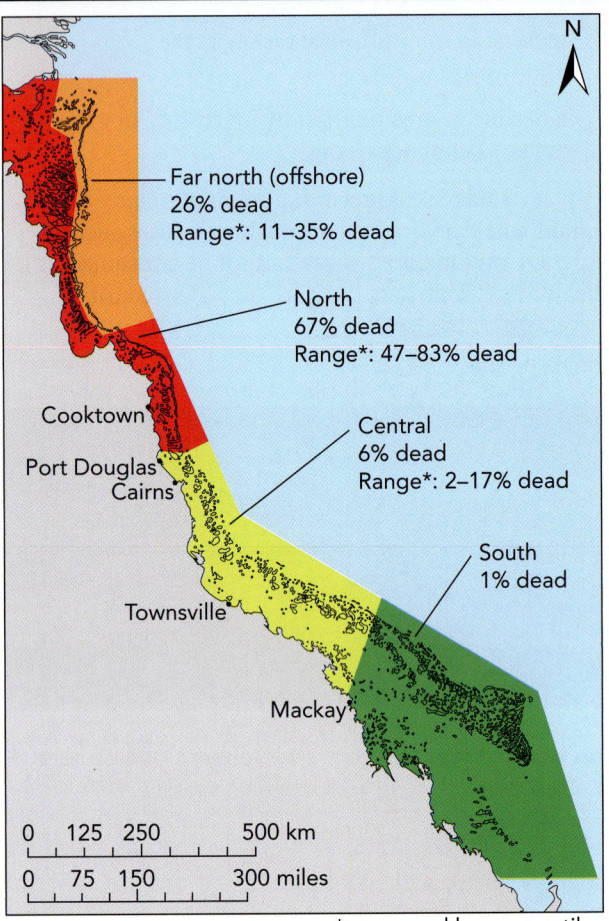

- Far north (offshore): 26% dead, Range*: 11–35% dead
- North: 67% dead, Range*: 47–83% dead
- Central: 6% dead, Range*: 2–17% dead
- South: 1% dead

*upper and lower quartiles

Figure 8.26: Mapping of the 2016 coral bleaching event shows higher rates of bleaching in the northernmost reaches of the reef.

KEY TERM

*euphotic zone: the upper layer in the ocean that sunlight penetrates, allowing for photosynthesis

CONTINUED		
Sea level rise	Sea level rise triggered by global warming reduces the amount of light reaching corals. This results in lower levels of photosynthesis by **zooxanthellae** which live in a **mutualistic relationship** with the corals. Zooxanthellae give corals their range of colours (Figure 8.25). The two organisms rely on each other for survival. A change in sea level also impacts the balance of erosion and construction processes that shape the reef. Increased wave frequency and energy damage coral structures. More severe storms caused by climate change increase the size and frequency of waves pounding on a reef and damaging it.
Coral bleaching	Warmer temperatures of oceans' upper surfaces increase the occurrence of coral bleaching (corals becoming white). When corals become stressed by environmental factors, such as the heating of ocean waters or pollution, which are above their tolerance range, they expel zooxanthellae. This weakens the coral, making it susceptible to disease, and reduces its growth rate. Bleaching is caused by corals becoming white as they expel zooxanthellae. Corals can recover from bleaching events that are not too frequent or too intense (too often or too hot). However, both the frequency and intensity of increased water temperature is increasing. This means that corals will ultimately die off. Major bleaching events have been occurring since 1998. They have resulted in a significant die off of portions of the Great Barrier Reef. In 2016, a 700 km area of reef in the northern region of the Great Barrier Reef was found to have lost on average 67% of its shallow-water corals (Figure 8.26).
Ocean acidification	The oceans absorb CO_2 from the atmosphere as part of the carbon cycle. An increase in the amount of the atmospheric CO_2 is resulting in oceans absorbing more CO_2. This increases the acidity of oceanic waters, leading to ocean acidification. Acidic water dissolves calcium carbonate which corals use to build their exoskeletons. Without skeletal growth, corals erode, and their growth rates slow. This prevents them from building the reefs in which they survive.
Habitat changes	As the water temperatures rises in northern reaches of the reef, many marine organisms migrate south into cooler waters. This increases the competition for food and shelter in reefs to the south, resulting in threat to the corals as carrying capacity is exceeded.

Table 8.1: The impacts of climate change of the Great Barrier Reef.

KEY TERMS

*zooxanthellae: photosynthetic, single-celled marine organisms that live in coral polyps

*mutualistic relationship: a relationship between two organisms where both species benefit from their interactions

CAMBRIDGE INTERNATIONAL AS LEVEL ENVIRONMENTAL MANAGEMENT: COURSEBOOK

CONTINUED

Figure 8.27: Staghorn corals show the effects of bleaching, after periods of higher than normal temperatures on the northern Great Barrier Reef, Australia.

Evaluation

If atmospheric CO_2 levels and temperatures continue to increase at their current rate, the coral reefs will be severely impacted. Reducing the rate of greenhouse gas pollution levels over the coming years and decades will have a significant impact on these marine organisms.

Extreme ocean temperatures (marine heatwaves) are likely to occur more regularly and will increase the frequency of coral bleaching events. Meanwhile, ongoing increases in CO_2 levels will continue to acidify the oceans, putting more stress on the corals.

The complex impact of global and local climate change, and the rate at which it is happening, means that the corals, their associated species and the food webs which they support are all at risk.

Fossil records show that corals have survived and adapted to changes in sea level and temperature in the past. Their long-term resilience and ability to adapt to change is not yet fully understood. However, they do potentially possess a mechanism that could allow them to mitigate the impacts of climate change and survive.

Case study questions

1. Describe the process of coral bleaching.
2. Explain the effect that sea level rise and ocean acidification is having on corals in the Great Barrier Reef.
3. In pairs, carry out research to find out the impact that damage to the coral reefs could have on the wider marine ecosystem. Create a spider diagram to show your findings.

8.3 Managing climate change

Strategies for managing climate change require a global effort. No individual country or person can, by themselves, make sufficient change to solve the climate problem.

Reducing the use of fossil fuels

Reducing carbon emissions in order to limit the amount of heat-trapping carbon molecules in the atmosphere is the primary action in the fight against climate change.

This is referred to as **mitigation**. The combustion of coal, oil and gas all result in the release of carbon from the slow carbon cycle into the atmosphere. This adds to the carbon load and increases the enhanced greenhouse effect.

> **KEY TERM**
>
> ***mitigation:** the action of reducing the severity or seriousness of something

8 Managing Climate Change

Figure 8.28: A surface coal mine showing coal deposits. Coal is mined to supply power stations with an energy source for the generation of electricity.

There are many ways that fossil fuel use can be reduced, such as:

- Individual actions such as conserving energy at home, using energy efficient appliances, and insulating homes effectively. Also, using public transport or sharing cars, choosing to walk or ride a bike where possible.
- Actions of governments and local councils could include developing clean energy policies, such as using public transport run on clean energy or low carbon fuels.

Switching to low carbon fuels

Low carbon fuels (LCF) are fuels with a lower carbon content than conventional petroleum fuels (petrol and diesel). These include energy sources such as natural gas or biofuel.

The main goal of LCFs is to decrease the amount of CO_2 emissions from vehicles with internal combustion engines. Between 2007 and 2011, the concept of low carbon fuels was applied in California, USA, then British Columbia, Canada, before being adopted by the EU.

One of the limitations of switching to LCFs is that they indirectly impact land use. Biofuels are produced by processing crops such as corn and sugar cane. Converting large areas of natural vegetation to sugar cane, for example, affects habitats and biodiversity.

Using alternative forms of energy

We use energy to run all our systems, from transport to heating, cooling and lighting. Moving from carbon based fuels to alternative energy sources is therefore key to stabilising increasing levels of CO_2 in the atmosphere caused by human activity.

Although renewable energies (Figure 8.29) do release some CO_2 into the atmosphere, this is far lower than fossil fuels. For example, for one kWh of electricity, coal emits approximately 820 g CO_2, while wind energy emits 11 g. Moving to renewable energy sources such as geothermal energy, HEP, solar and wind energy will reduce the daily volume of CO_2 being emitted into the atmosphere, and help to stabilise CO_2 levels in the atmosphere.

Figure 8.29: Alternative energy forms include renewable energy sources such as wind generated electricity.

Updating transport policies

Traditional transport technology uses fossil fuels as a source of energy. This means that transport is targeted when considering adaptations and policies to reduce CO_2 emissions. There is widespread agreement to achieve a 50% reduction in CO_2 emissions by 2050.

CO_2 **abatement** and improved fuel efficiency are key concepts when considering solutions to the problem of high carbon emissions from transportation.

- Fuel efficiency can be improved through new vehicle technologies including advanced engine management systems and efficient engine function.
- Vehicle emissions can be reduced through the use of technology such as catalytic converters.
- Sustainable biofuels, such as biodiesel and biogas, can be used to power engines.
- The development of reliable, cost effective public transport systems reduces the number of vehicles on the road.
- Improved transport infrastructure reduces congestion and encourages the use of public transport.
- Development of new technology, such as electric vehicles, reduces combustion of fuel vehicles.
- Legislation such as congestion charges. For example, in London, UK, drivers are charged a daily rate to enter the city centre. Money collected from the congestion charge is used towards maintaining the public transport system. This makes public transport a more viable option and reduces the number of vehicles on the road.

> **KEY TERM**
>
> ***abatement:** the reduction or removal of something

Reduction of global and individual carbon footprint

A carbon footprint is the total amount of greenhouse gas emitted as a result of daily activities. We can reduce our carbon footprint in many different ways. For example, we can change the way we build (to reduce energy consumption), what we eat, our electricity usage, and we can even change our shopping habits (buying local or imported goods, choosing between goods with a long or short lifespan). Usually, the most significant portion of an individual's carbon footprint comes from housing, transportation and food.

Figure 8.30: Alternative forms of transport, such as a bicycle, can reduce an individual's carbon footprint.

Food's carbon footprint is determined by how far it is transported and whether it is in season or not. If a product cannot be grown locally, it has to be imported, which creates CO_2 emissions through long-distance transport. Eating meat has a large carbon footprint associated with it, due to the water, food and energy required to raise livestock. Cattle in particular release huge amounts of methane, too, and creating farm land causes deforestation. Switching from eating beef to poultry decreases CO_2 emissions for an individual by approximately 700 kg of CO_2 per year. Following an entirely vegetarian diet decreases a person's footprint by an estimated 1360 kg per year.

Figure 8.31: Reducing meat consumptiom by eating vegan burgers made from plant products.

8 Managing Climate Change

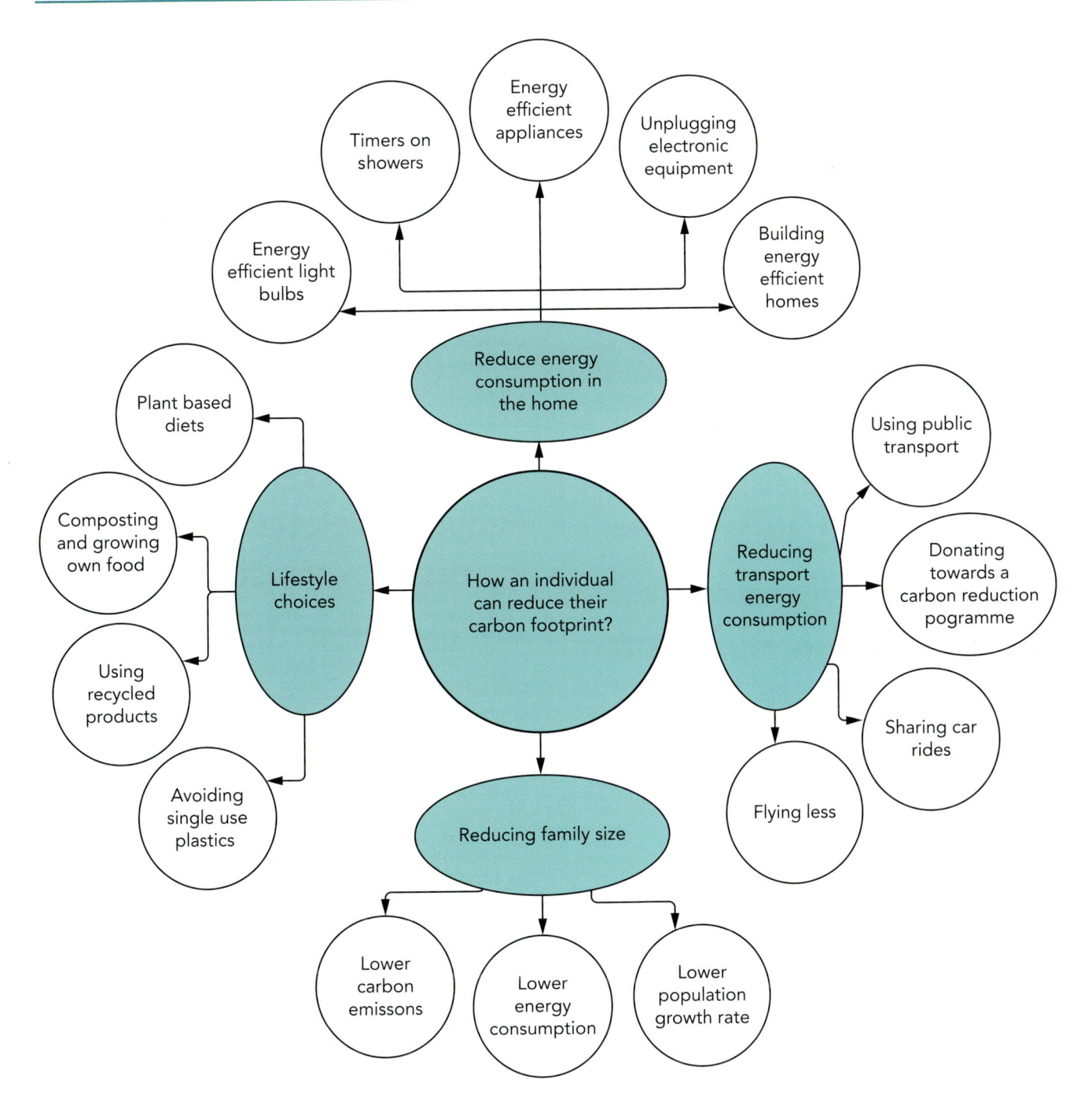

Figure 8.32: A spider diagram showing how individual choices can reduce a carbon footprint.

> CAMBRIDGE INTERNATIONAL AS LEVEL ENVIRONMENTAL MANAGEMENT: **COURSEBOOK**

> **ACTIVITY 8.3**
>
> **Carbon footprint**
>
> 1. Keep a diary of your habits for a week. Record items like:
>   - how you got to school
>   - the meals you ate
>   - if you recycled or used recycled products
>   - where your food came from
>   - if you switched off lights in empty rooms
>   - if you unplugged or switched off your appliances when not in use
>
> 2. Switch diaries with a study partner to compare answers.
>
>   a. What advice can you give them to help them reduce their carbon footprint?
>
>   b. What can you learn from them?
>
> 3. Consider your notes from the week, the discussion with your study partner, and the spider diagram in Figure 8.32. Write a short essay of approximately 300 words, explaining how you could reduce your personal carbon footprint.

Use of carbon capture and storage

Carbon storage is the process of capturing CO_2 before it is released into the atmosphere and storing it. The main methods of capturing it are pre-combustion, post-combustion and oxyfuel.

- Pre-combustion is carried out before fossil fuels are burned. It involves converting the fossil fuel into a mixture of CO_2 and hydrogen.
- Post-combustion technology removes CO_2 after the burning of fossil fuels from the flue gases.
- Oxyfuel technology involves burning fossil fuels with almost pure oxygen, producing CO_2 and steam. The CO_2 is then captured.

Post-combustion and oxyfuel technology can be used in new power plants or retrofitted (added after something was built - like air conditioning to a house built in the 1800s) to older power plants. Pre-combustion technology requires significant modifications to a power plant and is therefore more suited to newly built power plant.

Once CO_2 has been captured, it is compressed to form a liquid. It is then transported to the storage site by tanker, pipeline or ship. It is then pumped underground, usually at a depth of 1 km or greater. Old mined coalbeds or depleted oil reservoirs are often used for this purpose. Carbon storage has the potential to reduce the atmospheric carbon load.

The limitations of the technology are that it is still expensive. There is also the risk of CO_2 leakage from storage sites, although this risk is considered low.

Reducing deforestation and increasing reforestation

Carbon sequestration is reversed through the process of cutting down or burning of trees. During this process, CO_2 is released into the atmosphere, while the trees that take carbon dioxide from the atmosphere and store it as biomass are lost. This reduces the amount of CO_2 captured by forests. Furthermore, deforestation changes the ability of a surface to reflect insolation, decreasing the albedo effect. This results in more energy being absorbed, which further increases global warming.

Reforestation, **afforestation** and forest conservation allow for the growth of healthy forested areas. The greater the tree biomass and density, the greater the amount of carbon being stored.

This is a relatively cost effective way of combatting climate change, as the cost of planting a tree is low in comparison to other carbon management systems.

> **KEY TERMS**
>
> ***carbon sequestration:** a process by which CO_2 is removed from the atmosphere and stored in either solid or liquid form
>
> **afforestation:** the planting of trees in an area that was not previously forested

Figure 8.33: Reforestation and afforestation are terms which refer to the planting of trees. Here, trees are being planted in a savanna ecosystem to replace those lost during deforestation of the area.

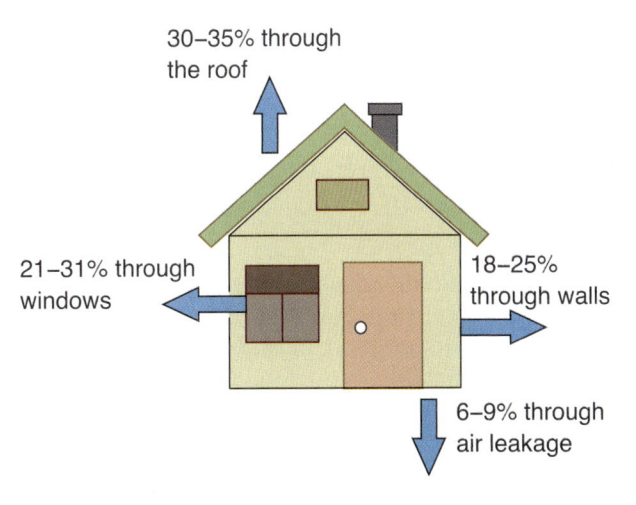

Figure 8.34: The percentage of heat lost from an insulated house in a temperate climate.

- 30–35% through the roof
- 21–31% through windows
- 18–25% through walls
- 6–9% through air leakage

Energy-efficient building and infrastructure

Cities create an estimated 75% of global CO_2 emissions, with buildings being the largest contributors (Figure 8.34). The 'greening' of buildings creates a more sustainable future. For example, in the EU all new buildings must be zero rated. Zero rated buildings are characterised by zero net energy consumption.

Buildings in the northern hemisphere need to be south facing to take advantage of the sun's energy (passive solar daylight and heating). In the southern hemisphere, buildings should be north facing to keep them warm.

Buildings can be built with deep overhangs (verandas) that give shade when it is hot, but allow sunlight in during the colder time of the year when the sun is low in the sky.

Buildings need to be well insulated. Windows can be double or triple glazed to limit the amount of energy flowing through them. In addition, low energy consumption features like LED lighting and star rated electrical equipment will reduce the amount of energy consumed.

Passive buildings

A passive house/building is energy efficient, comfortable, affordable and eco-friendly. It is a construction concept that considers the source and type of materials, and the building design. Passive houses/buildings consume low amounts of energy, and therefore have low carbon emissions, throughout their manufacture, building and use. Heating and cooling energy savings can be up to 90% compared to typical buildings. They also make use of energy sources inside the building such as solar heat, and they recover energy to contain it in the building.

In passive buildings, the building shell and windows provide good insulation. Roof and floor slabs keep in the heat during the winter and let it out during the summer. These buildings are also designed to ensure a consistent supply of fresh air.

Passive buildings significantly reduce greenhouse emissions. It is not only new builds that can be passive buildings, an existing building can be retrofitted to become a passive building.

> **ACTIVITY 8.4**
>
> **Managing the impacts of climate change**
>
> Use the storyboard you created in Activity 8.2 to complete this task. Consider the impacts you identified in Activity 8.2.
>
> 1 Explain the different climate change management strategies that could be employed to address the causes of these impacts. Write these out in a table, naming the impacts, and then describing the strategies to address them.

8.3a Questions

1. Define the terms reforestation and aforestation.
2. Explain how energy efficient buildings reduce CO_2 emissions.
3. Energy production is a significant contributor to climate change. Discuss the ways in which this impact can be reduced.
4. Explain how carbon capture and storage can reduce the impacts of rising CO_2 levels on global climate.

Adaptations to climate change

A community can make adjustments to survive the impacts of climate change. Adaptations made by a community depend on the climate change threats that it faces. There are a variety of different strategies, outlined in Table 8.2.

Sea-level rise	Sea levels are expected to rise between 20 cm and 200 cm by 2100, threatening low lying coastal areas. Miami, Florida, USA, is investing in hard engineering such as raising road levels, strengthening sea walls, and installing drainpipes and pumps to pump water out of areas. Other areas are resorting to soft engineering to reduce the risk of coastal erosion and damage from storm surges. Examples include the restoration of coastal wetlands and mangroves.
Wildfires	Preparing for longer, drier, warmer summer seasons and associated wildfires is important in arid areas. Educating the population about risks and evacuation plans minimizes risk to life. Training rapid response teams and developing infrastructure and technology to battle the fires that threaten populated areas is also key.
Severe weather	More intense storms and heatwaves threaten public health and safety. Many communities are updating infrastructure to cope with these changes, for example, installing higher sea walls and larger or better-designed storm water drains. Approaches to real estate development are also changing, as likely flood zones are considered, along with construction materials that protect the occupant in the event of strong winds, rising water or extreme heat.
Emergency education and planning	Emergency education and planning are vital for communities at risk. Understanding the scale and frequency of the risks that they face, evacuation procedures, and safe places are important. Individuals need this knowledge to protect themselves and their families. Governments and NGOs are making large scale, long-term plans and putting reserves in place. These include basic survival items such as medicine, water, food and shelter.
Protecting air quality	Changes to the climate can result in an increased number of microscopic particles in the air. These cause allergic reactions and increase the incidence of hay fever. They also harm respiratory and cardiovascular systems. Communities around the world are looking at ways to reduce emissions in their immediate surroundings to ensure cleaner air.
Managed retreat planning	When a community is damaged beyond recovery (from fires, coastal erosion, increased storm frequency or sea level rises), they can choose to move to a new, safer area. This might mean moving to higher land, or to areas at lower risk of climate related disasters.

Table 8.2: Adaptations to climate change.

8 Managing Climate Change

Figure 8.35: The Thames flood barrier, London, prevents storm surges from flooding the low-lying regions of the Thames River.

Figure 8.36: Firefighting equipment putting out a fire in France. The use of firefighting technology such as helicopters and aeroplanes minimises the risk to people on the ground and provides access to more remote areas.

Figure 8.37: Emergency supplies are delivered to the victims of the 2016 Japanese earthquake. Planning for a disaster makes this process easier.

National and international agreements

International agreements such as the Montreal Protocol (1987), Kyoto Protocol (1992), Paris Agreement (2015), and COP26 (2021) are designed to limit the global temperature increase to 2 °C by 2100 and now aim to limit that even further to only 1.5 °C.

The Montreal Protocol (1987) was established to facilitate international co-operation on research into the ozone layer and the impacts of CFCs (chlorofluorocarbons) on ozone gas in the stratosphere.

The Rio Summit (1992) resulted in an important agreement on climate change and the development of the Kyoto Protocol (1997). This international agreement was aimed at reducing Green House Gas Emissions based on the premise that:

1   Global warming exists
2   Human-caused emissions are the underlying cause.

The 2015 Paris Accord was an agreement within the UN addressing greenhouse gas emissions mitigation, adaptation and finance, starting in 2020. Annual conferences continue to raise global awareness and deal with the problems of high emissions. The latest conference was **COP26** (2021, Figure 8.38) which took place in Glasgow. This was seen as a turning point for action on fossil fuels, including the phasing out of governments subsidising fossil fuel production. This is a key development as it prevents governments from financially supporting the mining and use of fossil fuels that would otherwise not be profitable for the producer.

Figure 8.38: UN Climate Change Conference COP26 in Glasgow where international climate change discussion and agreements took place in 2021.

> **KEY TERM**
>
> ***COP:** Conference of the Parties. 'Parties' refers to the 197 nations that agreed to an environmental pact (UNFCCC – the United Nationals Framework Convention on Climate Change) in 1992

The launch of the 'Beyond Oil and Gas Alliance' means that the need for action against all fossil fuels is now on the political agenda. Governments will now find their climate leadership questioned if they continue to support the production of oil and gas.

Also of significant importance was the pledge from the USA and China to boost climate cooperation over the next decade. Both countries agreed to take steps on a range of issues including methane emissions, transitioning to clean energy and decarbonisation.

Geo-engineering strategies to counteract climate change

Geoengineering is the intentional manipulation of the natural environment using engineering to reduce the anthropogenic impacts on the global climate. Many of its concepts are still theoretical. However, they all consider ways of reducing the impacts of climate change. This can be by returning long-term climate stores back into Earth, or by reducing the amount of incoming solar radiation in order to limit the heating effect of greenhouse gases.

Solar radiation management (SRM)

Solar radiation management is a theoretical approach to managing climate change. If ever put into practice, it would operate by reflecting some of the insolation (incoming solar radiation). This would prevent energy from entering the atmosphere and would therefore limit the amount of energy trapped by greenhouse gases. SRM would effectively act as an artificial reflective shield to provide regional or global cooling.

At ground level, SRM proposals include using large pumps to introduce microbubbles into bodies of standing water. This would increase the albedo of the surface of the water. Another proposal is the spreading of reflective films on top of the ocean surface, or on the surface of vulnerable areas of ice. Even painting roads and the roofs of buildings white would help to cool local temperatures, as insolation is reflected rather than absorbed. If implemented widely, these methods could have a limited impact on global heating.

Three SRM proposals would operate in or above the atmosphere.

1   Marine cloud brightening involves introducing saltwater particles from the ocean up into the clouds. Theoretically, this would result in the formation of new whiter clouds, increasing the reflection of insolation. This is one of many different proposals being looked at in Australia to reduce the impacts of heating on the Great Barrier Reef. In 2020, an early trial used a turbine with hundreds of nozzles to spray salt crystals into the sky.

2   Stratospheric aerosol injection would involve putting reflective particles into the upper atmosphere.

3   Space mirrors are satellites which are placed outside of the atmosphere, and are designed to alter the amount of solar radiation entering it. The mirrors deflect a fraction of the amount of insolation coming into the atmosphere, and so reduce global temperatures (Figure 8.39).

SRM poses a number of ethical and political problems. Should these technologies be developed? Should they be used? Who decides which technologies are used, and based on which criteria? Who would be in control of the technologies? Who would pay for them? Could some communities suffer if technologies are implemented incorrectly? These questions need investigation, and the resulting international agreements would need to address areas of concern in order to ensure that SRM does not harm some while benefitting others.

Negative or unintended side effects are not always understood when making changes to energy flows on the complex system that makes up the world's energy budgets. SRM would require extensive modelling to try to ensure that it was beneficial and not detrimental.

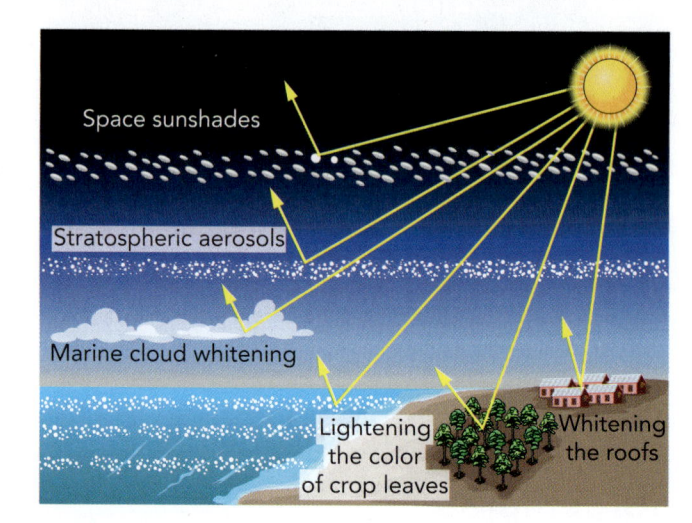

Figure 8.39: Methods of solar radiation management.

8 Managing Climate Change

8.3b Questions

1. Adapting to climate change is one approach to managing its impacts. Explain how humans can adapt to climate change.
2. With the use of examples, explain how international agreements play an important role in managing atmospheric pollution.
3. Explain the limitations in the use of geo-engineering strategies for counteracting climate change.

EXTENDED CASE STUDY

Island nations and sea level rise

Many Pacific nations made up of clusters of islands are directly threatened by climate change and the resulting sea level rise. Island nations such as Kiribati, Tuvalu and the Marshall Islands (Figure 8.40) are all low-lying islands in danger of disappearing under the sea in the coming decades.

With rising sea levels come the additional threats of more frequent and intense flooding and salt water intrusion into groundwater resources. This poses a risk to both food and water security.

The capital city of the Marshall Islands, Majuro, has lost a significant amount of land surface to the sea. Residents have had to build seawalls to protect their homes from the tides. On some islands, such as Tarawa of the Pacific Island nation Kiribati, residents have tried to reclaim land from the sea. However, the rate of sea level rise has overwhelmed these attempts.

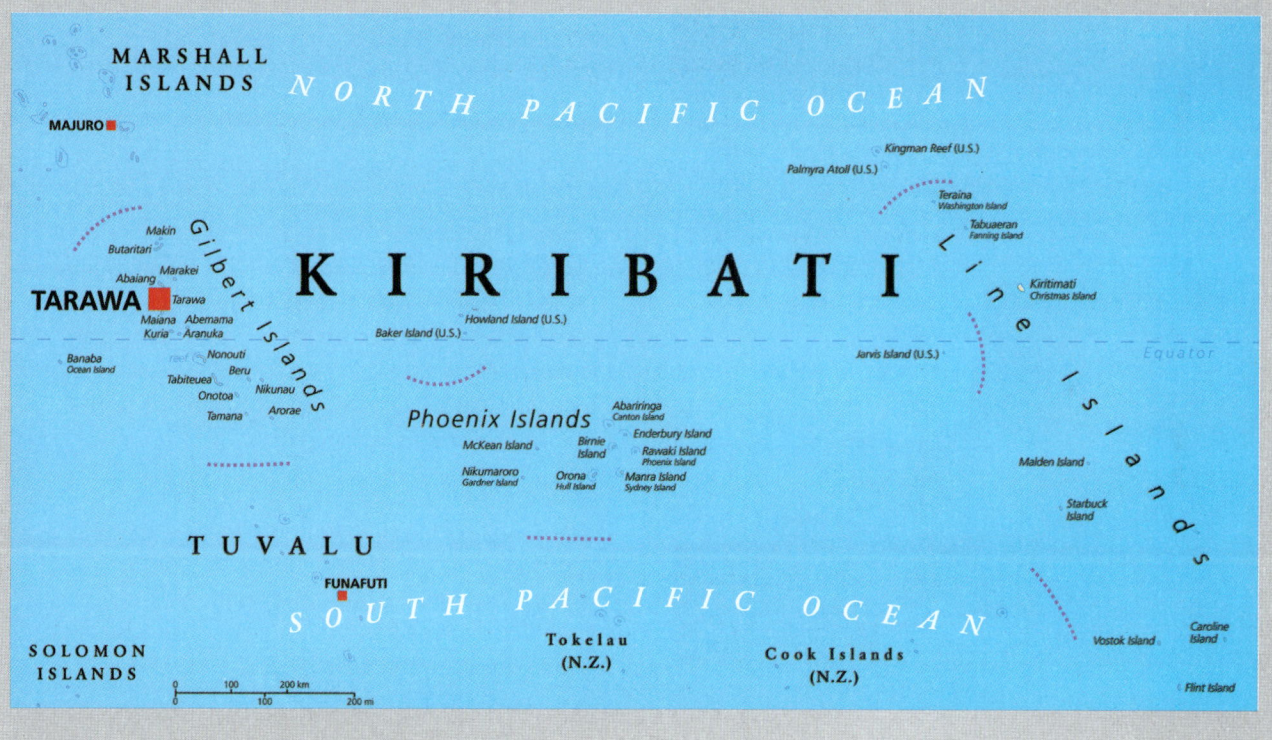

Figure 8.40: A map showing the location of Kiribati, Tuvalu and the Marshall Islands.

CONTINUED

Islands off the coast of Bangladesh have been equally impacted by sea level rise, and rice paddies have been inundated with sea water. Rice farming was the country's main economic activity, and so farmers are trying to adapt to the change by turning their flooded rice paddies into 'salt paddies', and mining salt instead.

Figure 8.41: Sea pounds along the sea wall, causing coastal erosion and threatening the homes of people living on the islands.

Figure 8.42: Damage caused to buildings by rising sea levels and stormy seas.

These impacts on developing island nations highlight the need for equitable social plans to assist with the migration driven by climate change. However, those migrants are not the people who caused the problem in the first place. Their population is now not only homeless, but landless, and they need an opportunity to thrive elsewhere.

The slow-moving disaster of sea level rise will cause people to move and never return. Where sea level rise is occurring along the coastlines of continents, residents are forced to move inland to survive. Island nations do not have this luxury. As their islands slip beneath the waves, they seek refuge with neighbouring nations.

Moving populations costs vast amounts of money. It is not only transport that needs to be considered. Receiving areas need infrastructure such as roads, septic waste removal, water and food security, housing, schools and medical care. Job opportunities need to be developed so that the new population has a chance to earn a living and become economically independent.

This is a global problem that requires global action. There are NGOs to support migrants who are 'climate displaced', such as the International Refugee Assistance Project (IRAP) to fight for their legal rights. The IRAP addresses the social, political and economic inequalities that these refugees face, and which impact their ability to move to a safer location. Migrants with fewer financial resources or those who care for others find it harder to start a new life in a new location.

Social inequality does not just lie in the fact that the wealthy can move while the impoverished cannot. It is also reflected in the fact that a disproportionate amount of climate change impacts are caused by wealthy countries, but have a more significant impact on low income countries.

People with the top 1% highest incomes globally (approximately 63 million people) are estimated to have emitted twice the amount of carbon emissions in the 25 years between 1990 and 2015 than the 3.1 billion people with incomes in the bottom 50%.

CONTINUED

In 2015, world leaders recognised the need to develop an integrated approach to solving these problems. At COP21 in Paris, a task force was established to start the important conversation on how to help climate refugees. The Global Compact on Safe, Orderly and Regular Migration and the Global Compact of Refugees was created. However, one of the greatest obstacles to progress is the lack of available funding for the mitigation of climate impacts.

Figure 8.43: Sandbags placed along the beaches in the Marshall Islands to prevent further coastal erosion.

Across the globe, the number of people displaced by sea level rise by 2017 was estimated at 26 million. This number could increase to 150 million by 2050. However, other estimates suggest that the number of people displaced by sea level rise by the year 2100 could be anywhere between 410 million and 2 billion people (if sea level rise reaches 2 metres above current levels).

Extended case study questions

1. Explain the causes of sea level rise.
2. Research and explain the concept of social equity.
3. The Marshall Islands frequently experience typhoons (hurricanes). Describe how sea-level rise worsens the impacts of a typhoon.
4. Sea level rise predictions for the Marshall Islands have been determined for three different levels of emissions.
   a. Look at the data in the table. Plot it on graph paper to create a line graph.
   b. Describe and explain the differences in the predictions.
   c. Calculate the percentage change for all three levels of emissions between the years 2030 and 2090.

Year	2030	2055	2090
	Predicted sea level rise in cm		
Low emissions	10	22	41
Medium emissions	15	30	63
High emissions	16	37	72

Extended case study project

Carry out research into the impact of sea level rise on the island nation of the Maldives.

Consider the following factors:

- The cause of sea level rise.
- The rate of sea level rise.
- The impacts of sea level rise on the population and environment.
- What is being done to try and combat the problem.
- The estimated date when the islands will no longer be inhabitable.

Task: Imagine you are a resident of the Maldives. Write a 300-word blog on the impact of the sea level rise. Give your personal viewpoint on how it is going to impact you.

SUMMARY

The greenhouse effect is natural, but the enhanced greenhouse effect is driven by human activities, including burning fossil fuels, livestock and rice farming and landfill sites.
Climate change is the increase in temperatures around the globe, causing changes in ice sheets, precipitation, ocean temperature, ocean acidity, wind patterns, species distribution and biodiversity.
Extreme weather events such as flooding and droughts are expected to increase in frequency and intensity.
Crop yields are expected to change, affecting global food security.
Climate models are limited as the relevant data is not yet totally understood and feedback mechanisms are still only being discovered.
Managing climate change includes factors that reduce greenhouse gas emissions.
Adapting to climate change involves the adjustments a community makes to survive climate change threats.
Geo-engineering strategies using SRM include increasing albedo at ground level, marine cloud brightening and stratospheric aerosol injection.

REFLECTION

Do you find the concept of climate change overwhelming? Have you tried to look at the solutions that humans are considering to manage the problem? Do you find that taking some time to problem solve helps with the anxiety that this topic and social media 'hype' may be causing?

8 Managing Climate Change

PRACTICE QUESTIONS

1. **Figure 1** is a model representing the greenhouse effect, and the amount of energy available in the atmosphere.

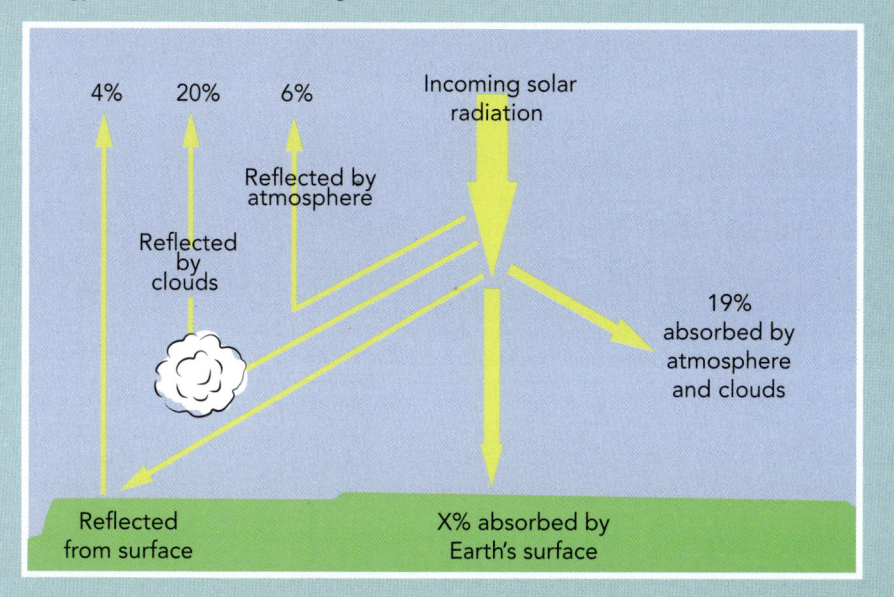

   Figure 1

   a Calculate the percentage of energy that is absorbed by Earth's surface. [2]

   b Describe and explain how the energy budget is expected to change due to the enhanced greenhouse effect. [3]

   c Describe the main causes of the enhanced greenhouse effect. [2]

   d Discuss some of the impacts of the enhanced greenhouse effect on the natural environment. [6]

2. Figure 2 shows the change in Arctic sea ice in March between the years of 1979 and 2021. March is when the Arctic sea ice typically reaches its maximum extent.

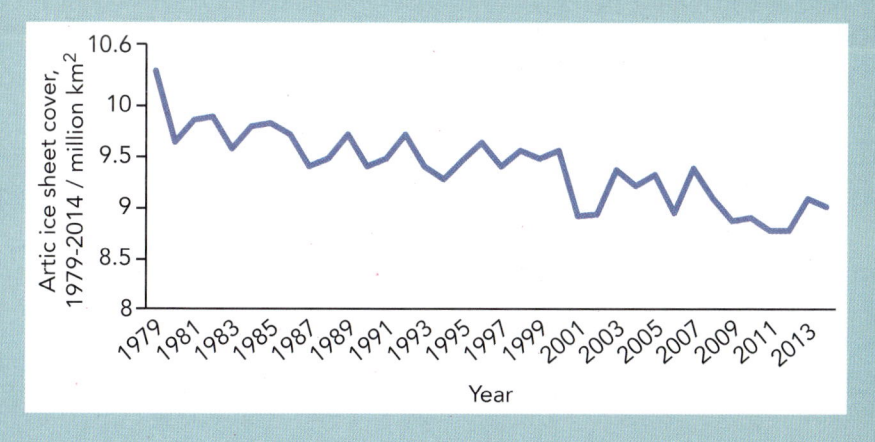

   Figure 2

CONTINUED

    a Describe the trend in sea ice cover between 1979 and 2014. [2]
    b Suggest an explanation for the trend observed. [3]
    c Describe the impact of melting land ice on coastal zones around the globe. [3]

3 This text discusses the potential decrease in food and water supplies by the year 2050.

> ### Decreasing food supplies and water scarcity
>
> The agricultural sector is the greatest consumer of water in the world, using around 70% of all the water consumed by humans. The possibility of crop yields failing is projected to be 4.5 times higher by 2030 and as much as 25 times higher by 2050. Water scarcity is considered to be one of the key causes of this problem. In Kenya it is estimated that the area suitable for growing wheat will decline from 101 666 km² (current time) to 23 792 km² in 2050 and as low as 15 435 km² in 2070.
>
> Global demand for food and water is also rising due to changes in consumption patterns and the ever-increasing global population. The main strategies for meeting this food demand are to increase cropland intensification and to expand current agricultural lands. However, climate change threatens available water resources in the food growing regions of the planet.

    a Calculate the percentage decrease in the area of land that will be suitable for growing wheat between the current time and the year 2050. [2]
    b Explain how climate change is expected to change water supplies around the globe, and therefore have an impact on crop yields. [3]
    c CO_2 emissions are believed to have a potentially positive impact for some crops. Explain why some crops are expected to have an increase in yield due to increased CO_2 in the atmosphere. [2]
    d Managing climate change is possible. Discuss some of the strategies that can be employed to help manage the impacts that humans are having on the atmosphere. [6]
    e Adapting to climate change is one of the options available to humans. Explain why this is not considered an option for many ecosystems. [3]

CONTINUED

4 **Figure 3** shows two suggested geoengineering strategies to address the impacts of global warming on the planet.

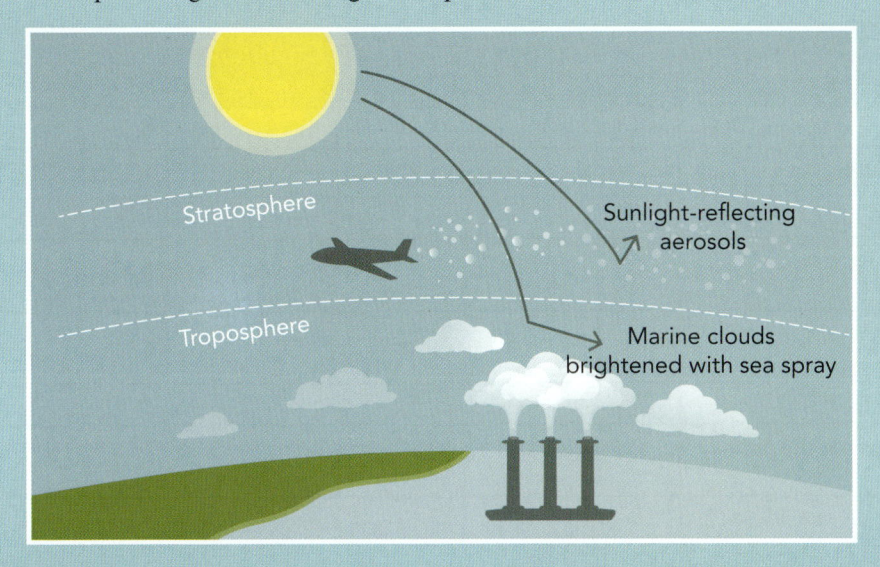

Figure 3

- a Explain what is meant by the term geoengineering. [2]
- b Describe how these two strategies would help to manage global climate change. [2]
- c Explain the current concerns associated with the use of geoengineering solutions in solving climate change. [3]

5 Evaluate the success of strategies to manage the impact of climate change in a location of your choice. [20]

Give reasons and include information from relevant examples to support your answer.

6 'The most significant impact of climate change is sea level rise'.

To what extent do you agree with this statement? [20]

Give reasons and include information from relevant examples to support your answer.

CAMBRIDGE INTERNATIONAL AS LEVEL ENVIRONMENTAL MANAGEMENT: COURSEBOOK

SELF-EVALUATION CHECKLIST

After studying this chapter, think about how confident you are with the different topics.
This will help you to see any gaps in your knowledge and help you to learn more effectively.

I can	Needs more work	Getting there	Confident to move on	See Section
Define climate change.				8.1
Outline the causes of climate change.				8.1
Outline the impacts of climate change.				8.2
State the impacts of climate change on the environment.				8.2
Describe the impacts of climate change on human populations.				8.2
Describe strategies for managing climate change.				8.3
Outline geo-engineering strategies to counteract climate change.				8.3
Evaluate strategies for managing climate change.				8.3

> Glossary

Command Words

Below are the definitions for command words which may be used in exams. The information in this section is taken from the Cambridge International syllabus (8291) for examination from 2025. You should always refer to the appropriate syllabus document for the year of your examination to confirm the details and for more information. The syllabus document is available on the Cambridge International website www.cambridgeinternational.org.

Analyse: examine in detail to show meaning, identify elements and the relationship between them

Assess: make an informed judgement

Calculate: work out from given facts, figures or information

Comment: give an informed opinion

Compare: identify/comment on similarities and/or differences

Consider: review and respond to given information

Contrast: identify/comment on differences

Define: give precise meaning

Demonstrate: show how or give an example

Describe: state the points of a topic / give characteristics and main features

Discuss: write about issue(s) or topic(s) in depth in a structured way

Evaluate: judge or calculate the quality, importance, amount, or value of something

Examine: investigate closely, in detail

Explain: set out purposes or reasons / make the relationships between things evident / provide why and/or how and support with relevant evidence

Give: produce an answer from a given source or recall/memory

Identify: name/select/recognise

Justify: support a case with evidence/argument

Outline: set out main points

Predict: suggest what may happen based on available information

Sketch: make a simple freehand drawing showing the key features, taking care over proportions

State: express in clear terms

Suggest: apply knowledge and understanding to situations where there are a range of valid responses in order to make proposals / put forward considerations

Summarise: select and present the main points, without detail

Key terms

*Abatement: the reduction or removal of something

Abiotic: climate, soil type, slope angle and non-living things or things without life are all abiotic factors that influence the structure of an ecosystem

*Abstraction: the process of taking water from a ground water source

Abundance: calculating abundance means counting the number of a specific organism present. Abundance can be low, with few individuals present. Where abundance is high many of the identified organisms are present

Acid deposition: a mix of air pollutants that deposit from the atmosphere as acidic wet deposition (with a pH <5.6) or acidic dry deposition

Aerobic respiration: the chemical reactions in cells that break down glucose molecules and release energy, carbon dioxide and water

Afforestation: the planting of trees in an area that was not previously forested

Ageing population: a population with a high percentage of old people (aged 65 years or older)

*Albedo: the ability of a surface to reflect light. White surfaces have a high albedo

Anomaly: data that is unusual, and which deviates from the patterns and trends that the rest of the data indicates

Antinatalist policy: a population strategy designed to discourage people from having children and to decrease birth rates

Aquaponics: soil-free farming system that uses the waste produced by aquatic organisms (fish) to supply nutrients to plants being grown hydroponically

Aquifer: an underground layer of permeable rock in which water is stored in the rock pores

*Aquitard: a zone within Earth that restricts the flow of groundwater from one aquifer to another; comprised of either clay or layers of non-porous rock

Artesian well: underground water that is under pressure. When punctured by a well or borehole the water will rise to the surface

Atmosphere: the envelope of gases, vapour and dust, that surround Earth

Bias: when a scientist knowingly or unknowingly incorporates systematic errors into sampling or testing by selecting or encouraging one outcome over another

Big Data: extremely large sets of numerical information collected using technology and analysed using computers

Bioaccumulation: the build-up of a toxin in the body of an organism

*Biodegrade: when organic material is decomposed by bacteria or other living organisms

Biodiversity: the number of different living organisms found within an ecosystem or region

Bioethanol: an alcohol produced from plant matter such a sugar cane or maize which can be used as an alternative to petrol

Biofuel: a fuel derived from biomass (plant or algal material, or animal waste)

Biogas: a gas such as methane that can be used as a fuel and is produced by fermenting organic matter

Biomagnification: the build-up of a toxin in a food chain, e.g. the concentration of mercury increases up the food chain as each consumer eats organisms that have mercury in their tissues

Biomass: the total quantity or weight of organic material in an ecosystem, or, plant material used as an energy source

Biome: a large ecological zone characterised by its soil, climate vegetation and wildlife. Tropical rain forest is a biome

*Biosphere: the zones of Earth where living organisms can survive

Biotic: living organisms (e.g. plants)

Birth rate: the number of live births per thousand people in the population, per year. Also known as the crude birth rate as it does not take age or gender into account

*Buttress roots: tree roots that develop above the ground and extend up the side of the trunk. They are large wide roots on all sides of shallowly rooted trees

Carbon cycle: the flow of carbon between various carbon stores

Carbon footprint: the amount of CO_2 released into the atmosphere as a result of the activities of an individual, organisation, community or activity

*Carbon sequestration: a process by which CO_2 is removed from the atmosphere and stored in either solid or liquid form

*Carbon sink: this is anything that absorbs more carbon from the atmosphere than it releases

*Carbon source: a carbon store that releases more carbon than it stores

Carbon store: carbon is stored in carbon sinks, as organic material in organisms, the soil, fossil fuels and the oceans

*Carnivore: an organism that only eats meat, also known as a tertiary consumer

*Carrying capacity: the number of a species which a region can support without environmental degradation

*Cascading change: a top-down process, where a change made at the top of a food web makes a change throughout the food web and the ecosystem

Cataracts: a cloudiness of the lens of the eye which makes vision blurry

*Child mortality rate: the number of children, per 1 000 live births, that die under the age of five in a population in a year

Chlorofluorocarbons (CFCs): nontoxic, nonflammable chemicals containing carbon, chlorine, and fluorine, that are used in the manufacture of aerosol sprays, foams and packing materials, solvents, and refrigerants

Chlorophyll: green pigment in the leaves of all green plants, which is responsible for the absorption of light to provide energy for photosynthesis

Climate change: detectable change in global temperatures. It is also referred to as global warming

Climate change prediction: to estimate future climate conditions, taking into account both time (months, years and decades) and place (global, regional or local)

Climate feedback mechanism: a process that speeds up or slows down the trend of climate warming. This is currently not fully understood. It is also referred to as a climate change loop

Glossary

Climate model: a computer simulation of Earth's climate system using mathematical equations. It seeks to simulate the outcomes of changes to factors that influence Earth's climate

***Closed question:** yes or no answers, or answers where the respondent can select an answer from tickable boxes

Combustion: the burning of an item, e.g. the burning of fossil fuels to use their energy

Community: the different populations that live together in an ecosystem

Competition: the relationships between organisms that need the same resource in the same space

Composting: decomposition of biotic/organic material that can be used as a fertiliser for plant growth

Condensation: the process by which a gas changes into a liquid duc to cooling

***Confined aquifer:** an aquifer below the land surface that is found between layers of impermeable material

***Confirmation bias:** when data that does not fit with the hypothesis is ignored. Data is then interpreted to support the hypothesis, even when some of it may not

Conservation: the protection and scientific management of natural areas to protect biodiversity in a sustainable manner

Contamination: the action of making or being made impure by pollution

Continents: the main continuous expanses of land found on Earth (Europe, Asia, Africa, North and South America, Australia, Antarctica)

**ial to the 197 nations that agreed to an environmental pact (UNFCCC – the United Nationals Framework Convention on Climate Change) in 1992

Consumer: an organism that cannot produce its own food, and must eat other organisms in order to obtain nutrients

**ain to allow it to float on the surface of water and stop the movement of an oil spill. The boom acts as a barrier to prevent the spread of an oil spill on the surface of the water

**ed or unexposed to the independent variable. The results from this group are then compared to the results of the test subjects

Control variable: any variable that is held constant in an experiment

Convectional rainfall: rainfall that occurs when the energy of the sun heats up Earth's surface and causes water to evaporate and become water vapour. This then condenses to form clouds at higher altitudes

Correlation: a relationship or link between two sets of data

**response to another argument, in order to support a viewpoint

Criteria: a characteristic that is considered important, and by which something may be judged or decided

Critical thinking: the objective analysis and evaluation of an issue in order to form a judgement

Cryosphere: parts of Earth's surface that are made up of ice, including ice, snow, glaciers, ice caps, ice sheets and areas of permafrost

Data: a set of information, in the form of facts, numbers, measurements, or statistics, that can be used for analysis

Data stream: the process of transmitting a continuous flow of data, typically via data processing software

Death rate: the number of deaths per thousand people in the population, per year

Decomposer: an organism that breaks down organic material

Deforestation: the action of clearing forested areas; the cutting down of trees

Dependency ratio: the measure of the dependent (non-working) portion of the population (age groups 0–14 and 65+) compared to the total independent (working) portion of the population (15–64 years). The ratio is expressed as the number of dependents per hundred people in the workforce.

Dependent variable: this variable depends on other factors. It is the variable being measured in the experiment

Desert: a hostile, barren landscape where less than 250 mm of precipitation occurs annually, and biodiversity is low

Discharge zone: the zone where water originating from an aquifer flows out into water courses such as lakes, rivers and wetlands

Diurnal temperature range: a variation between high and low air temperatures that occurs during the same day (e.g. changes between night and daytime temperatures)

DNA: the material in cells that carries information about how a living organism will look and function. Genes make up portion of the DNA

337

Dry acid deposition: atmospheric pollution deposited on Earth's surface in the absence of moisture, like dust or gases

*****Ecological demand:** the amount of water required by an ecosystem to continue functioning properly

*****Ecological niche:** the role and position that a species fills in an ecosystem, including the conditions and feeding needs necessary for the survival of the species

Ecological pyramid: a graphic representation of the relationship between organisms at different tropic levels in an ecosystem

*****Ecological succession:** the process by which the structure of a biological community changes over time

Ecosystem: a biological community of organisms interacting with each other and the physical environment

Ecosystem productivity: the rate of production of biomass for an ecosystem

*****Emigration:** people migrating out of a country

Energy security: the reliable availability of energy sources at an affordable price with a consideration of the environmental impacts

Enhanced greenhouse effect: an increase in the warming of the atmosphere, over and above the natural greenhouse effect, through gases produced by human activities. These gases increase the amount of infrared radiation being retained in the atmosphere, trapping heat from the sun

Environment: the surroundings or habitat in which an organism lives

*****Environmental impact assessment (EIA):** the evaluation of the environmental consequences of a plan, policy, program, or project before a decision is made to move forward with it

Evaporation: the process by which liquid turns to gas

*****Euphotic zone:** the upper layer in the ocean that sunlight penetrates, allowing for photosynthesis

*****Eutrophication:** an increase in nutrients in a body of water results in a rapid growth of algae. Algal blooms cover the surface of the water, forming a green layer. When the algae decay and die, a decline in oxygen level occurs, causing significant ecological degradation

*****Evolution:** the process by which living organisms have developed and adapted into different forms

*****Evolutionary tree:** a branching diagram or 'tree' showing the evolutionary relationships among various biological species based on similarities and differences in their physical and genetic characteristics

*****E-waste:** electronic waste

*****Exabyte:** a unit of information equal to one quintillion (10^{18}) bytes

*****Extensive farming:** a system of farming that uses a small amount of labour and capital investment relative to the area of land being farmed

Famine: the extreme scarcity of food

False reporting: the reporting of information that is false, fabricated or biased

*****Fauna:** the animal life characteristic of an area, region or environment

Fermentation: the chemical breakdown of substances by yeast or bacteria anaerobically to create an alcohol and biogas

Fertiliser: a chemical or natural product that can be added to soils to increase the nutrients available for plants

Food aid: help given to a country or region suffering from food insecurity

Food chain: the feeding sequence of organisms indicating the flow of energy as one species is consumed by the next, from the primary producer through to the apex predator

Food web: the connection of all the individual food chains within a community

Food security: when all people, at all times, have physical, social and economic access to sufficient, safe and nutritious food that meets their dietary needs and food preferences for an active and healthy life

Forced migration: the involuntary movement of a person or people away from their home or home region. Causes include natural disasters, wars, and food or water insecurity.

Fossilisation: the process through which organic material is replaced with mineral substances in the remains of an organism. It is a physical, chemical and biological process the preserves the plant and animal remains over time

*****Flora:** the plants of a particular area, region or environment

Fragmentation: an ecosystem that has been broken up into patches that are too far apart for species to properly interact and reproduce

Frequency: how often a specific species (e.g. plant) occurs in a sample

Genes: the basic unit of heredity (characteristics) passed down from parent to young. For example tall parents are more likely to have tall children as their genes carry that characteristic

Glossary

Genetically modified (GM) crops: foods derived from organisms in which DNA has been changed by humans

Geospatial/ geographic information systems (GIS): electronic mapping systems designed to capture, store, analyse and manage geographic information

Geothermal energy: energy generated from the heat under the surface of Earth

*Geyser: a hot spring in which water boils, periodically sending up tall columns of water and steam

Global warming: the increase in the average surface temperatures on Earth's surface due to rising levels of greenhouse gases

Grasslands: a biome with grassy plains and few trees, in the tropics and subtropics, typically referred to as savanna

Greenhouse gases: gases in the atmosphere that absorb infrared radiation

*Greenwashing: the process of presenting misleading information (often to consumers) about how a product is more environmentally friendly than it actually is

Gross primary productivity: the rate at which producers convert solar energy into biomass

Ground-level ozone: forms when NO_x and volatile organic compounds (VOCs) react in the presence of sunlight

Groundwater: the water found underground in cracks and spaces in the soil, sand and rocks

Groundwater flow: water which flows under the ground until it reaches the surface, often through boreholes or wells

*Groundwater seepage: when there are excessive amounts of groundwater and it pushes to the surface as a spring or an area of saturated soil

*Gross national income (GNI): the total amount of money earned by a nation's people and businesses. This is used to measure a nation's wealth

Habitat: the place that an organism makes its home. It meets all the environmental conditions that an organism needs for survival

*Hazardous waste: waste that has properties which make it dangerous or capable of harming the environment or human health

Herbicides and fungicides: chemicals used to control insects, unwanted plants and fungi in commercial food crops

*Herbivore: an organism that only eats plants, also known as a primary consumer

*Hibernation: a period of time when a plant or animal remains in a dormant or inactive state resembling sleep

High-income countries (HICs): countries that have strong, well-developed economies and a good standard of living, where the GNI per capita is more than US$13 205

Homogeneous: of the same kind, e.g. the crops produced by a farmer may be all the same kind

*Hot springs: a source of ground water that is heated by underground volcanic activity

Humidity: the percentage of water vapour in the air

Hydroelectric power (HEP): electricity that is generated using the energy of flowing water

Hydroponics: the growth of plants without soil. Instead, plants are grown in nutrient-rich water

*Hydrophilic: a water loving material which attracts water, and is capable of holding onto it

*Hydrosphere: all the water on Earth's surface, including lakes, seas, ground water and atmospheric water such as clouds

Hypothesis: a precise, testable statement that a researcher makes, predicting the outcome of a study that is designed to answer a specific question

*Ice caps: found in mountainous areas such as the Himalayas and Andes

Ice sheets: the mass of glacial land ice that covers the polar regions

*Ice shelves: the place where sea ice and ice sheets meet. They are platforms of ice that extend over the edge of the land onto the oceans

Independent variable: this variable stands alone and is not changed by other variables. It is the variable being changed in the experiment to test the hypothesis

Infant mortality rate: the number of infant deaths for every 1 000 live births, of children under the age of one

*Immigration: people migrating into a country

*Imply: to suggest or express something indirectly or without saying it

Incineration: the process of burning materials

*Indigenous: originating or occurring naturally in a specific area; a species that is native to an ecosystem

Infiltration: the movement of water into the soil from the surface

Infrared radiation: a type of radiation not visible to the human eye, but felt as heat

*In-migration: to move into an area or region in order to settle down and live

*Intensive farming: a system of farming that uses large amounts of investment and labour relative to the area of land being farmed

Interception: the blocking of rainfall by vegetation, preventing it from reaching the ground

Inter-specific: between individuals of different species

Interview: when people meet face to face, with one person asking questions and another answering them

Intra-specific: between individuals of the same species

Invasive species: a species that is able to outcompete other species, causing changes to an ecosystem's balance

Irrigation: the supply of water to land or crops to help plants grow

Landfill: a place where waste is disposed of by burying it

Leachate: a typically acidic fluid that has filtered through the waste in landfills; leaching results in the fluid becoming contaminated with heavy metals, toxic chemicals and biological waste

Leaching: when water soaks into soils, removing the minerals and nutrients and reducing their ability to support plant life

*Life expectancy: the average age that a new-born child is expected to live to

Limitations: shortcomings in a study that can influence the information collected. These include research design, methodology, materials and time constraints

Limiting factor: anything that may slow population growth, or constrain population size. The term limiting factor can also be used in other contexts to refer to any factor that can slow or reduce the chance of an event occurring

Lithosphere: the rigid outer layer of Earth

Long-term energy security: the supply of energy that is in line with economic developments and environmental needs

Low-income countries (LICs): countries that have the weakest economies and are least developed. The category is determined by the low GNI per capita of US$1 086 or less

Malnutrition: lack of adequate nutrition, caused by not having a balanced diet, or enough to eat

*Marginal: the trees found along the edges of a forest or cleared area

Mesosphere: the zone of the atmosphere above the stratosphere

*Microhabitat: a habitat that is small or limited in extent and that differs from the surrounding habitat

Microplastics: extremely small pieces of plastic waste in the environment. This results from discarded plastic breaking down into very small fragments

Middle-income countries (MICs): countries that have started to develop, with growing industry and GNI per capita increasing (more than US$1 086 but less than US$13 205)

Migration: the movement of peoples from one place in the world to another

*Mitigation: the action of reducing the severity or seriousness of something

Model: a scientific model is the production of a physical, conceptual or mathematical representation of a real occurrence that is difficult to observe

*Mucus: a slimy, sticky substance that coats, protects and moistens the surface it covers

*Mutualistic relationship: a relationship between two organisms where both species benefit from their interactions

Native species: a species that originated and developed in a specific ecosystem or region and has adapted to living in that area

Natural disaster: an event such as a flood, earthquake or hurricane, that causes great damage or loss of life

Natural greenhouse effect: the warming of the atmosphere by gases, found naturally in the atmosphere, trapping the heat from the sun

*Natural increase: the difference between the birth rates and death rates in a population; natural increase differs from overall increase

*Natural population change: the change in the size of a population due to birth and death rates

*Natural recharge: water that moves from the land surface or unsaturated zone into an aquifer. Where porous rock at the surface allows water to seep into an aquifer

*Net migration: the difference between the number of people entering a country (immigration) and the number of people leaving a country (emigration). Net migration is negative when more people leave a country than enter it

Net primary productivity: the rate at which producers convert solar energy into biomass minus the loss of energy through respiration

Non-renewable resources: resources that will run out and not be replenished for millions of years, for example oil, gas and coal

Glossary

Nuclear power: nuclear power uses radioactive materials such as uranium or plutonium. These materials undergo reactions and power is produced from the energy released

Numerical data: information that is expressed as numbers

Observation: to watch, view or note for scientific investigation

Oceanic circulation: the large-scale movement of waters in the ocean basins through ocean currents and the oceanic conveyor belt system

*Old age dependency ratio: the old-age dependency ratio is the number of older dependents (age 65+) in a population, in relation to the working-age population (15–64 years old). The ratio is expressed as the total number of older dependents per hundred people in the workforce

*Omnivore: an organism that eats both meat and plants, also known as a secondary consumer

*Open question: questions that allow the respondent to give a free-form answer with opinions and detailed information

*Out-migration: to leave one's community or area in order to settle in another area

*Overall population change: the change in the size of a population due to birth rates, death rates and net migration rates

Ozone: O_3, a colourless, odourless gas found naturally in the stratosphere and formed from oxygen by UV light

Ozone layer: a layer of the stratosphere rich in ozone (O_3) molecules, which absorb much of the incoming UV radiation

Ozone hole: an area where the average concentration of ozone is below 100 Dobson Units

Particulates: solid particles and liquid droplets in the air. Generally, these come from any type of burning or dust generating activities

*Pedosphere: the outermost layer of Earth that is made up of soils

Percentage cover: a measure of how much space an organism is taking up as a proportion of a specified area

*Perched aquifer: a perched aquifer occurs above a regional water-table. It is usually a relatively small body of water that lies above the large aquifer due to an impermeable rock layer blocking the downward flow of water

Permafrost: areas of permanently frozen ground

*Permeable: a material that allows liquid to pass through it (sandstone is a permeable rock type)

*Petabyte: a unit of information equal to one thousand million million (10^{15}) bytes

Photochemical smog: a mixture of air pollutants and particulates, including ground level ozone, that is formed when oxides of nitrogen and volatile organic compounds (VOCs) react in the presence of sunlight

Photosynthesis: the process by which plants synthesise glucose using carbon dioxide, water and energy from sunlight

*Phytoplankton: algae found in the upper parts of the ocean, the algae photosynthesise, capturing energy from the sun to live and grow

Pilot survey: a survey carried out prior to a full-scale study. Designed to identify areas of concern or areas for improvement before the full study is carried out

Pioneer species: a hardy species which is capable of being the first to colonise disturbed or newly formed environments

*PM$_{10}$: particulate matter of diameter of 10 micrometres or smaller. The particles can be inhaled. The number next to the PM indicates the size of the particulate matter being measured, e.g. PM$_1$ is 1 micrometre in diameter or smaller

Polar stratospheric clouds (PSCs): stratospheric clouds that form over the poles in winter at altitudes of between 15 000 and 25 000 metres. One of the main types of PSC is mostly made up of supercooled droplets of water and nitric acid

Polar vortex: a large, long-lasting rotating low-pressure system located over the north and south poles. It weakens in summer and strengthens in winter

Pollution: the presence or introduction into the environment of a substance which is harmful or has poisonous effects, for example polluted water is harmful to drink

Population: a group of organisms of the same species living within an ecosystem

Population density: the number of individuals of a species living in a specific unit of area (e.g. square metre or mile)

Population distribution: the way in which the population is spread out across a given area

Population dynamics: the study of how and why populations change in size and how they can be managed

Population size: the number of individuals in a population

Population structure: the number of males and females within different age groups in a given population

Precipitation: water that falls to the ground as rain, snow, hail and sleet

Prediction: a statement of the expected results of an experiment if the hypothesis is true

*Primary air pollutants: an air pollutant emitted directly from the source

*Primary data: information that is collected by the researcher (e.g. rainfall which is collected daily and recorded)

*Primary industry: industry such as mining, agriculture, fishing or forestry that involves harvesting raw materials

Primary producer: the organism within a food chain that produces its own food source through photosynthesis

*Primary Productivity: the rate at which energy is converted into organic material through photosynthesis by plants (producers)

Primary Succession: the gradual process by which an ecosystem develops and changes in a region that has not previously been colonised, for example new lava flows

Pronatalist policy: a population strategy designed to encourage people from having children and to increase birth rates

*Proxy data: preserved physical characteristics of the environment that stand in for measurements of the actual climate information

*Push Factors: these are the factors that cause people to leave an area. They include war, drought, floods, lack of housing, food, education, lack of jobs or a poor standard of living

Quadrat: typically one square metre samples, selected for assessing the local distribution of plants or animals

Qualitative data: data that are non-numerical, or descriptive. These data are collected through observations, interviews and focus groups

Quantitative data: data that is numerical, giving the quantity, range or amount of a variable. For example, monthly rainfall

Questionnaire: a form with a series of questions for respondents to complete, designed to seek data for an investigation

Random sampling: samples based on drawing names/numbers out of a hat or using a computer program to give a random list

Range: the difference between the upper and lower limits on a particular scale (e.g. temperature)

Rationing: to limit the amount of food each person or family is allowed to purchase

Recycling: the action or process of converting waste into reusable material. For example, glass is melted down and reused to form a new product

Reliable data: data that is reasonably complete and accurate, works towards answering the hypothesis in a clear and transparent manner and has not been inappropriately altered

Renewable energy source: any source of energy that can be naturally and quickly replenished, e.g. wind and solar power

Rewilding: restoring an area of land to its natural undisturbed state, specifically through the reintroduction of species of wild animals that have been driven out or hunted to extinction in the area

*Rhetorical question: a question that is asked in order to create a dramatic effect, or to make a point, rather than to get an answer

Runoff: the draining away of water as overland flow

Salinisation: an increase in salt content, usually of agricultural soils, irrigation water, or drinking water

*Salt water intrusion: the movement of salt water into fresh water aquifers

Sample: a set of data (number of plants, number of species, plant distribution) taken from a larger population for measurement

Sanitation: the provision of clean drinking water and sewage disposal

Scientific method: a procedure that involves systematic observation, measurement and experiment to test hypotheses

Scientific theory: an explanation of an aspect of the natural world that has been tested repeatedly to verify it through the use of the scientific method

*Sceptic: a person who doubts or does not believe in a concept or hypothesis

Sea ice: the ice that floats on the surface of the oceans and seas

*Secondary air pollutants: an air pollutant that forms when pollutants react in the atmosphere

*Secondary data: data that is collected by somebody else in a separate investigation (e.g. climate data from the local airport)

*Secondary industry: industry that converts raw materials such as farming or mining products into products for sale. The manufacturing industry

Glossary

Secondary succession: the gradual process by which an ecosystem develops and changes in a region that has previously been colonised, but which has been disturbed, damaged or removed

*****Sedentary:** organisms that do not move, such as plants or rocky shore species like the barnacle

Selective breeding: when humans grow plants and animals for specific characteristics, e.g. high yields or drought tolerance

Short-term energy security: systems that react promptly to sudden changes in the supply-demand balance

*****Smog:** a mixture of smoke and fog; smog intensifies atmospheric pollution

Soil moisture: the amount of water that is found in the soil

Solar power: electricity that is generated by utilising the energy of the sun

*****Solar variation:** fluctuations in the amount of radiation output from the sun

Species: a group of living organisms made of up individuals that can produce fertile offspring when they reproduce

*****Starvation:** suffering or death caused by lack of food.

*****Statistics:** the practice of collecting, analysing and interpreting numerical data in large quantities. This includes ways of reviewing and drawing conclusions from the data. Statistics are a way to see patterns in numerical data or to determine whether data shows a difference between two treatments

*****Stem-flow:** rainfall that reaches the ground in a forest by draining down the trunks of plants

Stockpiling: to store large amounts of good or materials, in this instance food stores

*****Stomata:** pores in the leaf or stem of the plant. These form a slit which allows the movement of gases in and out of the spaces between the cells; found mainly on the underside of leaves

Stratosphere: the zone of the atmosphere above the troposphere where the ozone layer is located

*****Surface mass balance:** the balance between the build-up of and loss of glacial surface

Sustainability: the ability to meet the needs of the present without compromising the ability of future generations to meet their own needs

Systematic sampling: choosing a sample based on regular intervals rather than random selection

Tectonic plates: a large slab of irregularly shaped solid rock that makes up part of the outer crust of Earth. These plates move in relation to each other over time

*****Terabyte:** a unit of information equal to one million million (10^{12}) bytes

*****Thesis statement:** a thesis statement is a sentence in a paper or essay which introduces the main topic or argument to the reader

*****Through-fall:** rain that falls through the leaves and branches of plants

*****Titanium dioxide:** titanium is a metal oxide commonly found in plants and animals and is the ninth most common element in Earth's crust. It is a white power that can be made into a bright white pigment. It is used in products such as paint, paper, plastic, ink, soap, food colouring and sunscreen

Total dependency ratio: the total dependency ratio is a measure of both young (age 0–14 years) and older dependents (age 65 and older) added together to show their total versus the independent population (15–64 year olds). The ratio is expressed as the total number of dependents (young and old) per hundred people in the workforce

*****Transnational corporation (TNC):** a company that has its head office in one country, and has branches or factories in other countries. It works across international borders (*trans*-national)

Transpiration: water lost through the leaves of plants

Trends: the general relationship between two sets of data

Trend line: a line indicating the general relationship between two sets of data

trophic level: a group of organisms within an ecosystem which fill the same level within a food chain

Troposphere: the lowest zone of the atmosphere that extends from Earth surface to a height of approximately 10 km

Tundra: a biome found far north in Asia and Alaska, characterised by long cold, dark winters, and short cool summers. Permanently frozen ground limits vegetation growth to short shrubs and grasses

Turbidity: the cloudiness or haziness of water; the lower the visibility the higher the turbidity

*****Unbiased:** not affected or influenced by a person's beliefs or opinions

*****Unconfined aquifer:** an aquifer in which the water table is at atmospheric pressure. There is no impermeable layer between the water table and the ground surface

*Upcycling:** reusing a discarded item in such a way as to create a product of a higher quality than the original. for example, using materials from discarded plastic bottles to make new shoes

Value: to have a use, or a worth

*Variable gases:** the concentration of the gas can differ either spatially or over time

Variables: a factor that can change in quality, quantity or size regarding the category of data that is being measured (e.g. rainfall)

*Variety:** diverse data

*Veracity:** the ability to verify or confirm data

*Vertical stratification:** the vertical layering of an ecosystem. Different species occupy different horizontal layers within the ecosystem

*Velocity:** speed with magnitude and direction

Volatile organic compounds (VOCs): compounds that have a high vapour pressure and a low water solubility; VOCs are emitted as gases from certain solids or liquids including paints, paint strippers, cleaning supplies, pesticides and building materials

*Volcanism:** any process associated with surface discharge of molten rock, hot water or steam from inside Earth

*Volume:** an amount or quantity of something

*Waste stream:** the flow of specific types of waste from their source through to recovery, recycling or disposal

Water security: the ability to access sufficient quantities of clean water to maintain adequate standards of food and manufacturing of goods, adequate sanitation and sustainable health care

*Water table:** the water table is an underground boundary between the soil surface and the area where ground water saturates the in the rocks

Wave and tidal energy: electricity that is generated using the energy of waves or the tides

Wet acid deposition: atmospheric pollution deposited on Earth's surface as precipitation, such as snow, rain, hail and fog

*Wetlands:** wetlands are areas where water covers the soil, or is present either at or near the surface of the soil all year or during periods of the year

Wind energy: electricity that is generated using the power of wind

*Youth dependency ratio:** a measure of the young dependents (age 0–14) in a population, in relation to the working–age population (15–64 years old)

*Zooxanthellae:** photosynthetic, single-celled marine organisms that live in coral polyps

› Acknowledgements

The authors and publishers acknowledge the following sources of copyright material and are grateful for the permissions granted. While every effort has been made, it has not always been possible to identify the sources of all the material used, or to trace all copyright holders. If any omissions are brought to our notice, we will be happy to include the appropriate acknowledgements on reprinting.

Thanks to the following for permission to reproduce images:

Cover Sue Cro/GI; *Inside,* **Key Skills** CasarsaGuru/GI; Billy Hustace/GI; Richard Bailey/GI; John Giustina/GI; Zuraisham Salleh/GI; JohnnyGreig/GI; Catherine Falls Commercial/GI; Phynart Studio/GI; **Chapter 1** Artur Debat/GI; Thorsten Milse/Robertharding/GI; Feng Wei Photography/GI; Guenterguni/GI; Ron Sanford/GI; Ashley Cooper/GI; Ugurhan/GI; Gallo Images-Dave Hamman/GI; Ghislain & Marie David de Lossy/GI; Anup Shah/GI; Artur carvalho/GI; Bernadette Schoeller/GI; Martin Harvey/GI; BarbAnna/GI; Shutterjack/GI; Ana Rocio Garcia Franco/GI; Darrell Gulin/GI; Figure 1.30-1.36 The Exploratorium, www.exploratorium.edu; Anup Shah/GI; Lou Coetzer/GI; **Chapter 2** Wolfgang Filser/GI; Greenshoots Communications/Alamy Stock Photo; SolStock/GI; Christopher Kimmel/GI; Paul Souders/GI; Figure 2.9 from climate.nasa.gov data credit: Luthi, D., et al. 2008; Etheridge, D.M., et al. 2010; Vostok ice core data/J.R. Petit et al.; NOAA Mauna Loa CO2 record; Jason Edwards/GI; Lukasz Larsson Warzecha/GI; Figure 2.12 adapted from www.climatecentral.org, data source: NADA GISS, NOAA NCEI, ESRL; Salvatore Virzi/GI; Ashley Cooper/GI; Fig 2.15 adapted from United States Environmental Protection Agency; Figure 2.17 climate model adapted from Ruddiman, W.F. Earth's Climate Past and Future, 2001, figure 3.26; Martyn F. Chillmaid/Science Photo Library; MShieldsPhotos/Alamy Stock Photo; Blickwinkel/Alamy Stock Photo; Martin Harvey/GI; Anadolu Agency/GI; 10'000 Hours/GI; Berkah/GI; Simon Landolt/GI; Giordano Cipriani/GI; somnuk krobkum/GI; Monticelllo/GI; Ralf Pollack/GI; kampee patisena/GI; PixeloneStocker/GI; Figure 2.39: Content is the intellectual property of Esri and is used herein with permission, © 2022 Esri and its licensors. All rights reserved; Christoph Burgstedt/GI; Figure 2.41 based on models by the Ocean Circulation Lab USF College of Marine Science; National Weather Service/GI; Dr T J Martin/GI; John M Lund Photography Inc/GI; Yin Wenjie/GI; Boonchai Wedmakawand/GI; Hinterhaus Productions/GI; Alessandro Biascioli/GI; **Chapter 3** Merrill Images/GI; Vivek Doshi/GI; Atlantide Phototravel/GI; Cinoby/GI; Stuart McCall/GI; Christian Fischer/GI; Jadwiga Figula/GI; Marco Longari/GI; Figure 3.8 data from World Population Prospects 2019; Tang Ming Tung/GI; Figure 3.11 from OurWorldInData; Figures 3.11-3.15 ©December 2019 by PopulationPyramid.net; Hadynyah/GI; Kali9/GI; Morsa Images/GI; Ariel Skelley/GI; Frans Lemmens/GI; Figure 3.23: https://www.un.org/sustainabledevelopment/ The content of this publication has not been approved by the United Nations and does not reflect the views of the United Nations or its officials or Member States; Marcus Clackson/GI; Peeterv/GI; Figure 3.27: Singapore Family Planning and Population Board Collection, courtesy of National Archives of Singapore; Andrew Aitchison/GI; David Sacks/GI; Figure 3.31 Japan population pyramids for 1950, 2005 and 2055 (Source: IPSS website (https://www.ipss.go.jp/index-e.asp)); **Chapter 4** Viktor Vichev/GI; Duncan1890/GI; Figure 4.2 from adapted figure in H.J de Blij & P.O. Muller, Physical Geography of the Global Environment, 1996; Patmeierphotography.com/GI; Anton Petrus/GI; Hillary Kladke/GI; Johner Images/GI; Bruce Yuanyue Bi/GI; Anders Blomqvist/GI; Michael Duff/GI; Paul Souders/GI; Figure 4.13 from source: Imhof ML, Bounoua L, Ricketts T, Loucks C, HArriss R, Lawrence WT (2004) Global patterns in human consumption of the net primary production. Nature 429:870-873. © 2008 The Trustees of Columbia University in the City of New York; Choksawadikorn/GI; Jason Edwards/GI; Shmelly/GI; EyeWolf/GI; David Pardoe/GI; Ivstiv/GI; Travelpix Ltd/GI; MediaNews Group/GI; Munib Chaudry/GI; Martin Harvey/GI; Zocha_K/GI; Philip Thurston/GI; Natalie Fobes/GI; Charles Van Zyl/GI; Tsuyoshi Kaminag/GI; Robin Bush/GI; Sara Grace/GI; Wokephoto17/GI; Pop_jop/GI; Kriswanto Ginting/GI; Goh Chai Hin/GI; Mike Simons/GI; Mischa Keijser/GI; Eduardo Fonseca Arraes/GI; Israel Sebastian/GI; Mike Powles/GI; Victor Moriyama/GI; Martin Harvey/GI; Patrick Pleul/GI; Tdub303/GI; Apomares/GI; O. Alamany & E. Vicens/GI; Gallo Images-Anthony Bannister/GI; UniversalImagesGroup/GI; Jim Julien/GI; Andrew Peacock/GI; Galen Rowell/GI; Roger Tidman/GI; Alison Wright/GI; Doug Marshall/GI; Deb Snelson/GI; George Lepp/GI; Michael Cummings/GI; **Chapter 5** Zhihao/GI; John Seaton Callahan/GI;

Figure 5.2 data from Statista; Jan Sochor/GI; John Carnemolla/GI; Martin Harvey/GI; Studio CJ/GI; Essa Ahmed/GI; Peter Summers/GI; Andia/GI; Kukiat Boontung/GI; Mmdi/GI; Figures 5.10-5.15 Tana Scott; Matteo Colombo/GI; Fotografixx/GI; SimonSkafar/GI; Javed Tanveer/GI; Andrew Merry/GI; View press/GI; Philippe Huguen/GI; DoraDalton/GI; Visoot Uthairam/GI; Imaginima/GI; Joao Abreu Miranda/GI; Eye Ubiquitous/GI; Brian Bumby/GI; Finnbarr Webster/GI; Hadynyah/GI; Mayur Kakade/GI; Yevgen Romanenko/GI; Alexander Hafemann/GI; Figure 5.35 from data Eurostat; Tina Stallard/GI; Montinique Monroe/GI; Dan_prat/GI; Emanuel Dunand/GI; Zhongguo/GI; John Downer/GI; Aleksandr Zubkov/GI; Tunart/GI; Narvikk/GI; George Rose/GI; Jackyenjoyphotography/GI; Chris Mansfield/GI; Grafner/GI; Peter Dazeley/GI; Arun Sankar/GI; Geraint Rowland Photography/GI; Songsak Rohprasit/GI; Gary Bell/GI; Figure 5.57 Raphaël Lagarde & Chris LeMoine/Brandon University; Denise Taylor/GI; K Neville/GI; Ashley Cooper/GI; Oscar Wong/GI; Tiziana Fabi/GI; PixeloneStocker/GI; vchal/GI; Xavierarnau/GI; Figure 5.67 from FAO - The State of Food Security and Nutrition in the World Report 2021; Chapter 5 Practice Questions Figure 5.2, source BP, The Economist; **Chapter 6** Kage Nesbitt/GI; Dewald Kirsten/GI; Morgana Wingard/GI; Ippei Naoi/GI; Figure 6.4 data World Bank; Ali Majdfar/GI; Nigel Hicks/GI; Giordano Cipriani/GI; Sinhyu/GI; Mario Tama/GI; The Washington Post/GI; Andrew Holt/GI; Ashley Cooper/GI; Hadynyah/GI; George Hammerstein/GI; Michael Milner/GI; Philartphace/GI; Hadynyah/GI; Suriyapong Thongsawang/GI; Tatsiana Volskaya/GI; Patrick T. Fallon/GI; Guido Dingemans, De Eindredactie/GI; Bob Sacha/GI; Hadynyah/GI; Karl Tapales/GI; TerryJ/GI; Mark Newman/GI; AscentXmedia/GI; Constantgardener/GI; Santiago Urquijo/GI; Miguel Sotomayor/GI; Getty Images/GI; Simon Maina/GI; Rajesh Jantilal/GI; Future Publishing/GI; Lonely Planet/GI; Roger Hutchings/GI; Figure 6.42 National Arsenic Mitigation Information Center, BAMwSP; Chapter 6 Practice Questions Figure 4 data from WHO and Unicef, 2006; **Chapter 7** JohnnyH5/GI; Qilai Shen/GI; Puneet Vikram Singh, Nature and Concept photographer/GI; Douglas Sacha/GI; Filo/GI; Dorling Kindersley/GI; CasarsaGuru/GI; Will & Deni McIntyre/GI; Figure 7.9 adapted from C.W. Stjern, 2011; Richard Baker/GI; Olivier Goujon/GI; Kontrolab/GI; Karl Tapales/GI; Dimitris_k/GI; Juanmonino/GI; Bojan Brecelj/GI; Nick Brundle Photography/GI; Socha/GI; Roslan Rahman/GI; Carsten Koall/GI; Yuwadee Singthong/GI; Patrick Lane/GI; Mateusz Brzezinski/GI; Mykhailo Polenok/GI; Figure 7.25 from Fahey W.D. and M.I. Hegglin (coordinating lead authors) Twenty Questions and Answers about the Ozone Layer: 2010 update, Scientific Assessment of Ozone Depletion, 2010, p33; Patrick J Endres-AlaskaPhotoGraphics/GI; JodiJacobson/GI; Chen, Huize & Han, Rong. (2016). Characterization of Actin Filament Dynamics during Mitosis in Wheat Protoplasts under UV-B Radiation. Scientific Reports. 6. 20115. 10.1038/srep20115; Stockbyte/GI; Album/Alamy Stock Photo; Hindustan Times/GI; **Chapter 8** Daniel Mingo/GI; Artur Debat/GI; Bloomberg/GI; Monty Rakusen/GI; Chadchai Ra-ngubpai/GI; Lucas Ninno/GI; Ashley Cooper/GI; Figure 8.9 adapted from www.environment.gov.au; Philippe Desmazes/GI; Gyro Photography/GI; Yuki Iwamura/GI; Michael Burkholder/GI; Majority World/GI; NurPhoto/GI; Paul Souders/GI; Figure 8.18 based on PNAS Vol 114, No. 46, 2017; Tony Arruza/GI; SOPA Images/GI; Georgeclerk/GI; Christopher Furlong/GI; Dneutral Han/GI; Figure 8.24 source EM-DAT: The OFDA/CRED International Disaster Database - www.emdat.net - Universite catholique de Louvain, Brussels, Belgium; Lea McQuillan/GI; Figure 8.26 based on ARC Centre of Excellence Coral Reef Studies; Brett Monroe Garner/GI; Monty Rakusen/GI; Silvia Otte/GI; Maskot/GI; Yagi Studio/GI; Martin Harvey/GI; FotoVoyager/GI; Sylvain Thomas/GI; Anadolu Agency/GI; Jeff J Mitchell/GI; PeterHermesFurian/GI; Hilary Hosia/GI; The Washington Post/GI; Jonas Gratzer/GI; Petmal/GI

Key GI= Getty Images

Index

abiotic components/factors, 33, 39–41
   humidity, 40
   light, 40
   limiting factors, 41
   oxygen, 40
   pH, 40
   salinity, 40–1
   water, 40
abstraction, 29, 237, 239, 245, 246
abundance estimation by ACFOR scale, 94
academic writing
   authority, 16
   good practice tips, 13–14
   see also evaluative essay
ACFOR scale, 94
acid deposition, 275–8, 296–7
   formation, 276
   impacts, 277–8
   types, 276
acidic dry deposition, 276
acidic wet deposition, 276
adaptations
   to climate change, 324
   to hot desert climate, 41–2
aerobic respiration, 43; see also photosynthesis
afforestation, 322, 323
African black rhinoceros, 157
African Pixie frog, 42
age gap see population age gap
ageing population
   challenges, 125–6
   defined, 125
   economic difficulties, 125–6
   Japan, 133–5
agreements see international agreements
agriculture, 112
   pests and diseases, 191
   water use, 254–6
   see also farming
agroforestry, 172
air pollutants
   abrasive, 281
   primary, 275
   secondary, 276

air pollution, 274–87
   acid deposition (see acid deposition)
   investigation, 279–80
   legislation, 286–7
   managing, 284–7
   photochemical smog, 275, 279–81
   pollutants, 275–6
   reducing emissions, 285–6
   sources, 275
air quality, 299
   protecting, 324
   real-time monitoring, 274
   variations (place to place), 275
Air Quality Index (AQI), 274
albedo, 312
alternative energy sources, 319;
      see also renewable energies
American bald eagle, 157
Andaman and Nicobar Islands, 109
anomalies, 3, 8
Antarctica
   human impacts, 175–9
   ozone depletion, 176, 288–9, 298
   research stations, 176
   sea ice, 177
   threats to, 175–6
Antarctic Treaty, 177–9
Antarctic Treaty Committee for Environmental
      Protection, 178
antinatalist policies, 126, 128–30
apex predators, 36
AQI see Air Quality Index
aquaponics, 194, 195
aquatic environments, acid deposition, 277
aquifers, 29, 238–40
   confined, 239, 240
   discharge zone, 238
   natural recharge, 238
   perched, 239, 240
   pollution, 245
   population pressure, 246
   unconfined, 239, 240
aquitard, 239; see also aquifers
arsenic pollution, 262–4
artesian well, 239; see also aquifers

atmosphere
- air pollution, 274–87
- Bhopal disaster, 293–5
- as carbon store, 49
- components, 30
- defined, 30
- gases, 30
- ozone depletion, 287–92
- structure, 30–2
- see also climate change

Bangladesh, 328
- arsenic contamination, 262–4
- population pyramid, 120

bar charts, 3
- divided, 4
- stacked, 4

Barisan Mountains, 164
Basel Convention, 224
beating trays/nets, 80–1
beavers, 180, 181; see also Yellowstone National Park
Beyond Oil and Gas Alliance, 326
Bhopal disaster, 293–5
bias, 63
big data, 99–100
biodegradable plastics, 221; see also waste management
biodegrades, 304
biodiversity, 25
- benefits from conservation, 155–6
- conservation, 153–69
- EDGE programme, 161–2
- habitat, 166–9
- indigenous species, 153, 154
- invasive species, 153–5
- legislation and protocols, 156–60
- native species, 153–5
- sustainable harvesting, 158

bioethanol, 201, 202
biofuels, 191
biogas, 201, 202
biomass, 38, 201–2
biomes, 33, 142–7
- soil, 143
- terrestrial, 142, 143–6

biosphere, 33
- as carbon store, 49

biotic components/factors, 33, 35–9
birth rate, 116–18
Black Triangle in Europe, 278
Brazil, 171
- Chico Mendez reserve, 167
- deforestation, 170

reducing emissions, 172

buildings
- acid deposition, 278
- deep overhangs, 323
- energy-efficient infrastructure, 323
- greening, 323
- insulated, 323
- northern hemisphere, 323
- passive, 323
- southern hemisphere, 323
- zero rated, 323

Cairo, Egypt
- air pollution, 282–4
- population, 282

Cape Town, South Africa
- water crisis, 235–6

captive breeding, 162–6
- Sumatran rhinoceros, 164–6

capture–mark–recapture technique, 82
carbon capturing and storage, 322
carbon cycle, 47–8
carbon dioxide, 29, 32, 35, 70–1, 206, 217
- aerobic respiration, 43
- alternative energy sources, 319
- atmospheric life time, 303
- buildings, 323
- capturing, 322
- carbon cycle, 47, 48, 49
- carbon footprint, 302
- carbon sequestration, 322
- changes in global amount, 70
- coral reefs, 318
- deforestation, 170–1, 304
- fossil fuel, 48, 71, 303
- increase, 71, 318
- landfill gas (LFG), 305
- low carbon fuels (LCF), 319
- ocean acidification, 175, 317
- photosynthesis, 43, 44, 47, 170
- storage, 322
- transport policies, 319–20
- as variable gas, 30
- see also climate change; greenhouse gases

carbon footprint
- defined, 302, 320
- food, 320
- reducing, 320

carbon sequestration, 322
carbon sink, 48, 49
carbon stores, 48–9
carnivores, 36

Index

carrying capacity, 21, 41
cascading changes, 169, 170
catalytic converters, 285
CCAMLR *see* Conservation of Antarctic Marine Living Resources
CFC *see* chlorofluorocarbons
charts
    analysis, 6–8
    bar, 3, 4
    divided bar, 4
    pie, 4
    stacked bar, 4
    see also graphs
Chernobyl Nuclear Power Plant, Ukraine, 111
Chico Mendez reserve, Brazil, 167
child mortality rate, 116
China, 260, 314
    climate cooperation, 326
    One Child Policy, 117, 129
    population policy, 117, 118, 129
    two-child policy, 129
chlorine gas, 289
chlorofluorocarbons (CFC), 32
    Antarctica, 176
    impact, 290–1
    Montreal Protocol, 292, 325
    Rowland-Molina hypothesis, 291–2
    stable synthetic chemical compounds, 288
    see also ozone depletion
chlorophyll, 43
cholera, 250
CITES *see* Convention of International Trade in Endangered Species of Wild Fauna and Flora
civil unrest, energy insecurity, 208
Clean Air Act, 287
climate, 110
climate change, 303–29
    adaptations to, 324
    Antarctica, 175
    biodiversity conservation help combating, 155
    data misuse, 71–2
    energy security, 206, 315
    extreme weather events, 110, 314, 315
    food security, 192, 314
    forced migration, 314
    historical data, 69
    impacts, 313–18
    managing (*see* climate change management strategies)
    melting of ice, 310, 312
    misleading data, 101–3
    monitoring and predicting, 306
    ocean circulation, 311
    precipitation, 310
    reporting on, 73–5
    scientific bias, 71–2
    scientific theory, 69–70
    sea level rise, 310, 312, 317, 324, 327–9
    technological advancement, 70–1
    temperature, 309–10
    water security, 243, 314–15
    wind circulation, 311–12
    see also greenhouse gases
climate change management strategies, 318–24
    adaptations to climate change, 324–5
    agreements, 325–6
    alternative energy sources, 319
    carbon capture and storage, 322
    carbon footprint, reducing, 320–2
    energy-efficient buildings, 323
    fossil fuels, reducing usage of, 318–29
    geo-engineering strategies, 326
    low carbon fuels (LCF), 319
    reforestation, 322
    transport policies, 319–20
climate change predictions, 303
climate feedback mechanisms, 308–9
climate models, 75
climax community, 149; *see also* ecological succession
closed questions, 88
Club of Rome, 126–7
coal, 303
    formation, 203, 204
    non-renewable energy, 203
    see also fossil fuels
combined graphs, 6
combustion of fossil fuels, 48, 275, 303
community, 34–5
competition, 35, 41
    inter-specific, 36
    intra-specific, 36
computer climate models, 308–9
    climate feedback mechanisms, 308–9
    data uncertainty, 309
    time delay on cause and effect, 309
computer modelling, 99
condensation, 26
Conference of the Parties (COP), 325
confined aquifers, 239, 240
confirmation bias, 71
congestion charging, 287
Conservation of Antarctic Marine Living Resources (CCAMLR), 178

conservation of biodiversity *see* biodiversity conservation
consumers, 36
containment booms, 67, 68
contamination
   arsenic, 262–4
   defined, 245
   illnesses, 250
   water, 245, 250, 262–4
continents, 22
control group, 64–5
control variables, 65–6
convectional rainfall, 143, 144
Convention of International Trade in Endangered Species of Wild Fauna and Flora (CITES), 157, 159
COP *see* Conference of the Parties
COP26, 325
corals
   bleaching, 317
   euphotic zone, 316
   habitat changes, 317
   ideal living conditions, 316
   ocean acidification, 317
   pH of water, 277
   sea level rise, 317
correlation, 5
counter-argument, 11
countries
   classifying, 23
   energy needs, 206
   see also high-income countries (HIC); low-income countries (LIC)
COVID-19
   food aid, 198
   food security, 88, 225–8
   rationing of food products, 199
crime, energy security and, 212
criteria, 10
critical thinking, 10
crop yields
   acid deposition, 277–8
   photochemical smog, 281
   see also agriculture; farming
crowd sourcing, 99
cryosphere
   as carbon store, 49
   defined, 49
cultural value, biodiversity and, 156

dams, 200; *see also* hydroelectric power (HEP)
data analysis, 67, 90–4
   abundance estimation by ACFOR scale, 94
   Lincoln index, 90–1
   percentage cover, 92
   percentage frequency, 92–3
   Simpson's index, 91–2
data collection
   interviews, 88–9
   questionnaire, 88–9
   sample/sampling, 76–88
   technology, 96–101
data streams, 100
DDT, 224
death rate, 116, 118
debt reduction, 171
decomposers, 37
Deep Water Horizon oil spill, 97
deforestation, 169–71
dependency ratio, 122–4
   calculating, 123
   defined, 122
dependent variables, 63, 64
desert, 142, 143, 146
   elephant, 20–1
desert frogs, 42
discharge zone, 238; *see also* aquifers
diurnal temperature range, 143, 144
divided bar charts, 4
DNA, 156
Dobson units, 287
domestic water use, 253–4
drip irrigation, 255
dry deposition *see* acidic dry deposition

Earthwatch Institute, 261
East African water crisis, 257–9
ecological demands, 253
ecological niche, 35
ecological pyramids, 36, 151–2
   energy transfer up the trophic levels, 151
   pyramid of biomass, 152
   pyramid of energy, 151
   pyramid of numbers, 152
ecological security, biodiversity and, 156
ecological succession, 148–50
   climax community, 149
   defined, 148
   intermediate community, 149
   primary, 148, 149
   secondary, 148, 149
   sequence, 149
   time scales, 149

Index

economic growth
    biodiversity conservation and, 155
economic recession, 208
ecosystems, 142–53
    defined, 25
    human impacts, 169–79
    sustainability, 24–5
ED *see* evolutionary distinctivness
EDGE *see* Evolutionarily Distinct and Globally Endangered species programme
education
    antinatalist policies, 129
    energy security, 211, 212
    waste management, 222
    water management, 253–6, 258, 260, 261
electric vehicles, 101–3, 285
electrostatic precipitators, 285
elephant, 20–1
emergency education and planning, 324
emigration, 119
emigration rates, energy security and, 212
Endangered Species Act, United States, 157
energy
    flow, 151
    pyramid, 151
energy-efficient buildings and infrastructure, 323
energy insecurity
    causes, 205–7
    energy needs of countries, 206
    global energy resource distribution, 205
    impacts, 207–9
    population growth, 205
    supply disruption, 206–7
    see also energy security
energy needs of countries, 206
energy prices, 208, 212
energy resources, 200–13
    geographical distribution, 205, 206
    non-renewable, 202–4
    renewable, 200–2, 319
    see also energy insecurity; energy security
energy security, 204–13
    challlenges/factors affecting, 210–13
    climate change, 206, 315
    defined, 204, 205
    long-term, 204, 205
    management strategies, 209–10
    short-term, 204, 205
    see also energy insecurity
energy sourcing, 209
energy transfer up the trophic levels, 151

enhanced greenhouse effect, 32
environment, 20, 21
Environmental Biodiversity Stewardship, 174
environmental factors, 110–11
Environmental Impact Assessment (EIA), 59–60
environmental investigation, 94–6
environmental management, 21
    scientific method, 59–68
EU CFP *see* European Union Common Fisheries Policy
euphotic zone, 316
European Economic Community, 292
European Union Common Fisheries Policy (EU CFP), 160
eutrophication, 155
evaluative essay, 10–18
    checklist, 17
    guidance, 13–14
    paragraph structure, 12–14
    peer assessment, 15
    planning, 11
    sentence length, 14
    structure, 12
    what to avoid, 15–16
    writing clearly, 15
evaluative writing, 10–18; *see also* evaluative essay
evaporation, 26
evolution, 161
Evolutionarily Distinct and Globally Endangered (EDGE) species programme, 161–2
evolutionary distinctivness (ED), 161
evolutionary tree, 161
e-waste, 224
exabytes, 100
extensive farming, 194, 195
extractive reserves, 167
extreme weather events, 110, 314, 315

false reporting, 71
famine, 192
farming
    energy security, 212
    extensive, 194, 195
    food security, 189
    inefficient, 189
    intensive, 194, 195
    livestock, 303–4
    rice, 303–4
    roof-top, 198
    subsistence, 197–8
fauna, 154
feedback mechanism, 304

fertilisers, 193
F-gases, 291, 292
	emissions, 291
	global warming effects, 291
FGD *see* flue-gas desulfursation
fieldwork techniques, 86–7
financial incentives, 223
fishing, as threat to Antarctic waters, 175
flora, 154
flue-gas desulfurisation (FGD), 285
food aid, 191, 198
food chain, 36, 38, 39
food insecurity
	agricultural pests and diseases, 191
	causes, 190–2
	diverting crops for biofuels, 191
	impacts, 192–3
	land degradation, 190–1
	managing, 193–9
	poverty, 191
	water shortages, 190
	see also food security
food security, 189–99
	biodiversity conservation, 155
	climate change, 192, 314
	considerations, 189
	COVID-19, 88, 225–8
	defined, 189
	inefficient farming methods, 189
	population growth, 191
	urbanisation, 189
	see also food insecurity
food web, 38, 39, 56
forced migration, 314
forest conservation, 322
fossil fuels, 318–19
	combustion, 303
	reducing usage, 284
fragmentation, 169, 170
frequency, 79
	percentage, 92–3
frozen water stores, 241–2
	heating and melting, 312
fuel desulfurisation, 285
fungicides, 194, 195

gases, atmosphere, 30
GE *see* global endangerment
genes, 156
genetically modified (GM) crops, 193
genetic diversity, 156

Geneva Convention on Long-Range Transboundary Air Pollution, 278, 287
geoengineering, 326
geographical distribution of energy resource, 205, 206
geospatial/geographic information system (GIS), 96–7
geothermal energy, 202
Germany, energy security, 211–13
geysers, 179
GIS *see* geospatial/geographic information system
Global Compact of Refugees, 329
Global Compact on Safe, Orderly and Regular Migration, 329
global endangerment (GE), 161
globalisation, 223
global warming, 303
GNI *see* gross national income
GPP *see* gross primary productivity
grain-based feeding of livestock, 199
graphs, 3–8
	analysis, 6–8
	combined, 6
	line, 5
	scatter, 5
	see also charts
grasslands, 142, 143, 145
	biodiversity conservation, 173–4
gravity-fed schemes, 253
grazing, 36
Great Barrier Reef, east coast of Australia, 316–18
Great Migration, 50–2
greenhouse effect, 306
greenhouse gases, 32, 69, 71, 203, 217, 242
	carbon cycle, 48
	electric vehicles, 101
	enhanced effect, 306
	F-gases, 291
	human sources, 303–6
	renewable energy, 200
	sustainable buildings, 302
	volcanic eruption, 72
	see also climate change
Greenland, 73–4
green revolution, 193
greenwashing, 102
gross national income (GNI), 23
gross primary productivity (GPP), 150, 151
ground-level ozone, 274
groundwater, 29, 238–40
	abstraction, 29, 237, 239, 245, 246
	flow, 29
	seepage, 237
	see also aquifers

Index

habitat, 35
    conservation and creation, 166–9
    extractive reserves, 167
    marine zones, 168–9
    national parks, 168
    nature reserves, 168
    protection, 167
    rewilding, 167, 179–81
hazardous waste, 213
    legislation, 224
HCFC *see* hydrochlorofluorocarbons
HEP *see* hydroelectric power
herbicides, 194, 195
herbivores, 36
HFC *see* hydrofluorocarbons
hibernation, 42
HIC *see* high-income countries
high-income countries (HIC), 23, 24
    ageing populations, 125
    Basel Convention, 224
    child mortality rates, 118
    dependency ratio, 122
    energy needs, 206
    energy security, 210–13
    food security, 189, 192
    irrigation, 256
    life expectancy, 119
    population age gap, 133–5
    population pyramids, 120–1
    pronatalist policies, 128
    transnational corporation (TNC), 293
    water availability, 247
    see also low-income countries (LIC)
homogeneous
    defined, 192
    food supplies, 192
hot springs, 179
HSBC Water Programme, 261
human impacts on ecosystems, 169–79
    Antarctica, 175–9
    tropical rainforests, 169–72
humidity, 40
hydrochlorofluorocarbons (HCFC), 291; *see also* F-gases
hydroelectric power (HEP), 200
hydrofluorocarbons (HFC), 291; *see also* F-gases
hydrophilic, 41
hydroponics, 194, 195
hydrosphere, 33
    as carbon store, 49
hypothesis, 61, 62–3
    testing, 63

IAATO *see* International Association of Antarctic Tour Operators
ice
    melting, 310, 312
    surface mass balance, 73
ice caps, 241
ice sheets, 241–2
    Greenland, 73–4
ice shelves, 175
ICUN Red List *see* International Union for Conservation of Nature (ICUN) Red List
illnesses, contaminated water and, 250
immigration, 119
implying, 11
imported energy sources, 208
independent variables, 63, 64
India
    air pollution, 281
    Bhopal disaster, 293–5
    tribal reserve, 109
indigenous species, 153, 154
industrial water use, 256
industry, energy security and, 212
infant mortality rate, 116, 117
infiltration, 29
infrared radiation, 303
in-migration, 112
insolation, 72
intensive farming, 194, 195
intermediate community, 149; *see also* ecological succession
international agreements
    Antarctica, 177–9
    climate change, 325–6
    hazardous waste, 224
    marine environment, 223–4
    ozone depletion, 292
    tropical rainforests, 171–2
    water use, 259–60
International Association of Antarctic Tour Operators (IAATO), 178
International Refugee Assistance Project (IRAP), 328
International Tropical Timber Organization (ITTO), 160
International Union for Conservation of Nature (ICUN) Red List, 160–1
International Whaling Commission (IWC), 157, 159–60
inter-specific competition, 36
interviews, 88–9
intra-specific competition, 36
invasive species, 153–5
IRAP *see* International Refugee Assistance Project

irrigation, 193, 255; *see also* agriculture; farming
irrigation mismanagement, 248–9
ITTO *see* International Tropical Timber Organization
IWC *see* International Whaling Commission

Japan
    ageing population, 133–5
    population pyramid, 121

Kaka'ako Waterfront Park, 188
kick sampling, 81, 85
Kigali Amendment, 292
Kiribati, 327
krill, 175, 178
Kyoto Protocol, 325

land degradation, food insecurity and, 190–1
landfill gas (LFG), 305
landforms, 110
LCF *see* low carbon fuels
leachate, 214
leaching, 111
legislation
    air pollution, 286–7
    biodiversity conservation, 156–60
    hazardous waste, 224
    marine, 223–4
    tropical rainforests, 172
    waste management, 223–4
LFG *see* landfill gas
LIC *see* low-income countries
life expectancy, 119
light, 40
light traps, 81–2, 85
limestone, acid deposition damaging, 278
limitations in scientific method, 66, 67
limiting factors, 38, 41
Lincoln index, 90–1
line graphs, 5
line sampling, 78
lithosphere, 33
    as carbon store, 49
livestock farming, 303–4
London Convention, 223–4
London Protocol, 224
long-term energy security, 204, 205; *see also* energy security
low carbon fuels (LCF), 319
low-income countries (LIC), 23, 24
    ageing populations, 125
    Basel Convention, 224
    child mortality rates, 118
    debt reduction, 171
    energy needs, 206
    energy security, 211–13
    food security, 189, 192
    irrigation, 256
    population age gap, 131–3
    transnational corporation (TNC), 293
    tropical rainforests, 171
    water availability, 247
    see also high-income countries (HIC)

Madrid Protocol, 178; *see also* Antarctic Treaty
malnutrition, 192
marble, acid deposition damaging, 278
marginal vegetation, 169, 170
marine cloud brightening, 326
marine conservation zones (MCZ), 168–9
marine environment
    legislation, 223–4
Marine Protection, Research and Sanctuaries Act (MPRSA), 223
Marshall Islands, 327
Masai Mara, 50–2
MCZ *see* marine conservation zones
MDG *see* Millennium Development Goals
'The Media is lying about Greenland', 73
medical care, energy security and, 212
medical resources, 155
Mekong River Commission in Southeast Asia, 260
melting of land ice and ice sheets, 310, 312; *see also* sea level rise
mesosphere, 31
methane, 32, 49, 70, 75, 214, 217, 320
    atmospheric life time, 303
    biogas, 200, 201, 222
    landfill gas (LFG), 305
    livestock farming, 304
    melting permafrost, 309, 312
    ozone, 288
    see also greenhouse gases
methyl isocyanate (MIC) gas, 293, 294
MIC *see* middle-income countries
microhabitats, 169
middle-income countries (MIC), 23
migration, 110
migration rates, 119
Millennium Development Goals (MDG), 261
mitigation, 318
models, 66
Molina, Mario, 291–2; *see also* ozone destruction hypothesis

Index

Mondi Forestry, 173
    Grasslands Programme, 173–4
monoculture crop/cropping, 190, 191
Montreal Protocol, 292, 325
MPRSA *see* Marine Protection, Research and Sanctuaries Act
Mt Gilboa reserve, South Africa, 174
mucus sac, 42
Murray-Darling River Basin, 260
mutualistic relationship, 317

naked produce, 220
Namib desert beetle, 41, 42
national parks, 168
native species, 153–5
natural disasters, 314
    water insecurity, 243
natural greenhouse effect, 32
natural population change, 119
natural recharge, 238; *see also* aquifers
natural resources, 111
natural threats, 111
nature reserves, 168
net primary productivity, 150, 151
Niger, 120, 124
    population pyramid, 120
nitric acid, 276, 277
nitrogen in acid deposition, 277
nitrogen monoxide, 277
nitrogen oxide, 276, 277
nitrous oxides, 32, 285, 303; *see also* greenhouse gases
non-renewable energy, 202–4
    coal, 203
    nuclear power, 203–4
    oil and gas, 202–3
North Sentinel Island, 109
Norway, 223
nuclear power, 203–4; *see also* non-renewable energy
numerical data, 3

Oahu, Hawaii, 188
observations, 61–4
    hypothesis, 61, 62–3
    qualitative data, 61, 62
    quantitative data, 61, 62
ocean acidification, 317
ocean circulation, 311
ocean currents, 72
oceanic circulation, 72
oceans, 22
oil and gas, 202–3

old age dependency ratio, 123, 124
omnivores, 36
open questions, 88
orbit of Earth, 72
out-migration, 112
overall population change, 119
over-cropping, 190, 191
oxyfuel technology, 322
oxygen, 40
ozone, 31, 32, 279, 281, 292, 303
    chlorine atoms, 288
    crop production, 281
    ground-level, 274, 279, 281
    measuring, 287
ozone depletion, 287–92
    Antarctica, 176, 288–9, 298
    causes, 288
    impact, 289–90
ozone destruction hypothesis, 291–2
    Rowland-Molina hypothesis, 291–2
    scientific investigation, 291
ozone hole, 176, 287, 288, 289, 292, 298; *see also* Antarctica
ozone layer, 31, 32, 288, 289, 291, 292

paragraph structure, 12–14
Paris Agreement, 325
particulates, 274
passive buildings, 323; *see also* buildings
pedosphere
    as carbon store, 49
peer assessment, 15
percentage cover, 92
percentage frequency, 92–3
perched aquifers, 239, 240; *see also* aquifers
permafrost, 146, 147, 242
permeable material, 238
persistent organic pollutants (POP), 224
pests and diseases *see* agriculture
petabytes, 100
pH, 40, 276
photochemical smog, 275, 279–81
    defined, 279
    impact, 281
photosynthesis, 43–7
phytoplankton, 31, 152
pie charts, 4
pilot survey, 88
pioneer species, 149
pitfall traps, 80, 84–5
Pixie frogs, 42

355

plastics
    biodegradable, 221
    deterioration/decay, 281
PM$_{10}$, 274
point quadrat, 79, 84
point sampling, 78
polar stratospheric clouds (PSC), 288–9
polar vortex, 288, 289
political stability, energy security and, 213
pollinating insects, protection, 197
pollution, 24
    air (*see* air pollution)
    Antarctica, 176
    arsenic, 262–4
    defined, 25
    water, 243–6, 262–4
POP *see* persistent organic pollutants
population, 33, 41
population age gap, 131–5
population change
    birth rate, 116–18
    child mortality rate, 116
    death rate, 116, 118
    defined, 116
    factors, 116–19
    impacts, 125–6
    infant mortality rate, 116, 117
    life expectancy, 119
    managing, 126–30
    migration rates, 119
    natural, 119
    overall, 119
population density, 87, 88
    calculating, 116
    economic factors, 112
    environmental factors, 110–11
    government policies, 112
    historical factors, 111
    social factors, 111
population distribution, 110, 112–15
    defined, 110
    patterns, 110, 112, 113
    South Sudan crises, 114–15
population dynamics, 110
population growth
    energy insecurity, 205
    food insecurity, 191
population pyramids, 119–22
    analysing, 121–2
population size, 87, 88

population structure, 119–22; *see also* population pyramids
post-combustion, 322
poverty
    biodiversity conservation help reducing, 155
    energy insecurity, 208, 212
    food insecurity, 191
    water management strategies, 256–7
    water supply and sanitation, 256–7
precipitation, 26, 28, 310
pre-combustion, 322
prediction, 62
prices
    energy, 208, 212
    food product, 192
primary air pollutants, 275; *see also* secondary air pollutants
primary data, 76, 77
primary ecological succession, 148, 149
primary industry, 112
primary producers, 36
primary productivity, 150–1
    gross, 150, 151
    net, 150, 151
pronatalist policies, 126, 128
PSC *see* polar stratospheric clouds
push factors, 115
pyramid(s)
    biomass, 152
    energy, 151
    numbers, 152
    population, 119–22
    see also ecological pyramids

Qatar, 124
quadrats, 78–9, 84
qualitative data, 61, 62
quantitative data, 61, 62
questionnaire, 88–9

radiation, 111; *see also* UV radiation
radio tracking, 98
random sampling, 76, 77, 83
rationing, 199
    water, 257
raw materials, 111, 112; *see also* natural resources
recreational value, biodiversity and, 156
reforestation, 172, 322
reliable data, 66
remote sensing *see* satellite sensing
renewable energies, 319

Index

air pollution control, 284–5
biomass, 201–2
geothermal energy, 202
hydroelectric power (HEP), 200
resources, 200
solar energy, 200, 201
wave and tidal energy, 201
wind energy, 201
research question, 62
research skills, 8–10
resources
biodiversity, 156
renewable energies (*see* renewable energies)
rewilding, 167
Yellowstone National Park, 179–82
rhetorical questions, 16
rice farming, 303–4
Rio Summit, 325
roof-top farming, 198
Rowland, Sherwood, 291–2; *see also* ozone destruction hypothesis
RRR (reduce, reuse, recycle), 220; *see also* waste management
rubber
deterioration/decay, 281
runoff, 28–9
Russia, 192

salinisation, 189
salinity, 40–1
salt water, 236, 237
sample/sampling, 76–7
beating trays/nets, 80–1
capture–mark–recapture technique, 82
kick sampling, 81, 85
light traps, 81–2, 85
line sampling, 78
pitfall traps, 80, 84–5
point quadrat, 79, 84
point sampling, 78
quadrat, 78–9, 84
random, 76, 77, 83
size, 77
strategies, 77
sweep nets, 80, 85
systematic, 76, 77, 83–4
techniques, 78–89
turbidity of water, 82–3, 86
see also data collection
sanitation, 243
sanitation mismanagement, 249
satellite sensing, 97–8

scatter graphs, 5
sceptics, 71
scientific bias, 71–2
scientific method, 59–68
data interpretation, 66–8
hypothesis, 63
observations, 61–4
variables, 64–6
scientific theory, 66
SDG *see* Sustainable Development Goals
sea ice, 73, 177
sea level rise, 310, 324
corals, 317
island nations, 327–9
melting of ice, 310, 312
secondary air pollutants, 276; *see also* primary air pollutants
secondary data, 76, 77
secondary ecological succession, 148, 149
secondary industry, 112
sedentary species, 78–9
selective breeding, 193
selective logging, 172
sentence length, 14
Sentinelese, 109
severe weather, 313, 324
shifting cultivation, 172
short-term energy security, 204, 205; *see also* long-term energy security
Simpson's index, 91–2
Singapore, 130
smog, 299
defined, 276
as secondary air pollutant, 276
social factors, 111
soil, 143
pioneer species, 149
terrestrial biomes, 144–6
type and quality, 111
soil moisture, 237
solar energy, 200, 201
solar radiation management (SRM), 326
solar variation, 72
South Africa
biodiversity conservation, 173–4
South Sudan crisis, 114–15
space mirrors, 326
species
defined, 21, 33
distribution, 20–1
sedentary, 78–9
stacked bar charts, 4

starvation, 192
statistics, 63
stem-flow, 28
Stockholm Convention on Persistent Organic Pollutants, 224
stockpiling
  defined, 199
  food resources, 198
stomata, 146, 147
stone carvings, acid deposition damaging, 278
stratosphere, 31
stratospheric aerosol injection, 326
subsistence farmer, 191
subsistence farming, 197–8
sub-surface fresh water, 237–40
sulfur dioxide (SO_2), 275, 276–7, 278
sulfuric acid (H_2SO_4), 276–7
Sumatran rhinoceros, 164–6
supply disruption, 206–7
surface fresh water stores, 237; *see also* sub-surface fresh water
surface mass balance, 73
surface water flow; *see also* runoff
sustainability, 24–5
sustainable architecture, 302; *see also* energy-efficient buildings and infrastructure
Sustainable Development Goals (SDG), 261
sustainable harvesting, 158
sustainable water extraction, 252
sweep nets, 80, 85
systematic sampling, 76, 77, 83–4

technology, 70
  data collection and analysis, 96–101
temperature, 309–10
  atmosphere layers, 30–1, 55
  diurnal, 143, 144
  precipitation, 310
  water and humidity, 40
terabytes, 100
terrestrial biomes, 142, 143–6
  desert, 142, 143, 146
  grasslands, 142, 143, 145
  tropical rainforest, 144
  tundra, 142, 143, 146
Thailand, 129
thermosphere, 31
thesis statement, 10
through-fall, 28
through-flow, 29
time delay on cause and effect, 309

time scales, ecological succession, 149
titanium dioxide, 59
TNC *see* transnational corporation
total dependency ratio, 122–3
tourism, as threat to Antarctic waters, 176
transborder water agreements, 260
transnational corporation (TNC), 293
transpiration, 26, 27
transport, 287, 319–20
trend line, 5
trends, 3, 6
trophic levels, 36–9
trophic pyramids *see* ecological pyramids
tropical rainforests, 144
  human impacts, 169–72
  sustainable management, 171–2
troposphere, 31
tundra, 142, 143, 146
turbidity of water, 82–3, 86
Tuvalu, 327

Uganda
  population pyramids, 132
  youthful population, 131–3
Ukraine, 192
UN Agenda 21, 127–8
unconfined aquifers, 239, 240; *see also* aquifers
UNEP *see* United Nations Environmental Programme
Union Carbide in Bhopal, India, 293–5; *see also* Bhopal disaster
United Arab Emirates, 124
United Nations Conference on Human Environment, 287
United Nations Environmental Programme (UNEP), 189
United States (US), 171
  Clean Air Act, 287
  climate cooperation, 326
  composting laws, 223
  Endangered Species Act, 157
  population pyramid, 120, 121
unsustainable production, 192
UV radiation, 31, 289, 290

value, data, 99
variable gases, 30
variables, 64–6
  control group, 64–5
  control variables, 65–6
  dependent, 63, 64
  independent, 63, 64

Index

see also scientific method
variety, data, 100
vegetation
    acid deposition, 277–8
    humans, 111
velocity
    big data, 100
    defined, 100
veracity, big data, 100
VOC *see* volatile organic compounds
volatile organic compounds (VOC), 279, 286
volcanism, 72
volcanoes, 72
volume
    big data, 99
    defined, 99

waste management, 213–24
    biodegradable plastics, 221
    composting, 222
    disposal at sea, 215
    education, 222
    exporting waste, 216
    fermentation, 222
    financial incentives, 223
    food waste for livestock, 221–2
    hierarchy, 220
    impact of disposal, 216–20
    incineration, 214
    landfill, 214
    legislation, 223–4
    methods, 214–16
    minimisation of waste, 220
    recycling, 213, 215
    storage, 214
    strategies reducing disposal impact, 220–4
    upcycling, 213, 214
waste streams, 213
water, 40
    abstraction, 29, 237, 239, 245, 246
    aquifers, 29, 238–40, 245
    contamination, 245, 250, 262–4
    groundwater, 29, 237, 238–40
    see also water cycle; water distribution; water insecurity; water security; water supply
WaterAid, 261
Water Convention of 1996, 260
water cycle, 25–9
    condensation, 26
    evaporation, 26
    groundwater, 29

infiltration, 29
precipitation, 26, 28
runoff, 28–9
stem-flow, 28
through-flow, 29
transpiration, 26, 27
water distribution, 236–42
    atmospheric water, 242
    sub-surface fresh water, 237–40
    surface fresh water stores, 237
water filtration, 240–1
water insecurity
    climate change, 243, 314–15
    competing demands, 246–7
    impacts, 249–51
    inequality of water availability, 247
    international competition, 247
    irrigation mismanagement, 248–9
    natural disasters, 243
    pollution, 243–6
    poverty, 250–1
    salinisation, 189
    sanitation mismanagement, 249
    sustainability solutions, 257–9
    urban *vs.* rural water access, 247–8
    see also water security
water management strategies, 251–61
    agreements and aid, 259–61
    improved supply, 252–3
    poverty reduction, 256–7
    reduced water usage, 253–6
    sustainable water extraction, 252
water-related aid, 260
water scarcity, 314–15; *see also* water insecurity
water security, 314–15
    defined, 243
    managing (*see* water management strategies)
    see also water insecurity
water shortages, food insecurity, 190
water supply
    gravity-fed schemes, 253
    hard engineering solution, 252
    human survival, 111
    reservoirs, 253
    soft engineering solution, 252
water table, 238
water vapour, 26, 30, 32, 70, 242, 303; *see also* greenhouse gases
wave and tidal energy, 201
weather, 110
wet deposition *see* acidic wet deposition

wetlands, 237
wildfires, 324
wildlife sanctuary *see* nature reserves
wind circulation, 311–12
wind energy, 201
wolves, 179–81; *see also* Yellowstone National Park
World Food Programme, 198
World Wildlife Fund, 157
writing clearly, 15
WWF, 261

Yellowstone National Park, 179–82
youth dependency ratio, 123, 124
youthful population
    Uganda, 131–3

Zimbabwe, energy security, 211–13
zooxanthellae, 317